Sustainability in Plant and Crop Protection

Series Editor
Aurelio Ciancio, Sezione di Bari, Consiglio Nazionale delle Ricerche Istituto per la Protezione delle Piante, Bari, Italy

More information about this series at http://www.springer.com/series/13031

Rajinder Peshin • Ashok K. Dhawan
Editors

Natural Resource Management: Ecological Perspectives

Springer

Editors
Rajinder Peshin
Division of Agricultural Extension
Education, Faculty of Agriculture
Sher-e-Kashmir University of Agricultural
Sciences and Technology of Jammu
Chatha Jammu, Jammu and Kashmir, India

Ashok K.Dhawan
The Indian Ecological Society
Ludhiana, Punjab, India

ISSN 2567-9805 ISSN 2567-9821 (electronic)
Sustainability in Plant and Crop Protection
ISBN 978-3-319-99767-4 ISBN 978-3-319-99768-1 (eBook)
https://doi.org/10.1007/978-3-319-99768-1

Library of Congress Control Number: 2019931545

This Springer imprint is published by the registered company Springer Nature Switzerland AG.
The registered company address is: Gewerbestrasse 11, 6330 Cham, Switzerland

This book is dedicated to the 608 delegates who attended and presented their work in the Indian Ecological Society International Conference 2016.

Preface

Modern input-intensive agriculture has increased the productivity of food and fibre crops, ensured food security for the 7.6 billion world population and provided food at affordable prices. The benefits of agricultural modernization overweigh the harms if we apply the consequential approach of utilitarian theory. However, modern agriculture has also led to certain undesired consequences, namely depletion of groundwater, soil erosion, pesticide resistance in pests, emergence of new insect pest and diseases, chronic health effects of pesticides, loss of soil fertility, environmental pollution and loss of biodiversity. Concern over global climate change, the energy crisis, pesticide-intensive pest management and new interest in the potential of biofuels have ushered in a new era of challenges and opportunities for agriculture and natural resource management.

To overcome harms of modern input-intensive agriculture and concerns of climate change, the emphasis is on ecologically sustainable agriculture to wisely manage natural resources. Both public and private sector organizations contribute to the development of agriculture. Development and transfer of proprietary technologies, namely hybrids, transgenic, agrochemicals and farm machinery as private goods by the private sector, has a predominant role. Private-sector research in germplasm improvements for food and non-food crops has been significant. Private-sector involvement in natural resource management research, however, has been very limited, constrained by high risk and non-exclusiveness of the research results. Public-sector research and extension are essential for natural resource management and development and diffusion of production management technologies. Sustainable use of natural resources requires environmental accountings to assess the damage to land and water resources, loss of biodiversity, pesticides inflicted damages to human health and beneficial organisms.

To address the issues of natural resource management from an intra and interdisciplinary perspective, an international conference was organized by the Indian Ecological Society in 2016 in which authorities from the disciplines of agriculture (entomology, plant pathology, crop production and improvement, extension education, resource economics), medical sciences, aquaculture, water management and engineering, private industry (pesticide, seed, forest products) and non-governmental organizations (World Wide Fund for Nature, Centers for International Projects Trust) presented papers. In this book, we have included selected keynote lectures to provide a holistic perspective of natural resource management issues. The book covers chapters on water management, the Himalayan environment, biodiversity, role of microbes in agriculture, impact of climate change on human health and crop pests, chronic health effects of pesticides, pesticide resistance, exploiting chemical ecology for pest management, integrated pest management, integrated farming systems, and the drivers of adoption of ecologically sustainable technologies.

Sher-e-Kashmir University of Agricultural Sciences and Technology of Jammu
Chatha Jammu, Jammu and Kashmir, India
Rajinder Peshin

Acknowledgements

I am grateful to Drs. Dinesh K. Marothia, Randeep Guleria, Keshav R. Kranthi, Kamal Vatta, Masud H. Wani, Surinder K. Gupta, Chanda Siddoo-Atwal, Bushan L. Jalali, Ambalal N. Sabalpara, Dharam P. Abrol and Kewal K. Sharma for their keynote lectures in the Indian Ecological Society International Conference 2016, on *Natural Recourse Management: Ecological Perspectives*, organized at the Sher-e-Kashmir University of Agricultural Sciences and Technology of Jammu, India, on Feb. 18–20, 2016. I thank our authors for agreeing to publish their interesting and valuable work presented in the conference in the form of book and Springer for agreeing to publish this volume.

Sher-e-Kashmir University of Agricultural Sciences
and Technology of Jammu
Chatha Jammu, Jammu and Kashmir, India

Rajinder Peshin

About the Book: Integrated Pest Management – Innovation Development Process, Vol. 1, Springer 2009

"*Peshin ... and Dhawan ... have ... produce an encyclopedic overview of fundamental concepts and recent advances in integrated pest management (IPM). Their ambitious undertaking is well executed, with a ... combination of breath and detail, and a truly global perspective. For those whose studies relate to IPM, this collection will be an invaluable resource. Summing up: Highly recommended. Upper-division undergraduates through researchers/faculty.*" [M. K. Bomford, *Choice*, Vol. 47 (3), November, 2009]

Contents

About the Editors and Authors

Editors

Rajinder Peshin is professor at the Sher-e-Kashmir University of Agricultural Sciences and Technology of Jammu (SKUAST Jammu), India. He received his Master of Science in extension education and Ph.D. from the Punjab Agricultural University, Ludhiana, India. He worked as agriculture extension officer in the Department of Agriculture, Government of Jammu and Kashmir, from 1989 to 1998 before joining SKUAST Jammu in 1998 as an assistant professor. His research expertise is dissemination and evaluation of issues associated with sustainable agriculture research and development programmes, especially for integrated pest management. He developed a model for predicting the adoptability of agricultural technologies when put on trial at farmers' fields and an evaluation methodology for integrated pest management programmes. Dr. Peshin has published over 65 scientific papers and 8 books. These include four books on integrated pest management series published by Springer in 2009 and 2014. Dr. Peshin is a fellow of the Indian Ecological Society.

Ashok K. Dhawan is former additional director and head in the Department of Entomology, Punjab Agricultural University (PAU), Ludhiana, India. At present he is the president of the Indian Ecological Society. He worked for more than 37 years on integrated pest management (IPM) in cotton. He was instrumental in devising and implementing insecticide resistance management strategy, mealy bug management in cotton and the transfer of technology for the management of cotton pests in more than 800 villages. He developed IPM models for rice, maize and vegetable pests and implemented these at the village level. He has authored 16 books, 430 research papers and 80 extension publications. He is a member of the executive council of various scientific bodies and a member of the advisory of various government and private sector organizations. He was awarded the Best Team Award by the Indian Council of Agricultural Research (ICAR) and the Department for International Development (DIFD-UK) for the transfer of technology.

Authors

Dharam P. Abrol is working as professor and dean of the Faculty of Agriculture, Sher-e-Kashmir University of Agricultural Sciences and Technology of Jammu, and Faculty of Agriculture, Chatha, India. He has been visiting scholar at ETH Zurich, Switzerland; Jagiellonian University, Krakow, Poland; Busan, South Korea, and Terengganu, Malaysia. His research addresses pollination biology, honeybee ecology and integrated pest management. He has been honoured by several national and international awards. Dr. D. P. Abrol has published more than 220 research papers, 20 chapters of books and 10 review articles and is the author of 15 books published by Springer, CABI, Academic Press and others.

Fatima Bano is a Ph.D. scholar in the Division of Agricultural Extension Education at Sher-e-Kashmir University of Agricultural Sciences and Technology (SKUAST) of Jammu, India. She is recipient of the National Fellowship for Higher Studies (NFST) granted by the University Grants Commission (UGC), New Delhi, and Ministry of Tribal Affairs, Government of India. She has qualified NET in agricultural extension education in 2016. In her Ph.D., she is working on farmers' knowledge and practices about agrochemical use in selected crops in Jammu and Kashmir State. She has received her M.Sc. degree from SKUAST of Jammu in 2014.

Vanya Bawa is currently a Ph.D. scholar working on *Brassica juncea* (oilseeds) in the Division of Plant Breeding and Genetics, Sher-e-Kashmir University of Agricultural Sciences and Technology of Jammu, India. She has obtained her degrees (B.Sc. and M.Sc. in Genetics and Plant Breeding) from the Sher-e-Kashmir University of Agricultural Sciences and Technology of Jammu. She is the recipient of merit scholarship in degree programmes and has her two publications in national and international (*PLOS One*) journal.

Ajanta Birah is working as principal scientist (agricultural entomology) at the ICAR, National Research Centre for Integrated Pest Management, New Delhi. She has worked as senior scientist in Central Island Agricultural Research Institute, Port Blair, Andaman and Nicobar Islands (2008–2013). For her contribution in the field of plant protection, she has been bestowed with Fakhruddin Ali Ahmed Award for Outstanding Research in Tribal Farming System. She has to her credit more than 64 research papers published in reputed peer-reviewed journals and 16 chapters in books. She is fellow of the Entomological Society of India and Society of Plant Protection Sciences, New Delhi, and life member of several professional societies and editorial board member/reviewer of many journals.

Chirantan Chattopadhyay is vice-chancellor at Uttar Banga Krishi Viswavidyalaya, Cooch Behar, West Bengal, since 2016. He joined Agricultural Research Service of the Indian Council of Agricultural Research in 1992 where he has served as scientist in different levels (IIOR, Hyderabad; DRMR, Bharatpur), head of the Division of Crop Protection (IIPR, Kanpur) and as director of ICAR, National Research Centre for Integrated Pest Management, New Delhi, during 2012–2016. For his contribution in the field of plant protection, he has been bestowed with various national-level prestigious awards including Lifetime Achievement Award by the Confederation of Indian Universities. His major area of research work is host resistance to plant diseases (oilseed crops), disease and insect-pest forecasting and bio-intensive integrated pest management.

Shilpa Chawla worked as a senior research fellow under the Technology Mission on cotton for more than 2 years at Central Institute for Cotton Research (CICR), Nagpur, India.

Ashutosh Dhanuka is a consultant pulmonologist and critical care specialist at Geetanjali Medical College and Hospital, Udaipur (India). He has done his D.M. in Pulmonology, critical care and sleep medicine from the prestigious All India Institutes of Medical Sciences, New Delhi. His areas of interest include interventional pulmonology, interstitial lung diseases and critical care. He was awarded "Young Indian Scholar – West Zone Topper" Award in 2009. He has presented posters in international conferences.

Ashok K. Dhawan is a former additional director and head of the Department of Entomology, Punjab Agricultural University (PAU), Ludhiana. At present he is the president of the Indian Ecological Society. He worked for more than 37 years on integrated pest management (IPM) of cotton. He was instrumental in devising and implementing insecticide resistance management strategy, mealy bug management in cotton and the transfer of technology for management of cotton pests in more than 800 villages. He developed IPM models for rice, maize and vegetable pests and implemented these at village level. He has authored 16 books, 430 research papers and 80 extension publications. He is a member of the executive council of various scientific bodies and a member of the advisory of various government and private sector organizations. He was awarded the Best Team Award by the Indian Council of Agricultural Research (ICAR) and Department for International Development (DIFD-UK) for transfer of technology.

Vikrant Dhawan is the deputy project director of the Agricultural Technology Management, Department of Agriculture, Punjab, India. He has obtained his M.Sc. in the discipline of Agricultural Economics from the Punjab Agricultural University, Ludhiana, India. He has worked on studying the impact of agricultural price policies on changing agrarian scenario in Punjab, India. He has published more than 15 original research papers in refereed journals, 12 conference papers, 1 book chapter and 8 extension papers.

Randeep Guleria is the director of All India Institute of Medical Sciences (AIIMS), New Delhi. He is professor in the Department of Pulmonary Medicine and Sleep Disorders. He is the first D.M. in Pulmonary Medicine in India and is the recipient of several awards and honours including the prestigious Padma Shri, the fourth highest civilian award of India in 2015 and Dr. B.C. Roy Eminent Medical Person in 2014. His research interest is on the effect of air pollution and pulmonary and sleep disorders with more than 450 publications, 49 chapters, 50 research grants and more than 300 invited lectures to his credit.

Surinder K. Gupta is currently working as professor and head/chief scientist (oilseeds) in the Division of Plant Breeding and Genetics, Sher-e-Kashmir University of Agricultural Sciences and Technology, Jammu, India, and holds a brilliant academic and service record. For almost two decades, he has devoted his research interest to the area of oilseed *Brassicas*. Dr. Gupta obtained his Post-Graduate degrees (M.Sc., Ph.D.) from Punjab Agricultural University, Ludhiana, India, in 1984 and 1987, respectively. He is the recipient of Post-doctoral Fellowship in plant biotechnology and has published more than 100 research papers in national and international journals, mostly on *Brassicas*. He has already developed five varieties of rapeseed-mustard. He has written two books on plant breeding and edited two volumes: one on recent advances in oilseed *Brassicas* and the second on rapeseed breeding – *Advances in Botanical Research*, Vol. 45, published by Academic Press, Elsevier. For his excellent scientific endeavours, he has been conferred with the Young Scientist Award 1993–1994 by the Jammu and Kashmir State Department of Science and Technology.

Bushan L. Jalali served as professor of plant pathology (1982–1992); head of the Department of Plant Pathology, Haryana Agricultural University, Hisar (1987–1990); director of ICAR, National Research Centre for Integrated Pest Management, New Delhi (1992–1993); visiting professor at the University of Hannover, 1987; University of Tubingen, Germany, 1994; University of Montreal, Canada, 1995; and Dongguk University, Seoul, South Korea, 2000; director of research at Haryana Agricultural University, 1998–2001; and member of the Commission on Agricultural Reforms, Research and Development, Government of Jharkhand, 2006–2008. He has been bestowed with various prestigious awards and also has been chairman/member of several national research management, planning and advisory committees/expert panels of ICAR, DBT, DST, NRDC, UGC and other national bodies.

Zeyaur R. Khan is a distinguished agricultural scientist and a professional entomologist working for the last 25 years with the International Centre of Insect Physiology and Ecology (*icipe*) in Kenya. He has dedicated his 35-year career as an international agricultural scientist to advancing the science and practice of agriculture by studying and applying chemical ecology, behaviour, plant-plant and insect-plant interactions to improve farm productivity to combat poverty and food insecurity in Africa. He is responsible for the discovery and wide-scale implementation of the Push-Pull technology (www.push-pull.net) and its subsequent adaptation to climate change for enhancing food security and environmental sustainability for small-holder cereal-livestock African farmers. His goal is to end hunger and poverty for ten million people by extending the Push-Pull technology to one million households in sub-Saharan Africa by 2020.

Keshav R. Kranthi is the head of the Technical Information Section, International Cotton Advisory Committee, Washington DC, USA. He is the former director of the Central Institute for Cotton Research Nagpur, India. He earned his Ph.D. from the Indian Agricultural Research Institute. His research areas are on genetics and stochastic modelling of bollworm resistance to insecticides and biotech cotton, and he has developed several diagnostic kits to detect biotech cotton, insecticide-resistant insects and substandard insecticides. Kranthi has two patents in India and four in South Africa, South Korea, China and Uzbekistan. He has published 35 papers in peer-reviewed journals. He has many national and international awards and was recognized as the "International Cotton Researcher of the Year Award 2009" by the International Cotton Advisory Committee, Washington, and was felicitated at Bremen, Germany, by the cotton fraternity in 2010 for his research contributions. Dr. Kranthi is a fellow of the National Academy of Agricultural Sciences, India.

Sandhya Kranthi is principal scientist and head (incharge) of the Crop Protection Division and has more than 25 years of experience in cotton crop protection research at the Central Institute for Cotton Research, Nagpur, India. She has published more than 50 papers in peer-reviewed journals and has a citation of 1767 and an H index of 15. Her basic degree in agriculture and a Master's degree in entomology were received from Acharya N. G. Ranga Agricultural University (ANGRAU), Rajendranagar, Hyderabad, India, and she obtained her Ph.D. from the Indian Agricultural Research Institute, New Delhi, India, in 1991 before joining the Agricultural Research Service.

Raj Kumar is a Ph.D. scholar in the Division of Agricultural Extension Education at the Sher-e-Kashmir University of Agricultural Sciences and Technology of Jammu. His Ph.D. research is on the impact evaluation of rice integrated pest management programme in subtropical North India. He received his Master's degree in extension education from Chaudhary Charan Singh University, Meerut, India.

Lalit Mahatma is an associate professor of plant pathology at N.M. College of Agriculture, Navsari Agricultural University, Navsari, India. His Ph.D. is from Anand Agricultural University, Anand, Gujarat, India. His research span the fields of molecular plant virology, plant disease epidemiology, soil biology, biological control, molecular plant breeding, plant biotechnology, organic farming, organic agriculture input production, bio-fertilizers, waste management and environment. Dr. Mahatma has published over 70 scientific papers and 10 book chapters. He has been the vice president (2013–2014) and fellow of India Society of Mycology and Plant Pathology, Udaipur. He has been awarded the Best Teacher Award (2014–2015) by Navsari Agricultural University, Navsari; Prof. H.C. Dube Outstanding Young Scientist Award (2015); and PP Singhal Memorial Pesticides India Award-II by the India Society of Mycology and Plant Pathology, Udaipur.

Dinesh K. Marothia earned his Ph.D. in 1981 from the University of Alberta, Edmonton, Canada. He is a recipient of the Canadian Commonwealth Scholarship and Fellowship, a member (non-official) of the State Planning Commission, Chhattisgarh, and president of the National Institute of Ecology. His main subject is agricultural and natural resource economics. He held many distinguished positions in academic institutions, federal and state governments and professional societies including president of the Conference of the Indian Society of Agricultural Economics and chairman of the Editorial Board (Indian Journal of Agricultural Economics). He had published extensively and authored books on natural resources economics and common pool resources. He had worked as team leader/advisor in many national and international projects and visited several countries for academic assignments

Vartika Mathur is an assistant professor in the Department of Zoology, Sri Venkateswara College, University of Delhi. She is the recipient of NUFFIC Fellowship from the Dutch Ministry to pursue Ph.D. from Wageningen University, Netherlands. Her research interest is on animal-plant interactions, environment and chemical and microbial ecology, with 12 research grants, 16 publications and an ecology undergraduate book. She is assistant editor for biology journal and reviewer in international journals of high repute. She has received "Young Scientist of the Year Award 2015" in the field of chemical ecology by the International Foundation of Environment and Ecology.

Charles A. O. Midega is a senior scientist at the International Centre of Insect Physiology and Ecology (*icipe*), Kenya, and holds a Doctoral degree (2005) in Agricultural Entomology from Kenyatta University (Kenya). His primary research interest is on building ecological frameworks for the management of insect pests in smallholder cropping systems through understanding the underlying mechanisms of plant-herbivore-natural enemy interactions. He is directly involved in the implementation of the innovative push-pull technology in Africa while adapting it to different agro-ecologies, cropping systems and farmer profiles. He also evaluates the technology's impacts on ecological integrity of agroecosystems, in addition to chemical ecology research on plant-to-plant communication and plant-insect interactions and their exploitation in managing pests of cereal crops and cotton. He is also an expert in agricultural development, including sustainable intensification of agricultural systems.

Sarita Nehare worked as a research associate in the NMITLI (New Millennium Initiatives in Technologies Leading India) in the area of lectin evaluation against sucking pests for more than 3 years at the Crop Protection Division, Central Institute for Cotton Research (CICR), Nagpur, India. She is currently working as a lecturer in the Anand Niketan College of Agriculture, Warora, Maharashtra, India.

Rajinder Peshin is a professor of agricultural extension education at the Sher-e-Kashmir University of Agricultural Sciences and Technology of Jammu (SKUAST Jammu), India. He received his Master of Science in Extension Education and Ph.D. from the Punjab Agricultural University (PAU), Ludhiana, India. He worked as agriculture extension officer in the Department of Agriculture from 1989 to 1998 before joining SKUAST Jammu in November, 1998, as an assistant professor. His research expertise is dissemination and evaluation issues associated with sustainable agriculture research and development programmes especially integrated pest management. He has developed a model for predicting the adoptability of agricultural technologies when put on trial at farmers' fields and an evaluation methodology for integrated pest management programmes. Dr. Peshin has published over 65 scientific papers and has authored/edited 8 books. These include four books on integrated pest management series published by Springer in 2009 and 2014. Dr. Peshin is a fellow of the Indian Ecological Society.

John A. Pickett is a professor and is originally an organic chemist (B.Sc. 1967, Ph.D. 1971, D.Sc. 1993) recognized worldwide (e.g. Wolf Prize in Agriculture, 2008) for investigations into a group of volatile natural products and semiochemicals, which affect the behaviour and development of animals and other organisms. He was the first to identify sex-related pheromones of aphids, sand flies and mosquitoes. Recent successes include laboratory and field research into GM-based release of the aphid alarm pheromone and a practical programme for controlling stem borer pests and striga weeds collaboratively in Africa. He headed the Department of Biological Chemistry at Rothamsted Research from 1984 to 2010 and now holds the first Michael Elliott Distinguished Research Fellowship.

Jimmy Pittchar is a social scientist at the International Centre of Insect Physiology and Ecology (*icipe*). He holds a B.Sc. degree in mathematics, statistics and computer science and Master's degree in social sciences. As part of an interdisciplinary research team, he implements the push-pull technology among smallholder farmers in Africa and helps them to intensify and sustainably improve their cereal crop and livestock productivity, agroecosystem resilience and capacity in order to achieve better health, food security, incomes and conservation of their natural environments. His current research interests include social and technological innovation systems, technology diffusion processes, rural smallholder farmers' knowledge bases, livelihood strategies, options for scaling up best practices and interactions between the push-pull scientific innovation, agroecosystems and socio-economic factors. He also evaluates the impacts of push-pull on farmers and their production systems.

Aditya Pratap is a principal scientist in the Crop Improvement Division, ICAR-Indian Institute of Pulses Research, Kanpur. Holding a brilliant academic and service record, he has worked on the genetic improvement of crop plants including *Vigna* species, wheat, triticale, rapeseed-mustard and chickpea. He has been associated with the development and release of 11 crop varieties including 6 in green gram, 3 in rapeseed-mustard and 1 in facultative winter wheat. He has developed one of the first marker-assisted breeding products in pulses by developing *Fusarium* wilt-resistant chickpea. To his credit, he has >150 publications including research papers in several high-impact journals. He has edited five books published by Springer and CABI. His research interests include distant hybridization, doubled haploid breeding, plant tissue culture and molecular breeding. He is the recipient of the prestigious Norman E. Borlaug International Agricultural Science and Technology Fellowship, USA, and also the ICAR-Lal Bahadur Shastri Outstanding Young Scientist Award.

Sunil K. Rai is currently working as scientist and project leader at the All India Coordinated Research Project on Rapeseed-Mustard at the Division of Plant Breeding and Genetics, Sher-e-Kashmir University of Agricultural Sciences and Technology, Jammu, India. As a *Brassica* breeder, he has developed three varieties each of *Brassica rapa*, *B. Juncea* and *B. napus.*

N. Ravisankar is a principal scientist at the ICAR- Project Directorate for Farming Systems Research, Modipuram, India. As an outcome of the research and extension work done, Dr N. Ravisankar has published 58 research papers, 28 popular articles, 11 books, 34 book chapters, 9 technical bulletins, 7 training manuals and 2 Video films. Based on the farmer participatory research, 81 geo-referenced farming systems' success stories have been documented by involving 23 state agricultural universities of India. As a national principal investigator of Network Project on Organic Farming (NPOF), he published *Organic Farming: Crop Production Guide* for organic production of crops for 79 cropping systems suitable to 14 states of India.

Chetali Rodge is a technical assistant at the Central Institute for Cotton Research (CICR), Nagpur, India, and had worked with Dr. K.R. Kranthi under the Insecticide Resistance Management Project for more than 5 years, before which she was associated as senior research fellow/research associate in various funded projects of the Division of Crop Improvement at CICR Nagpur, India.

Ambalal N. Sabalpara retired as director of research and dean of post-graduate studies from Navsari Agricultural University, Navsari, Gujarat, India, and presently engaged in as an advisor of industries manufacturing and marketing biofertilizer and biopesticides. He obtained Ph.D. from the College of Agriculture, Junagadh, India. He is a fellow of the Indian Society of Mycology and Plant Pathology, Phytopathological Society, India; Mycological Society, Kolkata; and Society for Recent Development in Agriculture, Meerut. He has been the president of the Indian Society of Mycology and Plant Pathology for two consecutive terms in 2013–2014 and 2014–2015. He has been credited for the development of integrated disease management strategies for all the major crops of South Gujarat. He is specialist of biofertilizers, biopesticides, waste management and organic farming and on farm input production technologies. He has been actively associated with the teaching, research and extension programme for more than 35 years.

Uma Shankar is working as an associate professor at the Division of Entomology, Sher-e-Kashmir University of Agricultural Sciences and Technology of Jammu, India, and Faculty of Agriculture, Chatha, India. He has expertise in integrated pest management and pollination ecology of non-*Apis* bee pollinators. He has published 65 research papers and authored 3 books, 3 manuals and 20 book chapters of national and international repute. He has handled several research projects on IPM of fruits and vegetables and Network Project on Insect Biosystematics, Onion and Garlic and on Linseed.

Kewal Krishan Sharma did his M.Sc. in Zoology (Entomology) from Kurukshetra University, Kurukshetra, and Ph.D. from Chaudhary Charan Singh Haryana Agricultural University, Hisar, India. He joined the then Indian Lac Research Institute (rechristened as Indian Institute of Natural Resins and Gums in 2007) as scientist in 1986 and is presently working as director of the institute. He is also project coordinator of Network Project on Conservation of Lac Insect Genetic Resources. His major areas of professional interest are biodiversity conservation, insect-host plant interaction, environment simulation, gene bank and plantation management in lac culture. He has developed a number of improved lac insect breeds, new lac insect-host plant combinations and technologies of lac cultivation; discovered new lac insect species, host plants, parasitoids and fungi associated with lac insect; and developed field gene banks of lac insects and host plants. He has published more than 71 research papers in national and international journals, 28 books/technical reports, 31 book chapters, 21 research/extension bulletins and 15 popular article and guided two Ph.D. and three M.Sc. students. Dr. Sharma is fellow of *Entomological Society of India*, IARI, New Delhi.

Chanda Siddoo-Atwal completed her B.Sc. in Biochemistry from the University of London. Her Ph.D. was taken in Applied Sciences from Simon Fraser University in Burnaby (her research conducted at the BC Cancer Research Center), and she did a post-doctoral fellowship in the Biochemistry Department at the Medical College of Wisconsin. Cancer is her subject of specialization, and she started with research on ataxia-telangiectasia (AT) patients, who are genetically prone to lymphomas. Since then, she has focused mainly on the mechanisms of carcinogenesis in various models including the AT disease state and ionizing radiation-induced cancer, mycotoxin-induced and chemically induced cancers and ultraviolet radiation-induced skin cancer. She is the president and primary biochemist of Moondust Cosmetics, Ltd., and her current research activity includes the formulation of a novel sunscreen to combat apoptotic sunburn that has been associated with skin cancer and other products for the repair of skin damage related to sun exposure. She has been invited to present her research papers at conferences around the world and is the author of a recent book called *A New Approach to Cancer Risk Assessment: An Overview*, published by Lambert Academic Publishing.

Kamal Vatta is a professor and director of Centers for International Projects Trust, New Delhi, India. He is an agricultural economist specializing at the interface of economics with the natural resource degradation (specifically water), agriculture and rural development processes like labour markets. He has completed Ph.D. in Agricultural Economics from Punjab Agricultural University, Ludhiana, India. He also has an advanced diploma in Agricultural Research for Development from the International Center for Development-Oriented Research in Agriculture (ICRA),

Wageningen, Netherlands. Dr. Vatta is recognized for his contributions in the fields of research on sustainable agriculture, development economics and resource sustainability and has been endowed with several awards and distinctions throughout his career. He has over 70 peer-reviewed publications in agriculture, economics and development and has led more than 20 projects in various capacities in these fields both nationally and internationally.

Sohan S. Walia is a senior agronomist in the Department of Agronomy, Punjab Agricultural University, Ludhiana, India. He has earned his M.Sc. and Ph.D. in the discipline of Agronomy from Punjab Agricultural University, Ludhiana. He has published 125 research papers in refereed journals, 190 conference papers, 100 extension papers and more than 24 book chapters. Dr. Walia published three books, namely, *Science of Agronomy*; *Contingent Crop Planning* and *Crop Management*. He has guided seven M.Sc. and three Ph.D. students. He along with his co-workers has developed technologies on resource conservation cropping systems, 6 organic farming-based cropping systems, production technology of cultivation of direct seeded rice and consortium bio-fertilizer application in sugarcane, maize, wheat, potato, onion and turmeric, and 62 of from these have included in the *Package of Practices* of the Punjab Agricultural University. Based on his significant contributions in the field of agronomy, he has received P.S. Deshmukh Young Agronomist Award (2005) and conferred with the fellow of the Indian Ecological Society (2016).

Masaud-ul-Haq Wani is working as professor, Rajiv Gandhi chair and registrar at the Sher-e-Kashmir University of Agricultural Sciences and Technology of Kashmir, India. Prof. Wani has published more than 260 scientific papers in national/international journals, books, reviews, etc., and has organized conferences/seminars/workshops/capacity building programmes/orientation courses for faculty and other stakeholders and has completed 30 major research projects funded by various national and international financing agencies.

Saima Masood Wani is working as junior research fellow in the Division of Biotechnology at the Sher-e-Kashmir University of Agricultural Sciences and Technology of Kashmir, India. She has obtained her Ph.D. from IISC, Bangalore. The author has six national and international publications to her credit.

Chapter 1
Managing Wetland Ecosystems: A Polycentric Perspective

Dinesh K. Marothia

Abstract India is endowed with extensive wetlands of different kinds. Multi-use water bodies (MUWBs) or multiple-use water systems are the most important and valuable category of wetlands. MUWBs include small water storage bodies, village ponds, and irrigation and multipurpose tanks. MUWBs constitute an important component of community assets in India. These water bodies have been used as traditional commons by village communities for centuries to meet their domestic needs, for irrigating crops, and for practicing fish farming in many Indian states. The landscape of Chhattisgarh state is dotted by numerous age-old and recent MUWBs. The issues related to MUWB management are complex due to the different categories and characteristics of these de facto common water bodies; the scale, size, and coverage of fisheries, agriculture, and domestic, sociocultural, and religious activities; and the multiple agencies involved in water use. From this example of Chhattisgarh, the extent of completion of and conflicts over MUWBs and mechanisms to resolve conflicts can be understood. Data used in this study have largely been taken from the author's earlier published and unpublished work (between 1985 and 2015) pertaining to small-scale wetlands (multi-use village ponds and tanks). Competition and conflicts over MUWBs have been observed at four levels, i.e., within user groups (irrigators), across user groups (irrigators, fishermen, and villagers), interinstitutional [in fisheries cooperative societies, water user associations (WUAs), *panchayat* (a democratically elected village council in India), the irrigation department, the fisheries department], and between stakeholders and institutions (irrigators vs the water resource department, fishers vs WUA, or *panchayat*). Governing MUWBs is complicated because it can create disproportionate spatial and temporal externalities due to technical, socioeconomic, cultural, political, and environmental interdependencies. Issues regarding the role of traditional authori-

Keynote Address delivered at the Indian Ecology Society International Conference organized by Sher-e-Kashmir University of Agricultural Sciences and Technology, Jammu, Feb 18–20, 2016.

D. K. Marothia (✉)
National Institute of Ecology, New Delhi, India

Raipur, Chhattisgarh, India
e-mail: dkmarothia@gmail.com

R. Peshin, A. K. Dhawan (eds.), *Natural Resource Management: Ecological Perspectives*, Sustainability in Plant and Crop Protection,
https://doi.org/10.1007/978-3-319-99768-1_1

ties, user groups, stakeholder committees, different departments of the state, and *panchayat* and the relationships between multiple authorities with overlapping working zones are highly complex. Given such a complexity, we have advocated in this chapter distributed governance or polycentric approaches to manage MUWBs. We have also suggested that the policies of different departments of the state and *panchayat* need to be examined critically in view of the fact that many departments are loosely linked and intensely compete for water. A synthesis of traditional institutional mechanisms and the components of current policies can appropriately be made for the distributed or polycentric governance of MUWBs.

Keywords Wetland ecosystem · Multi-use water bodies · Water use · Water use policy

1.1 Introduction

In the recent past, several wetland[1] ecosystems have been altered, destroyed, or lost in different parts of India. The degradation process is also occurring in most countries of the world as a direct consequence of the use of the wetland ecosystem for urban development or its use for waste disposal by homes and industries or through its continuous exhaustion due to hydrological disturbances; pollution; recreational pressure; human, livestock, and fishing activities; and policy and institutional failures (Pearce and Turner 1990). Wetland ecosystems undeniably represent a valuable environmental resource with consequently high preservation, conservation, and utilization value. Despite this, existing evidence strongly suggest that wetlands in India are still not being managed sustainably. While a few attempts have been made in India to understand the causes and consequences of wetland ecosystem degradation, there is little explicit research conducted to resolve the complexities of interdependencies between economic, social, institutional, and environmental attributes and processes (Marothia 1995, 1997a, b, 2004a, b, c, d, 2015). Furthermore, user's perceptions have also not been adequately analyzed in designing wetland ecosystem policies and strategies (Marothia 2001, 2004a, b, c, d). Such understanding is extremely important as wetland ecosystems in India are basically common pool resources (CPRs). Wetland ecosystems are multipurpose in nature and have multi-use, with technical, socioeconomic, cultural, political, and environmental interdependencies. As a result, wetlands suffer from spatial and temporal externalities due to pollution, congestion, in-fillings, encroachment, natural and created use conflicts,

[1] Ramsar Convention, 1971 (Ramsar Convention Secretariat 2013) definition of wetlands: "areas of marsh, fen, peat land or water, whether natural or artificial, permanent or temporary, with water that is static or flowing, fresh, brackish or salt, including areas of marine water that depth of which at low tide does not exceed six meters" and "many incorporate riparian and coastal zones adjacent to the wetlands, and islands or bodies of marine water deeper than six meters at low tide lying within the wetlands."

misspecification of property rights regimes, entitlement systems, institutional hierarchy, and mechanisms for adjusting the allocation of wetland water resources (Marothia 2004a, b, c, d, 2015).

India is endowed with extensive wetlands of different kinds.[2,3] Multi-use water bodies (MUWBs) or multiple-use water systems are the most important and valuable category of wetlands. MUWBs include small water storage bodies, village ponds, irrigation, and multipurpose tanks. MUWBs constitute an important component of community assets in India.[3] These water bodies have been used as traditional commons by village communities for centuries to meet their domestic needs and for irrigating crops and practicing fish farming in many Indian states. The landscape of Chhattisgarh is dotted by numerous age-old MUWBs. Construction of ponds and tanks was traditionally undertaken by kings, jagirdars (feudal land grant holders), religious bodies, and rich and affluent classes in Chhattisgarh. The aim was to create community water assets and the management of these resources by means of well-defined informal codes to be used for different purposes and stakeholders, for repair, and for rejuvenation. Control over water bodies was exercised by the owner. In each village, MUWBs have traditionally been allotted for different uses, such as for tending cattle, washing clothes, irrigation, fish culture, nutrient-rich soils, fodder grass collection and brick making, micro-biodiversity resource base, social and cultural rituals (funeral, worship, temple, or sacred ponds), and social groups (Marothia 2015).

After 1952, most of the ponds and tanks were transferred to *panchayats*, municipal bodies, and the state irrigation departments based on their size, water spread areas, and the location of water bodies. New technical and institutional arrangements, entitlements, and multiple authority systems governed usufruct rights for different users and nonusers. MUWBs are being administered and controlled under different property rights regimes or levels of institutional hierarchy, namely, the *Panchayat* Raj Institution at Village, Janpad and district levels, State Department of Water Resource Development (SDWRD), irrigation/public works departments, soil and water conservation, wings of State Department of Agriculture, State Department of Fisheries (SDF), and private owners. These resources can be managed sustainably under state or common or private property regimes but may be also subject to degradation. There are many overlaps and combinations of state (public), community, and private management systems or governance structures in managing MUWBs. In other words, these resources are managed under different property regimes. Enough

[2] Broadly, wetlands include different kinds of water bodies: glacial lakes, peat bogs, shallow lakes, deep lakes, flood plains, marshes, swamps, oxbows, lagoons, reservoirs, tanks, temple tanks, fish ponds, village ponds, and paddy (paddy is synonymous with rice) fields (Brij Gopal 2015).

[3] According to the latest estimate of wetlands in India, the human-made inland wetlands cover about 37% area (3,841,832 ha), and the remaining 63% are natural wetlands (6,623,067 ha).There are also 4,140,116 ha of coastal wetlands (of which the intertidal mudflats of Kutch alone contribute about 51%) and 555,557 ha of wetlands smaller than 2.25 ha each. It is noteworthy that the paddy fields were included as wetlands in this inventory (Brij Gopal 2015).

evidence is available in India when MUWBs managed under common property regimes have been degraded into open-access resources due to weak property rights regimes, inadequate or poorly conceived institutional arrangements, and the breakdown of local authority system. Examples are also available of these resources being degraded under an open-access system brought under a state, private, or community management regime through changes in institutional arrangements (Marothia 2015).

Issues related to the management of MUWBs are complex due to the different categories and characteristics of these de facto common water bodies; the scale, size, and coverage of fisheries, agriculture, and domestic, sociocultural, and religious activities; and the multiple agencies involved in governing the water resource. In many cases, the completion of and conflicts over MUWBs have been observed at different levels, i.e., within users, across user groups, interinstitutional, and between stakeholders. From this example of Chhattisgarh, the extent of completion of and conflicts over MUWBs and mechanisms that to resolve conflicts can be understood. This study makes an effort to sketch path(s) of technical and institutional arrangements evolved over three decades to govern different categories of MUWBs administered under different property rights regimes in Chhattisgarh (some of the institutional interventions were inherited from undivided Madhya Pradesh: Chhattisgarh was part of Madhya Pradesh until Nov 1, 2000) (Marothia 1988, 1992a, b, 1993, 2006, 2007, 2010, 2012, 2014; Marothia et al. 2010). The study has documented workable institutional arrangements of decentralization, using a polycentric perspective, for minimizing water conflicts across users and the sustainable management of MUWBs.

1.2 Polycentric Approach

The institutional analysis framework of polycentricity developed by Ostrom and her co-authors (Kiser and Ostrom 1982; Oakerson 1986, 1992; Ostrom 1986, 1988, 1990a, b, 1992a, b, 1995, 1999a, b; Schlager and Ostrom 1992; Ostrom and Gardner 1993; Ostrom et al., 1994; Tang 1992; Townsend and Polley 1995) was used in the last 15–20 years to assess the performance of various CPR-based development programs, including wetlands, following a decentralized management approach. Institutional theories of polycentricism can effectively explain the relationships between multiple authorities with overlapping working zones. The emphasis on multilevel dynamics is extremely important in managing wetland ecosystems and designing wetland governance policies.[4]

[4]To analyze the process of decentralized governance and management of natural resources, researchers have developed an institutional framework of polycentricity that identifies key attributes of typical situations facing resource users, local communities, and decision-makers at different administrative hierarchies (Kiser and Ostrom 1982; Oakerson 1986,1992; Ostrom 1986, 1988, 1990a, b, 1992a, b, 1995, 1999a, b; Schlager and Ostrom 1992; Ostrom and Gardner 1993;

In general, among all the attributes of the institutional framework or derivatives of the framework for analyzing the institutional design of natural resources, property rights structures are the most important. Property rights regimes are part of the institutional arrangements through which resource users convert natural resources and environmental services into human-made capital or inputs of production (Folke and Berkes 1995). The nature of institutional arrangements defines the extent of property regime over land, water, and related resources. A property regime is a system or a set of institutional arrangements, or working rules of rights and duties, characterizing the mutual relationships of co-users with respect to a specific natural resource.[5] Property rights or resource management regimes can be classified under four categories: state property, private property, common property, and open-access system (Bromley 1989; Ostrom 1990a, b; Bromley (1992); Gibbs and Bromley (1989); Gibbs and Bromley 1989). These four categories of resource governance or property rights regimes have been extended by Townsend and Polley (1995) to recognize that governance can be shared between states, communities, and private interest groups in various ways at different decision-making levels. Distributed governance or polycentricity involves external institutional arrangements[6] between government and local communities or resource users, as well as internal institutional arrangements[7] within local community institutions or resource users.

Ostrom et al. 1994; Tang 1992; Townsend and Polly 1995; Marothia 2009a, b). The principal attributes of institutional analysis and development framework include various physical and technical attributes of a resource, characteristics of resource users and community, social and ecological context, and internal and external institutional arrangements that may affect the outcome directly or through collective action and patterns of interactions. The interrelationships among these variables can ensure efficient, equitable, and sustainable outcomes in managing a large group of renewable resources, including CPRs (plus wetlands, see Kiser and Ostrom 1982; Oakerson 1986,1992; Ostrom 1986, 1988, 1990a, b, 1992a, b, 1995, 1999a, b; Schlager and Ostrom 1992; Ostrom and Gardner 1993; Ostrom et al. 1994; Tang 1992; Townsend and Polly 1995; Marothia 2009a, b). "Polycentric is a broad type of a governance regime that possesses a number of specific institutional attributes capable of providing and producing essential collective goods and services to the citizens, in that regime. It seeks to unleash the ingenuity, and stimulate the creativity, of political entrepreneurs, structured so that actors within the system are given opportunities for institutional innovation and adaptation through experimentation and learning" (Ostrom et al. 1994). "A key aspect of all proposals for increased polycentricity (as opposed to just centralization or just decentralization) is the effort to enable institutions of multiple scales to more effectively blend local, indigenous knowledge with scientific knowledge" (quoted from Andersson and Ostrom 2008; see also Berkes and Folke 1998; McGinnis 1999, V. Ostrom 1991, 1997, 2007; V. Ostrom et al. 1961).

[5]See also classic work of Commons (1934) and North (1990) on role of institutions, property rights, and collective action.

[6]An external governance structure has essentially three alternatives of management systems (Townsend and Polley 1995), namely, *rights-based management* (the government grants usufruct rights to individual resource users under well-specified constraint conditions and assumes the role of monopoly over the resource base and retains all responsibility/authority for conservation decision), *co-management* (the government and the local communities share ongoing responsibility for decision-making over all or most of the resource management decisions), and *contracted management* (to transfer large part of the decision-making process to local bodies).

[7]The four alternative internal institutional arrangements that have been closely associated with the concept of distributed governance (Townsend and Polley 1995) include *self-organizing institutions*

Government, local communities, and private parties utilizing CPRs bring different interests, capabilities, and understanding to the resource management process (Marothia 2009a, b).

1.3 MUWB Status in Chhattisgarh

Estimation of areas under various wetland categories for the state of Chhattisgarh has been done by the Space Applications Centre, Indian Space Research Organisation (2010), using GIS layers of wetland boundary, water spread, aquatic vegetation, and turbidity (Fig. 1.1). A total of 7711 wetlands have been mapped at 1:50,000 scale in the state. In addition, 27,823 wetlands (smaller than 2.25 ha) have also been identified and delineated as point feature. Total area estimated is 337,966 ha, that is, around 2.5% of the geographic area concerned. The major wetland types are rivers/streams, accounting for about 53% of wetlands (179,088 ha), reservoirs (90,389 ha), and tanks and ponds (40,226 ha). The small wetlands (<2.5 ha) account for about 8.2%, assuming that each surface is equivalent to 1 ha. Out of the total wetland area, the extent of open water is 243,814 ha in post-monsoon season and 173,678 ha in pre-monsoon. There is a significant reduction in the extent of open water from post- to pre-monsoon. It is reflected in all categories of wetlands (Table 1.1). Turbidity is observed to be dominantly moderate in post-monsoon (183,025 ha) out of 243,814 ha of open water, followed by high (31,804 ha) and low turbidity (28,985 ha). Aquatic vegetation in Chhattisgarh accounts for about 0.6 and 5.8% of total wetland area in post-monsoon (2123 ha) and pre-monsoon (19,600 ha), respectively (see SAP, ISRO, Wetland Atlas Chhattisgarh, p. 19).

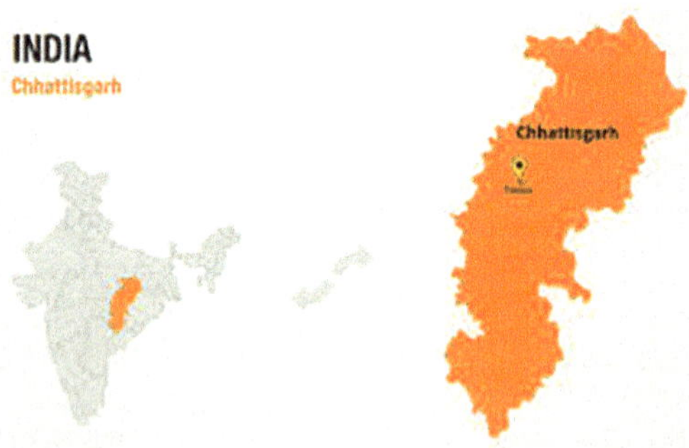

Fig. 1.1 Map of Chhattisgarh, India

(institutional and organizational decisions remain with local communities and the government may use the institutional building capacity to support and gain strength from self-organization), *communal management* (to reduce the existing authority of state and vest more localized interest), *cooperative management* (membership is limited with well-defined working rules for collective governance), and *corporate* (under the corporate governance, the owners and shareholders of the corporation would operate under governance rules typical of private corporations).

Table 1.1 Area estimates of wetlands in Chhattisgarh

Area in ha					
			% of	Open water	
Wetland category	Number of wetlands	Total wetland area	wetland area	Post-monsoon area	Pre-monsoon area
Inland wetlands – natural					
Lakes/ponds	–	–	–	–	–
Oxbow lakes/cutoff meanders	6	26	0.01	26	13
High-altitude wetlands	–	–	–	–	–
Riverine wetlands	8	174	0.05	83	76
Waterlogged	–	–	–	–	–
River/stream	156	179,088	52.99	124,712	93,095
Inland wetlands – man-made					
Reservoirs/barrages	604	90,389	26.74	85,148	54,012
Tanks/ponds	6906	40,226	11.90	33,671	26,366
Waterlogged	31	240	0.07	174	116
Salt pans	–	–	–	–	
Subtotal	7711	310,143	91.77	243,814	173,678
Wetland (<2.25 ha), mainly tanks	27,823	27,823	8.23	–	–
Total	35,534	337,966	100.00	243,814	173,678

Source: Indian Space Research Organisation, Space Applications Centre (2010). National Wetland Atlas: Chhattisgarh, Ahmedabad, Table 1.4, p. 19

MUWBs in the form of village ponds, irrigation, and multipurpose tanks are extensively distributed in all the villages of the state. MUWBs cover 52,211 village ponds and 1,616 irrigation tanks with 70,000 and 83,873 ha of water spread area, respectively, in the state. Of the total water spread area (153,873 ha) available in the state, 79 and 87% have been developed into 40,967 village ponds and 1,462 irrigation tanks, respectively, for fish culture (Table 1.2).

1.4 Database and Coverage

Data used in this study largely proceed from the author's earlier published and unpublished works, pertaining to multi-use village ponds and tanks in different parts of Chhattisgarh (between 1988 and 2015; see reference list). Management of the four categories of MUWBs, based on physical and technical attributes, ownership/tenure arrangements, multi-use, and multifunctional and multiple stakeholders, has been critically examined in this chapter. The first category deals with village irrigation tanks (VITs). In 1977–1978, VITs were constructed in Chhattisgarh (when it was part of Madhya Pradesh state) under a micro-minor irrigation tank

Table 1.2 MUWB availability in Chhattisgarh

Category of MUWBs	Available water area		Area under fish farming	
	Tank/reservoir (no.)	Water area (ha)	Tank/reservoir (no.)	Water area (ha)
Village ponds/tank				
Village panchayat	32,060	41,958	26,122	35,929
	(59.56)	(27.27)	(61.57)	(26.91)
Janpad panchayat	227	2394	141	1873
	(0.42)	(1.56)	(0.33)	(1.40)
Others	19,924	25,648	14,704	17,433
	(37.01)	(16.67)	(34.66)	(13.06)
Subtotal	52,211	70,000	40,967	55,235
	(97.00)	(45.49)	(96.55)	(41.36)
Irrigation reservoir				
Village panchayat	723	4232	645	3510
	(1.34)	(2.75)	(1.52)	(2.63)
Janpad panchayat	825	20,123	751	17,989
	(1.53)	(13.08)	(1.77)	(13.47)
District panchayat	34	20,719	32	18,002
	(0.06)	(13.47)	(0.08)	(13.48)
Fisheries Mahasangh	15	35,350	15	35,350
	(0.03)	(22.97)	(0.04)	(26.47)
Departmental	19	3449	19	3449
	(0.04)	(2.24)	(0.04)	(2.58)
Subtotal	1616	83,873	1462	78,300
	(3.00)	(54.51)	(3.45)	(58.64)
Grand total	53,827	153,873	42,429	133,535
	(100.00)	(100.00)	(100.00)	(100.00)

Note: Figures in parenthesis are percent to respective attributes
Source: Technical Report, State Department of Fisheries, Raipur, Chhattisgarh 2003–2004

(MMIT) scheme, or what is known as *Dabri Yojna*, through its soil conservation wing to provide life-saving irrigation facilities with provision for fish culture. The states had transferred these tanks to *panchayats* for management. For the purpose of this study, ten and eight perennial VITs and seasonal VITs were selected. Under the second category of MUWBs, seven canal-fed perennial and nine seasonal ponds and tanks (adequate water year-round in perennial and up to December–January in seasonal ponds and tanks) were chosen to understand water-sharing mechanisms between farmers, fishers, and villagers for household uses. Water user associations (WUAs), *Krishi Samittees*, fisheries cooperative societies (FCSs), *panchayat*, and SDWRs are collectively responsible for the allocation of water. The third categories of MUWBs have 11 seasonal/rainfed ponds and tanks, and primary users are fishers and village households. These are managed by village *panchayat*. In the fourth category, five privately owned rainfed ponds are also selected to compare their performance with ponds under the MUWB category. Water of these ponds is shared between the fisher and owner for family use (see Table 1.3).

Table 1.3 General attributes and institutional hierarchy in managing different categories of MUWBs

	Category of MUWBs	Classification based on seasonal/ perennial/ rainfed water bodies	Source of water	Availability of water	Major uses of water in order of priority	Local institutions involved in water uses	Administrative ownership and authority for water allocation	Leasing authority for fish culture	Lease rent Rs./ HWSA ($/ HWSA)	Restriction/ conditions to use water for fish culture
A.	Village ponds under MMIT scheme	Seasonal (5 ha)	Rainfed 40 ha*	Up to December	Domestic use, fish culture, irrigation,	Village *panchayat*, FGs/fish contractor	Village *panchayat*	Village *panchayat*	2600	Limited restrictions
		Perennial (4.90 ha.)	Rain/ canal-fed 48 ha*	Up to Feb–March	Irrigation, fish culture, domestic use	Village *panchayat* FGs/fish contractor	*Panchayat*	Village *panchayat*	857.00	Restriction on feed and manure
B.	Irrigation tanks	Perennial irrigation tank (101.05 ha.)	Rain/ canal-fed 836.50 ha*	Adequate, year round	Irrigation, domestic use, fish culture	FCS, WUAs, *panchayat*	SDWRD	Janpad *panchayat*	240.00	Feed and manure can't be used by FCS, other users can't be excluded to use water
		Seasonal (28.52 ha.)	Rain/ canal-fed 161.10 ha*	Up to Feb–March	Irrigation, fisheries, and domestic use	*Panchayat*, *Krishi Samittee*, FCS	SDWRD	Janpad *panchayat*	243.00	Feed and manure can't be used by FCS, other users can't be excluded to use water

(continued)

Table 1.3 (continued)

	Category of MUWBs	Classification based on seasonal/ perennial/ rainfed water bodies	Source of water	Availability of water	Major uses of water in order of priority	Local institutions involved in water uses	Administrative ownership and authority for water allocation	Leasing authority for fish culture	Lease rent Rs./ HWSA ($/ HWSA)	Restriction/ conditions to use water for fish culture
C	Panchayat-owned village ponds for fish culture	Seasonal (1.256 ha)	Rainfed	Up to December	Domestic use, fish culture	SHG,FG, and village *panchayat*	Village *panchayat*	Village *panchayat*	2229.00	No restriction
D	Privately owned ponds for fish culture	Perennial (1.60 ha.)	Rainfed	Adequate water for fish culture up to February	Fish production	Private	Private	Private (lease out to fisherman)	5625.00	No restriction

Note: Figures in brackets indicate water spread area of respective tanks. Irrigation charges for Kharif and Rabi crops are Rs. 230 and Rs. 600, respectively. HWSA = water spread area in hectare. * Command Area. 1US $ = Rs. 62 at 2014 exchange rate. $ rounded up to the nearest whole number
Source: Based on author's work listed in references of this paper

1.5 Institutional Arrangements for Water Sharing

Regarding water-sharing mechanisms in MUWBs, based on the author's earlier studies, this synthesis provides some meaningful lessons. Competition and conflicts over pond/tank water have been observed at different levels, i.e., within user groups (irrigators), across user groups (irrigators, fishermen, and villagers), interinstitutional (FCSs, WUAs, *panchayat*, the irrigation department, the fisheries department), and between stakeholders and institutions (irrigators vs SDWRD, fishers vs WUA, or *panchayat*). Outcomes of major categories, mentioned above, are briefly summarized in Table 1.4.

Table 1.4 Institutional interventions designed/initiated/strengthened for water sharing of MUWBs

Categories of MUWBs	Potential competition within users and across users over water sharing	Potential interinstitutional conflicts and conflicts between multi-users	Institutional mechanism to resolve completion and conflicts
Village ponds under MMIT scheme	Conflict within farmers located at head, middle, and tail end between farmers and fisher group or fish contractor	*Panchayat*-farmers-fisher group/contractor	*Panchayat* designed institutional arrangements for irrigation scheduling for farmers and ensured minimum water availability for fish culture and domestic uses. Irrigation fees and fish leasing funds were used to maintain MUWBs.
Irrigation tanks	Between farmers and fishers	WUA/SDWRD-FCS-*panchayat*-*Krishi Samittees*	Due to the protective nature of irrigation systems and field-to-field methods of irrigation, there were some conflicts observed between head-, middle-, and tail-end farmers during low rainfall years. Since SDWRD and panchayat ensure minimum level of water required for fish culture, there is no conflict between the fisherman community (FCS) and farmers. In case of decline of water level in the tank, which may affect fish culture, FCS collectively forces village and Janpad panchayat and SDWRD to release additional water. With the release of additional water, fish growth period continues up to June. The FCS cannot use fish feed and manure in the tank; the *panchayat* resolves conflicts, if any, between domestic users of tank water and FCS. The *panchayat* ensures minimum level of water required for fish culture and coordinates between *Krishi Samittee* (responsible for maintaining irrigation water use) and FCS. With lease money, the *panchayat* repairs bunds of the ponds to stop the outflow of fish. In case of water stress condition in rainfed ponds, FCS transfer fishes to perennial ponds interinstitutional.

(continued)

Table 1.4 (continued)

Categories of MUWBs	Potential competition within users and across users over water sharing	Potential interinstitutional conflicts and conflicts between multi-users	Institutional mechanism to resolve completion and conflicts
Panchayat-owned village ponds for fish culture	Between village households using water and fisher groups	Panchayat and fisher group	These MUWBs are leased largely by SHGs, and since there is no restriction on the application of fish feed and manure, sometimes water users for domestic uses have serious conflict with SHGs. Panchayats now keep a few ponds exclusively for domestic use to minimize conflicts.
Privately owned ponds for fish culture	No conflict between owner and fisher	Well-defined leasing arrangement	Well-designed institutional arrangements for use of water for fish rearing and use of water by owner.

Source: Based on author's work listed in references of this paper

The MMIT program was introduced with the assumption that contributions of farmers in different types of resources, labor, materials, and capital will be forthcoming once it is determined that the MMITs are useful to farmers and *panchayats*. This assumption has largely been proved wrong. Due to lack of legislative and administrative powers, *panchayats* could not manage the MMITs under a common property regime. MMITs had extremely poor excludability, i.e., it was difficult to exclude noncontributors in labor or capital resources from taking advantage of the water use in the absence of a well-defined structure of rights and duties for users. The *panchayats* could not evolve rules for the use of water and collection of fees and enforce their authority. As a result, the common property MMITs ultimately degenerated into open access. After the formation of Chhattisgarh, the state government has empowered the *panchayats* to be legitimate local authority, and MMITs are now managed using well-defined institutional mechanisms for water sharing between/among different users.

Perennial canal-fed MUWBs are primarily for irrigation. Due to adequate water availability throughout the year, conflicts between fishers and irrigators have not been observed. However, conflicts between head, middle, and tail-end farmers due to field-to-field and protective irrigation systems (irrigation is provided at critical crop growth stages) are quite common. Occasionally, conflicts also have surfaced between SDWRD and farmers/WUA. For fish culture, minimum water level is ensured by SDWRD. Even in canal-fed seasonal MUWBs, irrigation is invariably the first priority. In case of inadequate water supply during critical crop growth stages in *kharif* season (synonymous with the wet season/summer season) (in some cases one irrigation is provided in *rabi* season: *rabi* is dry/winter season), farmers put collective pressure on *panchayats* through *Krishi Samittees* (formal local body of cultivators recognized by *panchayats* that regulates water distribution and availability of water) to release water. If fish growth is affected due to lower level of water, the amount of lease gets an exemption from the *panchayat*. FCS gets desirable support from SDF to settle the issue. In case of water stress condition, FCSs transfer fish from rainfed/seasonal to perennial ponds/tanks. The conflict is more common in rainfed/perennial ponds that are located in/near villages and caters domestic, fish culture, and irrigation needs. Therefore, FCSs prefer to lease in ponds located 2–3 km away from the main village periphery, and water is basically shared between irrigators and fisherman communities.

It was observed in the study area that productivity levels, investment, and net benefits per ha of fisheries under MUWBs were much lower than private ponds. Two reasons are evident. First, while private ponds are used solely for fish production, MUWBs are not. Different users cannot exclude each other, and this factor affects the viability of MUWBs, particularly seasonal ponds/tanks for fish production, which declines as other uses ascend. Second, villagers often complain about health problems associated with feed and manure used for intensive fish culture. As a result, fishermen are constrained from applying profit-maximizing levels of inputs, even in individually leased ponds owned by the *panchayat*, particularly in the dry season when water for cattle and human population is most needed. These constraints have created inequalities of income, yield, and employment distribution in the management of MUWBs. Owners of private ponds also sometimes impose certain restrictions on input use that reduce productivity.

It is essential to note that in almost every village of Chhattisgarh, MUWBs have traditionally been allotted by local *panchayats* for different uses, such as for tending cattle, washing clothes and baths, irrigation, fish culture, and social rituals (funeral, worship, etc.). These water bodies were managed through collective work under a common property regime. There is invariably one or two temple(s) or sacred ponds/tanks in most of the villages. In almost every village, ponds/tanks were separately allotted for women and social groups. As an exception to the traditional property rights arrangements, a few ponds may be left out for domestic uses and social rituals in a village, and the rest can be exclusively used for culture fisheries, without any restrictions on the application of growth-promoting inputs to achieve potential yield. Fortunately, in several villages, a good number of community ponds are still existing. Close-in ponds can represent a reserve for common use and more distant ponds for fish culture. Such simple reorientation in property rights arrangements can substantially reduce intercommunity conflicts. Similarly, irrigation ponds/tanks can be exclusively used for irrigation and fish culture, if some of them can be kept aside for catering the needs of villagers. Further, a feasible solution can be worked out between fishers, irrigation groups, and SDWRD for desilting the tanks, as one of major concerns of SDWRD is the increasing silt load and reducing water intake capacity due to use of fish feed and manure. Such institutional arrangements can increase fish yield and the total productivity of MUWBs by many folds besides minimizing social conflicts. It has been observed in some parts of the state that FCSs have a tendency to lease in all the common ponds/irrigation tanks within an 8 km periphery (a norm prescribed in leasing policies of fisheries) in a particular village to strategically eliminate fisherman groups, thereby increasing individual fishers' chance to lease in these common water resources. Further, in some cases, all the leased out ponds/tanks are not used for fish culture. Similarly, in many cases, the lessee has to be a member of the fisherman community to lease in *panchayat*-owned ponds. This puts restrictions on other poor members of the village.

1.6 Ways Forward

MUWBs are multipurpose in nature and have multi-use, with technical, socioeconomic, cultural, political, and environmental interdependencies. Their management is difficult because it increases the level of conflicts among a range of stakeholders and can create disproportionate spatial and temporal externalities. Issues regarding the role of traditional authorities, user groups, stakeholder committees, different departments of the state, and *panchayat* and the relationships between multiple authorities with overlapping working zones are highly complex. Given such a complexity, we advocate for pluralistic and polycentric or distributed governance approaches to manage MUWBs. We also suggest that the policies of different departments of the state and *panchayat* need to be examined critically in view of the fact that many departments are loosely linked and intensely compete for MUWBs. It is also equally important to learn, from the traditional property rights arrangements/soft institutional arrangements, how to substantially reduce the intercommunity and multi-department conflicts. A synthesis of traditional institutional mechanisms and the components of current policies can appropriately be made for the distributed or polycentric governance of MUWBs.

Recognizing the importance of common pool resources (CPRs) in general and water bodies in particular, the Supreme Court of India passed on January 28, 2011, a landmark judgment on the restoration of these resources. The Supreme Court noted in its ruling: "*Our ancestors were not fools. They knew that in certain years there may be droughts or water shortages and water was also required for cattle to drink and bathe. Hence they built a pond attached to every village, a tank to every temple.*" The apex court directed all state governments to prepare schemes for the eviction of those occupying water bodies and other village commons and restore them to the community. In the state of Chhattisgarh, an overarching policy direction on commons, exemplified by MUWBs; recognizing the various social-cultural, economic, and ecological functions of CPRs; clarifying rights of access to and the use and management of these resources; and devolving and decentralizing the governance of CPRs would be crucial to the ecological security and total welfare of the society.

Natural resources in general and wetlands in particular have political and economic dimensions that may have a strong bearing on governance policies. Further, governance of MUWBs is complicated because it increases the level of conflicts among a range of stakeholders and can create disproportionate spatial and temporal externalities. Issues regarding the role of traditional authorities, user groups, stakeholder committees, NGOs, state and national governments, private organizations, and international donors and the relationships between multiple authorities with overlapping working zones are highly complex. Given such a complexity, researchers and policy makers would require suitable analytical frameworks to evaluate the effectiveness of wetland governance. It is here that some of the emerging frameworks advocating pluralistic and polycentric or distributed approaches to management of natural resources, including wetlands, have shown considerable potential from a policy angle. However, it is important to examine the layers or components of the institutional framework that is essential to put into operation the concept of distributed or polycentric governance, particularly where many institutions are loosely linked and intensely compete for the same space. Can the wetland conservation and development programs initiated by the state graduate into polycentric governance? If so, what are the alterations needed and additional conditions to be ensured? More research is needed on these newer frameworks before they can be validated for policy applications (restructured from Marothia 2009a, b in the context of wetland governance).

References

Andersson, K., & Ostrom, E. (2008). Analyzing decentralized natural resource governance from a polycentric perspective. *Policy Sciences, 41*(1), 1–23.

Berkes, F., & Folke, C. (1998). *Linking social and ecological systems: Management practices for social mechanisms for building resilience*. Cambridge: Cambridge University Press.

Bromley, D. W. (1989). Property relations and economic development: The other land reform. *World Development, 17*, 867–877.

Bromley, D. W. (Ed.). (1992). *Making the commons work*. San Francisco: ICS Press.

Commons, J. R. (1934). *Institutional economics*. New York: Macmillan Company.

Folke, C., & Berkes, F. (1995). Mechanism that link property rights to ecological system. In S. Hanna & M. Munasinghe (Eds.), *(1995) Property rights and the environment – Social and ecological issues* (pp. 121–138). Washington, D.C.: The Beijer International Institute of Ecological Economics and World Bank.

Gibbs, C. J. N., & Bromley, D. W. (1989). Institutional arrangements for management of rural resources: Common property regimes. In F. Berkes (Ed.), *Common property resource: Ecology and community-based sustainable development* (pp. 22–32). London: Belhaven Press.

Gopal, B. (2015, September). *Wetland conservation for biodiversity and ecosystem services :Needs a shift in land and water resources policies*. Policy Brief, National Institute of Ecology, p. 16.

Indian Space Research Organization, Space Applications Centre. (2010). National Wetland Atlas: Chhattisgarh. 138 pp.

Kiser, L. L., & Ostrom, E. (1982). The three worlds of action, a met theoretical synthesis of institutional approaches. In E. Ostrom (Ed.), *Strategies of political inquiry* (pp. 179–222). Beverly Hills: Sage.

Marothia, D. K. (1988). Tank irrigation in Chhattisgarh region: Management and policy issues. *Indian Journal of Agricultural Economics, 43*(3), 305–309.

Marothia, D. K. (1992a, September 18–20). *Village ponds and aquaculture development: Issues of inequality in common property regimes*. Contributory paper: Third conference of the International Association for the Study of Common Property, Washington, DC, USA.

Marothia, D. K. (1992b). Village irrigation tanks: Institutional design for sustainable resource use. *Agriculture Situation in India, 47*, 479–484.

Marothia, D. K. (1993). Property regimes and institutional arrangements: Concepts and their relevance in managing the village commons. *Indian Journal of Agricultural Economics, 48*(3), 557–565.

Marothia, D. K. (1995, July 4–7).*Village ponds and aquaculture development: Is conflict inevitable?* Paper presented at the National workshop on poultry, fisheries and food processing organized for the Members of Parliament at the National Academy of Agricultural Research Management, Hyderabad.

Marothia, D. K. (1997a). Agricultural technology and environmental quality : An institutional perspective – Keynote paper. *Indian Journal of Agricultural Economics, 52*(3), 473–487.

Marothia, D. K. (1997b). Property rights, externalities and pollution. In A. Agrawal (Ed.), *The challenge of balance*. New Delhi: Center for Science and Environment.

Marothia, D. K. (2001). Valuation of a day use recreation site: Application of alternative estimation techniques. *Indian Journal of Agricultural Economics, 56*(3), 313–324.

Marothia, D. K. (2004a). Institutional structures for governance of freshwater aquaculture: Lessons from Chhattisgarh. *Indian Journal of Agricultural Economics, 59*(3), 421–434.

Marothia, D. K. (2004b, January). *Minor irrigation water storage tanks-a case study of Bastar District*, Chhattisgarh, research report IWMI-TATA Water Policy Programme, Anand, Gujarat, India, p. 44.

Marothia, D. K. (2004c). Restoration of lake ecosystems: An environmental economics perspective-keynote paper. *International Journal of Ecology and Environmental Sciences, 30*(4), 1–11.

Marothia, D. K. (2004d). Transforming institutions for sustainable development of natural resources programmes. *Agricultural Economics Research Review, 17*, 163–190.

Marothia, D. K. (2006). Managing culture fisheries in multi-use water bodies: A property rights paradigm. In K. K. Vass, K. Mitra, V. R. Suresh, P. K. Katiha, & N. P. Shrivastava (Eds.), *River fisheries in India: Issues and current status* (pp. 227–240). Barrackpore: Inland Fisheries Society of India.

Marothia, D. K. (2007). Fish farming in small multi-use water bodies: Water sharing mechanism and its implications. In K. K. Vass, N. Sarangi, K. Mitra, J. K. Jena, V. R. Suresh, N. P. Shrivastava, & P. K. Katiha (Eds.), *Water management in fisheries and aquaculture* (pp. 151–159). Bhubaneswar: Inland Fisheries Society of India, Barrackpore, W.B. and Association of Aquaculturists.

Marothia, D. K. (2009a). Decentralization of natural resource management in India: An institutional perspective. Presidential address published in Indian. *Journal of Agricultural Economics, 65*(1), 1–34.

Marothia, D. K. (2009b, February 4–7). *Governance and institutional change in traditional commons: Lessons from Chhattisgarh, India.* Keynote paper presented in **4th** World Congress on Conservation Agriculture, New Delhi. New Delhi: Published in the Proceeding of 4thWCCA. ICAR.

Marothia, D. K. (2010). Technological and institutional options for common property resource management in rainfed areas. *International Journal of Ecology and Environmental Sciences, 36*(I), 45–57.

Marothia, D. K. (2012). Performance of culture fisheries under alternative property rights regimes in Chhattisgarh. *International Journal of Ecology and Environmental Sciences, 38*(4), 163–207.

Marothia, D.K. (2014, August 27). *Urban and peri-urban wetlands of Chhattisgarh: The dwindling ecological heritage*. Keynote Address delivered at the joint workshop organized by INTACH and Pt RSU on traditional water harvesting system and its importance in urban perspective.

Marothia, D. K. (2015, Feb 3-6). *Improving water sharing through fostering institutional creativity: Lessons from multi- use water bodies management*. Invited paper presented in Policy Challenges for Protecting Smallholders Session, 12th Agricultural Science Congress 2015. Karnal: NDRI.

Marothia, D.K., Pandy S. and Choudhry, V.K. (2010). Irrigation tanks in Chhattisgarh: Traditional technology for sustaining rainfed agriculture. In: Singh,S and Rathore, M.S. (Eds.).Rainfed agriculture in India: Perspective and challenges, Rawat Publications, Jaipur, India.

McGinnis, M. (Ed.) (1999). *Polycentricity and local public economies: Readings from the Workshop in political theory and policy analysis*. Ann Arbor: University of Michigan Press) (Ramsar Convention Secretariat 2013).

North, D. C. (1990). *Institutions, institutional change and economic performance*. Cambridge: Cambridge University Press.
Oakerson, R. J. (1986). *A model for the analysis of common property problems*. In Proceedings of the Conference on Common Property Resource Management, National Academy of Sciences, Washington, DC, USA.
Oakerson, R. J. (1992). Analyzing the commons: A framework. In D. W. Bromley (Ed.), *Making the commons work* (pp. 41–62). San Francisco: ICS Press.
Ostrom, E. (1986). An agenda for the study of institutions. *Public Choice, 48*, 3–25.
Ostrom, E. (1988). Institutional arrangements and the commons dilemma. In V. Ostrom, D. Feeny, & H. Picht (Eds.), *Rethinking institutional analysis and development* (pp. 101–139). San Francisco: ICS Press.
Ostrom, E. (1990a). *Governing the commons: The evolution of institutions for collective action*. New York: Cambridge University Press.
Ostrom, E. (1990b). *Governing the commons: The evolution of institutions for collective action*. New York: Cambridge University Press.
Ostrom, V. (1991). *The meaning of American federalism: Constituting a self-governing society*. San Francisco: ICS Press.
Ostrom, E. (1992a). *Crafting institutions for self-governing irrigation systems*. San Francisco: ICS Press.
Ostrom, E. (1992b). The rudiments of a theory of the origins, survival, and performance of common property institutions. In D. W. Bromley (Ed.), *Making the commons work* (pp. 293–318). San Francisco: ICS Press.
Ostrom, E. (1995). Designing complexity to govern complexity. In S. Hanna & M. Munasinghe (Eds.), *Property rights and the environment: Social and ecological issues* (pp. 33–45). Washington, DC: Beijer International Institute of Ecological Economics and the World Bank.
Ostrom, V. (1997). *The meaning of democracy and the vulnerability of democracies: A response to Tocqueville's challenge*. Ann Arbor: University of Michigan Press.
Ostrom, E. (1999a). Coping with tragedies of the commons. *Annual Review of Political Science, 2*, 493–535.
Ostrom, E. (1999b). Institutional rational choice: An assessment of the IAD framework. In P. Sabatier (Ed.), *Theories of policy process*. Boulder: Westview Press.
Ostrom, V. (2007). *The political theory of a compound republic: Designing the American experiment*, 3rd ed. Lanham: Lexington Books. First published in 1987.
Ostrom, E., & Gardner, R. (1993). Coping with asymmetries in the commons: Self-governing irrigation systems can work. *Journal of Economic Literature, 7*(4), 93–112.
Ostrom, V., Tiebout, C., & Warren, R. (1961). The organization of government in metropolitan areas. *American Political Science Review, 55*, 831–842.
Ostrom, E., Gardner, R., & Walker, J. (1994). *Rules, games, and common pool resources*. Ann Arbor: University of Michigan Press.
Pearce, D.W., & Turner, R.K. (1990). *Economics of natural resources and the environment*. The Johns Hopkins University Press, Baltimore 21, pp 320-335.
Ramsar Convention Secretariat. (2013). *The Ramsar convention manual: A guide to the convention on wetlands (Ramsar, Iran, 1971)* (6th ed.). Gland: Ramsar Convention Secretariat.
Schlager, E., & Ostrom, E. (1992). Property rights regimes and natural resources: A conceptual analysis. *Land Economics, 68*(3), 249–262.
Tang, S. Y. (1992). *Institutions and collective action: Self-governance in irrigation*. San Francisco: ICS Press.
Townsend, R. E., & Polley, S. G. (1995). Distributed governance in fisheries. In S. Hanna & M. Munasinghe (Eds.), *Property rights and the environment – Social and ecological issues* (pp. 33–40). Washington, DC: The Beijer International Institute of Ecological Economics and the World Bank.

Chapter 2
Sustainability of Groundwater Use in Punjab Agriculture: Issues and Options

Kamal Vatta

Abstract Punjab agriculture is a fitting example of intensive cultivation with the dominance of rice-wheat monoculture. The state of Punjab has one of the highest levels of cropping intensity, assured irrigation network and exhaustive input use. Intensive agriculture has led to the widening gap between the demand for groundwater and sustainable levels of its supply over time. The present paper outlines the trends in depletion of groundwater in Punjab, identifying the major factors governing the depletion and giving broader solutions for long-term sustainability. The groundwater table in central Punjab has fallen annually by about 17 cm and 25 cm during the 1980s and 1990s, respectively. The decline has been the sharpest during the last about one and a half decades, averaging at 91 cm per annum during 2000–2005 and about 75 cm during the last decade. The decline in average annual rainfall from 754.6 mm in 1990 to 365 mm in 2012 had an adverse impact. There are many drivers of the demand-supply mismatch for groundwater use in Punjab. The major reason for increase in groundwater over-exploitation is a sharp increase in the area under rice, as well as the cropping intensity. Another important reason for demand-supply mismatch is the provision of free or highly subsidized supply of power for agriculture. Even the recent climate variability, which is translating into an increased frequency of droughts in Punjab, is not helping the state favourably in terms of groundwater sustainability. The depletion of groundwater resources has adverse implications for crop profitability, especially for small and marginal farmers. It has also increased the burden of electricity subsidies and affects the output of other sectors of the economy. Promoting crop diversification and large-scale use of low-cost conservation technologies can help in achieving long-term groundwater sustainability in Punjab agriculture.

Keywords Intensive cultivation · Monoculture · Cropping intensity · Irrigation · Groundwater depletion · Crop diversification

---

K. Vatta (✉)
Centers for International Projects Trust, New Delhi, India
e-mail: kamal@cipt.in; kmlvatta@yahoo.com

R. Peshin, A. K. Dhawan (eds.), *Natural Resource Management: Ecological Perspectives*, Sustainability in Plant and Crop Protection,
https://doi.org/10.1007/978-3-319-99768-1_2

2.1 The Groundwater Situation in Punjab

Punjab agriculture is a fitting example of intensive cultivation with the dominance of rice-wheat monoculture, with almost the entire area under assured irrigation and with one of the highest levels of cropping intensity and input use in India. The NPK use in Punjab is 243 kg/ha of net area sown, cropping intensity is 189% and area under assured irrigation is 98%, which are much above the national averages (135 kg/ha, 138% and 45%, respectively) (PAU 2015). Almost three-fourths of the cultivated area is irrigated by using groundwater resources. Assured procurement, high productivity and remunerative pricing of rice and wheat crops favoured the dominance of rice-wheat cultivation. The combined area under these crops was less than 40% during 1966–1967, which has touched almost 80% in recent times (Vatta et al. 2013).

The water-intensive nature of rice cultivation and the fast pace of urbanization and industrialization led to increasing dependence on groundwater resources, which has widened the gap between the demand for groundwater and sustainable levels of its supply over time. This shortage is visible from the net availability of groundwater in Punjab declining from 0.30 million ha m in 1984 to 0.03 million ha m in 1999, turning negative from 0.99 million ha m in 2004 to 1.48 million ha m in 2011 (Fig. 2.1).

Such a negative balance between demand and supply of water after 1999 implies over-exploitation of groundwater resources and hence a fall in the groundwater table. Furthermore, this decline peaked during the last decade (Fig. 2.2). However, it increased to 26.7% in 1990 and reached 91.6% in 2010. The area under water table depth of more than 15 m was only about 0.6% in 1973, remained still low at 2.9% in 1990 and finally jumped to 75.1% in 2010.

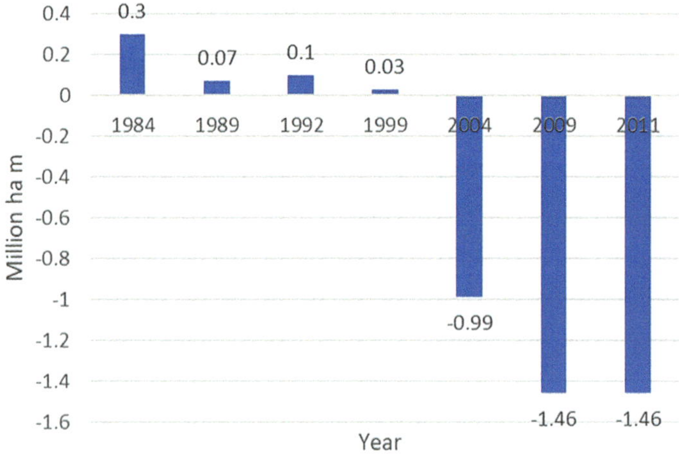

Fig. 2.1 Net groundwater availability in Punjab, million ha m during 1984–2011. (Source: Central Ground Water Board, Punjab)

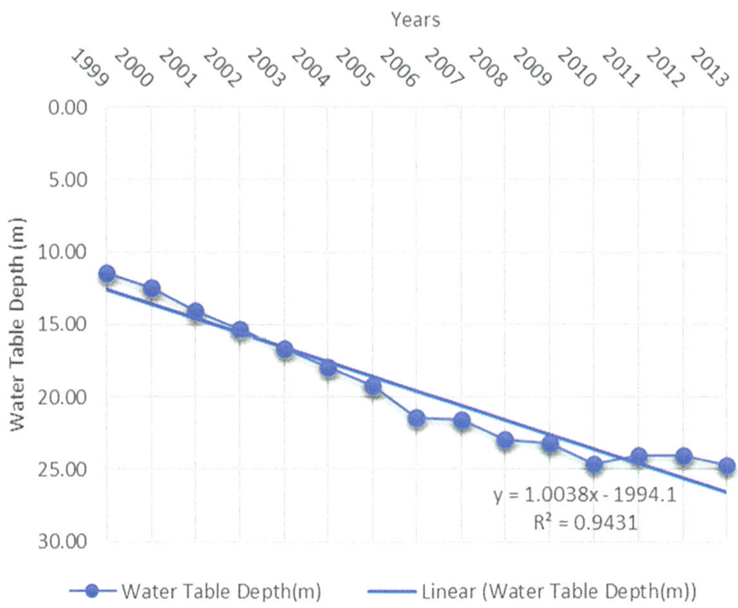

Fig. 2.2 Fall in groundwater table in central Punjab, 1999–2013. (Source: Central Ground Water Board, Punjab)

The overdependence on groundwater has been cited as a primary driver for groundwater depletion along with subsidized power and relatively higher profitability of water-intensive crops such as rice (Pandey 2016). Therefore, it is essential to examine the major causes of groundwater depletion and the related adverse impact of groundwater depletion on the agricultural economy of Punjab.

The availability of water for agriculture varied between 2.7 million ha m and 4.7 million ha m (it was exceptionally high at 5.96 million ha m during 1988, which was the year of flood). For about 6 years, during 1997–2012, the availability of water remained around three million ha m or less, indicating the rising incidences of drought in Punjab. Thus, owing to these drought conditions coupled with water-intensive rice-wheat cropping system resulted in the demand for water in agriculture, which started surpassing the supply after 1977 (year 1988 was an exception owing to the floods in Punjab). The gap is widening sharply, especially after the beginning of the 1990s. This has serious implications for long-term sustainability of water resources in the state (Fig. 2.3).

There has been a significant change in the components of demand and supply of irrigation water in Punjab. Rainfall, surface water and groundwater are the three major components of supply of irrigation water. The average annual rainfall for the state is 435.6 mm (Envis Centre, Punjab 2013). Punjab receives most of the rainfall (75%) from south-west monsoon, in the months of June to October, but it is highly variable in time and space. While the average rainfall in the state increased from around 600 mm to 700 mm during 1970–1997, it has considerably declined to

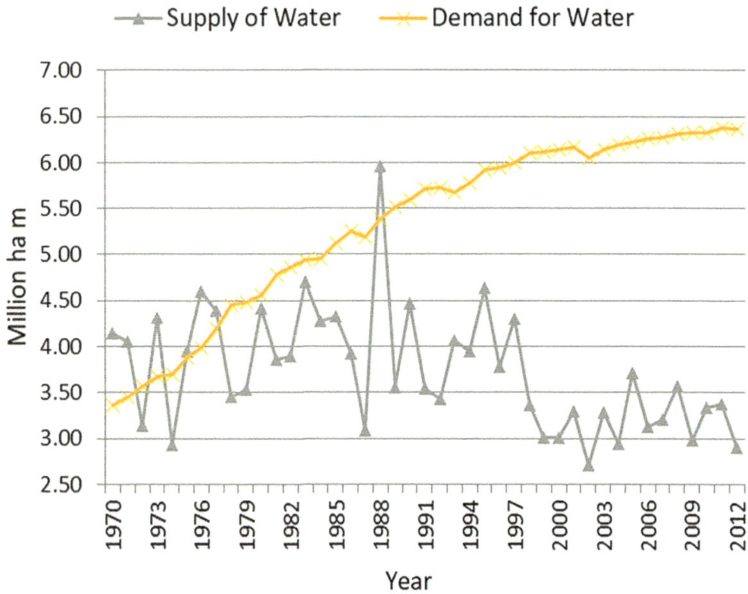

Fig. 2.3 Agricultural water supply and demand in Punjab. (Source: Author's calculations based on the rainfall and surface water data)

around 400 mm after 1997, which has repercussions on the availability of water resources in Punjab (Fig. 2.4).

Further to rainfall, there have been changes in the pattern of irrigation through canals and tube wells. Owing to the problem of salinity of groundwater, the area irrigated by canals in Punjab also declined sharply by about 33% since 1995–1996. It is worth noting that the current area irrigated by canals in most of the districts is substantially lower than the maximum canal area irrigated in these districts and significantly lower than the initial levels of irrigation by canals (TE 1974–75) during 1970–1971 to 2012–2013. In most of the districts, maximum area under canal irrigation was achieved during the mid-1990s (Fig. 2.5).

The cropping pattern in Punjab has seen a significant change since the 1970s. Owing to assured irrigation, increased use of modern inputs such as fertilizers and pesticides, higher productivity and effective public procurement at remunerative prices, the rice crop started dominating the cropping pattern after successful introduction of high-yielding varieties, since the early 1970s. Besides this, during a span of 40 years, during 1970–2010, cropping intensity increased from about 140% to 190% adding further to the water use in agriculture.

The decomposition of changes in water demand in the agriculture sector in Punjab into the above two components has been carried out in two steps. To estimate the impact of cropping intensity on demand for water in agriculture, the cropping pattern of the triennium ending in 1972–1973 was adjusted with the total cropped area of the triennium ending in 2012–2013. As the cropping pattern was

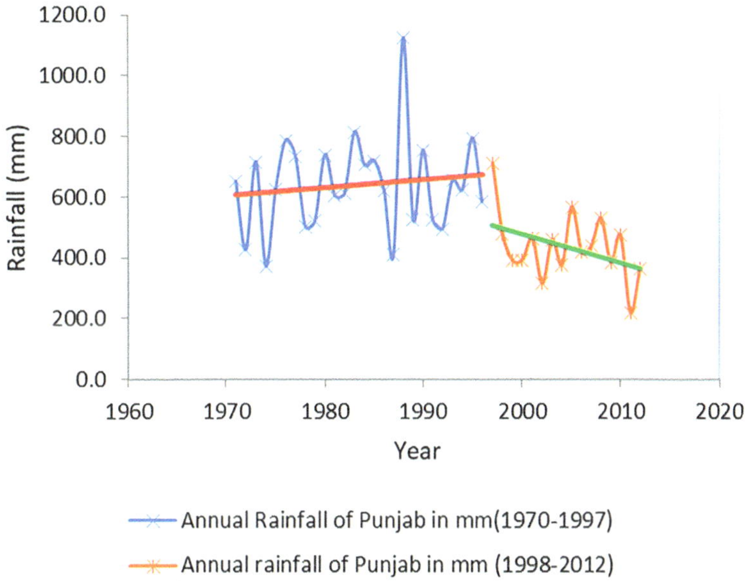

Fig. 2.4 Temporal variation in average annual rainfall of Punjab. (Source: Statistical Abstract of Punjab)

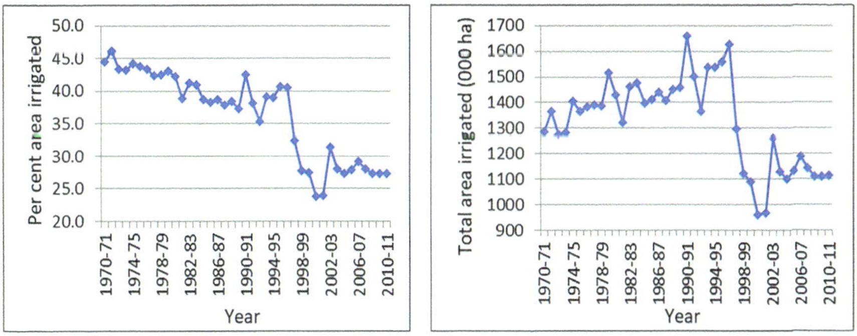

Fig. 2.5 Canal irrigation trends in Punjab (per cent area and total area irrigated by canals). (Source: Statistical Abstracts of Punjab)

adjusted to be the same as that during TE 1972–1973, the rise in demand was assumed to be due to the rise in cropping intensity. In the second step, the difference in water demand for current and adjusted cropping patterns was worked out, reflecting the increase in water demand due to changes in the cropping pattern. Total water use in agriculture during TE 1972–1973 to TE 2012–2013 increased by around 84%. While the increase in cropping intensity accounted for 45% rise in water use, changes in cropping pattern accounted for 39% rise in water use, during this period.

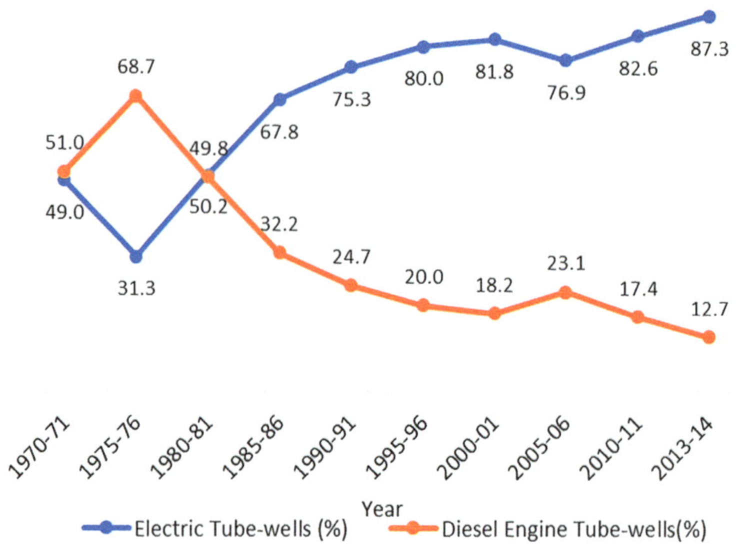

Fig. 2.6 Changing proportions of electric and diesel tube wells

Rising farm incomes due to increased productivity of major food crops encouraged the farmers to make private investments on their farms. The government policy of village electrification also favoured such investments. Figure 2.6 depicts that the proportion of electricity operated tube wells, which was just about 47% in 1970–1971, jumped to the level of almost 83% in 2010–2011. The overdependence on groundwater sources for irrigation in agriculture is leading to the over-exploitation of the groundwater resource and poses serious challenge for long-term sustainability of such resource.

Another important reason for demand-supply mismatch is the provision of free or highly subsidized supply of power for agriculture. The electricity was made free for agriculture since 1992, and the average cost recovery is much less. There is evidence that the water-use efficiency declined to minimal levels in Punjab due to the provision of free electricity. There is no need to keep the rice fields flooded after 15 days of initial flooding and further irrigations after longer durations will not have any adverse impact on productivity. However, despite the recommendation of the Punjab Agricultural University (PAU), most of the farmers do not follow this recommendation and keep the water flooded even for more than a month. Such profligate use of water happens at the cost of increased use of electricity, which is available free of cost and adds to the subsidy burden of the state government.

2.2 Implications of Groundwater Depletion

The over-exploitation of groundwater resources in Punjab had multiple serious implications which varied from an increased consumption of power and power subsidies in agriculture to increased investments and indebtedness, adverse impacts on smallholders, quality of water, etc.

1. Often the corollary of the faster depletion of groundwater resources in Punjab compelled farmers to invest frequently on deepening for installation of bore wells and on buying centrifugal motors of larger horsepower. The expenses on deepening of the borewells, motors of larger horsepower and subsequent shifts to the submersible pumps contributed significantly towards rising indebtedness amongst the Punjab farmers (Sidhu 2002; Singh et al. 2008a, b).
2. Mostly small (<2 ha landholding) and marginal (<1 ha landholding) farmers, who cultivate less than 5 acres (2 ha) of land, had to bear the brunt of groundwater depletion. Majority of small and marginal farmers have been found to share the electric motors with their relatives or neighbouring farmers, which has some negative impact on their access to groundwater resources (Kaur and Vatta 2015).
3. There has been a tremendous increase in the power consumption that increased six times between 1974–1975 and 1990–1991, due to land intensification. The consumption has almost doubled since the 1990s, which is mainly due to the depletion in groundwater table.
4. In addition to the increased power consumption, the provision of highly subsidized or almost free supply of electricity to the agriculture sector for a long time resulted in a substantial increase in electricity subsidy, which has grown significantly over time. While the consumption of electricity has almost doubled since the 1990s, the electricity subsidy has become almost nine times. Such a huge increase in power subsidies seems highly unsustainable, and it is believed to adversely affect the productive investments in agriculture by the government (Kaur and Vatta 2015).

The situation worsens during drought years, when the stress on groundwater resources increases due to the reduced supply of rainwater. Increased pumping of groundwater becomes necessary to sustain the productivity levels. The electricity consumption in Punjab during drought years, particularly in the agricultural sector, is diverted from the industrial and the domestic sectors for sustaining agricultural outputs. On the other hand, the growth in electricity consumption in the industrial sector was just 1.3% and 1.9%, which is much lower than the usual annual growth. It shows a significant diversion of electricity during the drought years, with a limited success to sustain the agriculture sector output.

The overdependence on groundwater resources has also impacted the quality of groundwater resources in recent times. Punjab is emerging as an important hot spot of groundwater pollution, and there are many sources of such pollution such as industry, distilleries and agriculture. As per the data from the Ministry of Drinking Water and Sanitation, Government of India, almost 15% of the water samples tested during 2012–2013 and 2013–2014 have been found to be contaminated.

There are a couple of interventions which have significant implications for long-term sustainability of groundwater use in Punjab agriculture. These include crop diversification and delayed transplantation of rice. Diversification of crop aims at diverting a significant area away from rice to less-water-consuming crops, and the delayed transplantation of rice has the potential to bring a significant reduction in water use in the rice crop itself.

Crop diversification in Punjab: the concern on over-exploitation of groundwater resources or unsustainability of intensive cropping system in Punjab urged to explore potential alternatives to the rice-wheat monoculture. Another potential reason for such an attempt was to stabilize food grains in the markets and avoid the environmental issues of resource depletion, owing to the dominance of these two crops. The upshot of this attempt led to the formation of committees which suggested ways to promote growth and productivity of agriculture in Punjab. Johl Committee in 1985 recommended 20% diversion in the area from rice and wheat crops. The diverted area was supposed to be shifted under oilseeds, pulses, sugarcane, fodder, fruits and vegetables. The Alagh Committee, constituted in 2005, also considered the potential of crop diversification and suggested a three-pronged strategy aimed at (1) promoting the production of commodities of mass demand such as milk, pulses, oilseeds and high-value commodities like fruits, vegetables and floriculture, (2) building value chains for such commodities and (3) enhancing productivity of traditional crops to offset decline in production, owing to the area shifts. Kalkat Committee, in its Draft Agricultural Policy of 2013 (PSFC 2013), also recognized the depletion of groundwater (especially in the central Punjab), climate change and global warming with its repercussions on the pace of water depletion, increasing investments on tube wells, rising power consumption and power subsidy for the agriculture sector as major challenges to the Punjab agriculture. It suggested the limiting of area under rice to just 1.6 million ha and shifting to other crops such as maize, pulses, etc.

Despite the concern for crop diversification, the area under rice increased. The reasons for the failure of the diversification efforts may be many. There was a severe drought in the country in 1986, a year after the Johl Committee submitted its report on diversification in 1985. The Punjab state again came to the rescue of the nation with a significant contribution to the national food security and the focus on diversification got loosened. There was no adequate financial support from the government (as was emphasized), which led to the failure of crop diversification strategies as emphasized in the Johl Committee Report of 2002 and the Alagh Committee Report of 2005. There was a clear economic advantage in following the rice-wheat rotation and all other crop alternatives lagged far behind in terms of profitability.

However, looking at the rate of fall in the water table in central Punjab in the recent years, there is an urgent need to look at all the possible options for sustainability of natural resources in agriculture, especially groundwater. While a single solution will not address the problem completely, there is a need to look at the water-energy-agriculture nexus in a holistic manner. The success of the act has given an important message to the policymakers, administrators, scientific community and development officials. It reflects that issues of common property resources can be more effectively addressed by effective policy instruments, without any adverse political fallouts.

2.3 Options for Groundwater Sustainability

The options for long-term sustainability of water use in agriculture have been broadly classified into two categories, namely (1) water-saving technologies and practices and (2) crop diversification, as envisaged in the Draft Agricultural Policy of 2013. After discussing the potential benefits of the technologies, practices and the diversification targets of the Draft Agricultural Policy, three different scenarios have been outlined with the possible impact on groundwater use and long-term sustainability.

2.4 Technologies and Practices for Water Conservation

Many technologies and practices have the potential to reduce water use in agriculture and improve water-use efficiency without any adverse impact on farmers' income. These technologies and practices include laser levelling of land, planting on permanent raised beds, zero tillage in wheat, use of tensiometers in rice and direct seeding of rice.

- The *laser-guided precision levelling* of fields became popular amongst farmers in the last few years. It was also facilitated by the effective functioning of Agro Machinery Service Centres (AMSCs), being largely owned and managed by the Primary Agricultural Cooperative Societies. The water saving translates to reduction in electricity consumption by 213.35 kwh/ha, with corresponding cost saving of Rs 1022/ha (US$ 15.7/ha, at 1$=Rs 65 at 2016 exchange rates).
- Apart from this, the *permanent raised bed* planting technique also helps to maintain the natural texture of soil. Without any adverse effect on the rice and wheat productivity, this technique saves 60 cm/ha of water in rice and 8 cm/ha in wheat, as compared to the conventional methods of planting/sowing.
- Another practice relates to improving organic matter and soil fertility. *Zero tillage in wheat* is reported to reduce the water use in wheat crop by 8.5 cm/ha, with a corresponding electricity saving of 50.11 kwh/ha.
- Also, *use of tensiometers* in rice improves water-use efficiency, thereby saving a significant amount of water without any adverse impact on productivity.

Experimental evidence has established a water saving of 37 cm/ha with the use of tensiometer, with a corresponding power saving of 218.13 kwh/ha.

- *Direct seeding of rice* is also being promoted as a potential alternative for water saving. It is estimated that water requirement of rice is reduced by about 45 cm per ha, with a corresponding power saving of 265.29 kwh/ha.
- *Mulching* is another practice to conserve the soil moisture and is estimated to improve the productivity of maize and wheat crops by about 14% and 23%, respectively (Singh et al. 2006).

2.5 Scenarios of Technologies, Practices and Crop Diversification and Their Impact

Three scenarios and their impact on water use have been examined. The first is the Business as Usual (BAU) Scenario, the second assumes the successful implementation of the diversification efforts in the state and is called the Crop Diversification (CD) Scenario. The last scenario assumes that the diversification efforts succeed along with increased adoption of water-saving technologies and practices. This scenario is called Crop Diversification plus Technologies and Practices (CDTP) Scenario.

Business as Usual (BAU) Scenario: No change in the cropping pattern and water requirements is assumed. It is the current scenario of agriculture in the state with relatively less water-use efficiency. There is a significant trade-off between the enhanced food production and groundwater sustainability. In this case, 1% increase in the rice production would increase the water use by $44.8 \cdot 10^3$ ha m-. This will place additional strain on groundwater resources.

Crop Diversification (CD) Scenario aims to divert 1.2 million ha area from rice to other crops, as envisaged by the Draft Agriculture Policy for Punjab, 2013. This will bring about reduction in water use in agriculture by 1.58 million ha m, which is quite sufficient to address the problem of groundwater depletion in Punjab (Table 2.1).

Crop Diversification plus Technologies and Practices (CDTP) Scenario is intended to introduce crop diversification along with the adoption of water-saving technologies and practices in Punjab. The CDTP scenario has the total potential to save 2.3 million ha m of water in Punjab agriculture. Under this scenario, the compulsion to divert 1.2 million ha area from rice to other crops will reduce by 0.55 million ha. It means that not 1.6 million ha of rice area but 2.15 million ha can be cultivated without endangering the groundwater sustainability in the long run.

Table 2.1 Targets of diversification, water use, and water savings in Punjab under crop diversification scenario

Crop	Targeted change in the area (thousand ha)	Water requirements (mm)	Total water savings (thousand ha m)
Rice	−1200	1600	–
Maize	425	350	531.25
Cotton	200	200	280.00
Sugarcane	175	850	192.50
Pulses	70	200	98.00
Fodder	100	300	130.00
Fruits and Vegetables	85	350/450	111.25
Agro-forestry	145	350	232.00
Total	–	–	1575.00

Source: Author's calculations

2.6 Summary and Conclusion

The intensive agriculture and dominance of rice crop in Punjab has led to severe depletion of groundwater resources. The net availability of groundwater resources started turning negative since 1999, with a significantly large number of development blocks witnessing the over-exploitation of the resource and turning critical. Currently, more than half of the area falls under the water table depth of more than 20 m.

The rise in cropping intensity and changes in the cropping pattern have led to a significant increase in the demand for groundwater. Also, declining rainfall by almost 50% in about one and a half decades and decline in surface water irrigation were two important factors responsible for the decline in supply levels. Both these factors contributed almost equally towards increase in demand for water in agriculture.

The fall in groundwater table necessitated the deepening of tube wells; however, this involved huge investments from the farmers in the form of fixed costs and increasing incidences of farmers' indebtedness with relatively larger burden on the smallholders. The use of electricity in agriculture also increased tremendously with a parallel increase in the power subsidies, the upshot of which was deterioration of groundwater quality.

Therefore, sustainability of groundwater resource can be achieved by following a two-pronged strategy, successful diversion of area from rice to less-water-consuming crops and promotion of water-saving technologies and practices. A comprehensive effort aimed at crop diversification as well as promoting the adoption of water-saving technologies can ensure long-term sustainability of groundwater resources in Punjab agriculture.

References

Alagh Committee. (2005). *Report of the expert committee on the World Trade Organization for Punjab*, Government of Punjab.
Envis Centre: Punjab. (2013). *Annual average rainfall by districts in Punjab (in mm)*. http://punenvis.nic.in/index3.aspx?sslid=3870&subsublinkid=3599 &langid=1&mid=1. Accessed on 11April 2018.
Johl, S. S. (1986). *Diversification of agriculture in Punjab, report of expert committee*, Govt. of Punjab, Chandigarh, India.
Kaur, S., & Vatta, K. (2015). Groundwater depletion in Central Punjab: Pattern, access and adaptations. *Current Science, 108*(4), 485–490.
Pandey, R. (2016). Groundwater irrigation in Punjab: Some issues and a way forward. In L. Singh & N. Singh (Eds.), *Economic transformation of a developing economy: The experience of Punjab, India* (pp. 97–117). Singapore: Springer.
PAU. (2015). *Statistics of Punjab agriculture*. Ludhiana: Punjab Agricultural University.
PSFC. (2013). *Draft agricultural policy for Punjab*. Punjab State Farmers Commission, Mohali, Punjab.
Sidhu, H. S. (2002). Crisis in Agrarian economy in Punjab: Some urgent steps. *Economic and Political Weekly, 37*(30), 3132–3138.
Singh, C. B., Chhibba, I. M., & Brar, J. S. (2006). Technologies for efficient management of soil, water and nutrients in Punjab. *Technical Bulletin, 1*.
Singh, R. P., Dhaliwal, H. S., Singh, T., Sidhu, H. S., Singh, M., Singh, Y., & Humphreys, E. (2008a). A financial assessment of the happy seeder for rice-wheat systems in Punjab, India. In E. Humphreys & C. H. Roth (Eds.), *Permanent beds and rice-residue management for rice- wheat systems in the Indo-Gangetic Plain, ACIAR Proceedings No. 127* (pp. 182–190). Canberra: Australian Centre for International Agricultural Research.
Singh, S., Kaur, M., & Kingra, H. S. (2008b). Indebtedness among farmers in Punjab. *Economic and Political Weekly, 43*(26/27), 130–136.
Vatta, K., Sidhu, M. S., & Kaur, A. (2013). Pattern of growth and emerging challenges for Punjab agriculture. *Report submitted to the United Nations Environment Programme*. Punjab Agricultural University, Ludhiana, India, 46.

Chapter 3
Sustainability of Himalayan Environment: Issues and Policies

Masaud-ul-Haq Wani and Saima Masood Wani

Abstract The Himalayan region has three major areas of interest, mountain environment, forest resources and fresh glacial water. Mountain people mostly follow agricultural livelihood system for their sustenance and economic wellbeing. However, contraction of resources and climate changes demand a shift from the conventional to improved management systems. The process of economic growth is reaching the unreached mountain societies, which are seen as a transition from total dependence to partial dependence on farming, through improved education, skill development and exploitation of potential niche areas for development. The unpredictable weather, fast depleting resources, inadequate infrastructure, low productivity of animals and heavy disasters witnessed during the recent past are some of the challenges to be met in the Himalayas. However, opportunities exist for resilience to these challenges through cultivation of high-value off-season vegetables, medicinal/aromatic plants/flowers and a path of reversing the shift from desirable to undesirable ecology. To enable the Himalayan people to promote sustainable farming, achieve sustainable livelihood and maintain mountain ecosystem services, these people need access to natural resources and empowerment of their women, which are supposed to be the key determinants for development and are required to be extended to them under specific legislations. Public investments in education, health, transport, research, extending credit, extension services, compensation for watershed management and conservation of biodiversity are some of the policy options suggested to achieve sustainable Himalayan development.

M.-ul.-H. Wani (✉)
Rajiv Gandhi Chair in Contemporary Studies on Livelihood and Food Security, Sher-e-Kashmir University of Agricultural Sciences & Technology of Kashmir, Srinagar, Jammu and Kashmir, India
e-mail: rgckashmir@gmail.com

S. M. Wani
Division of Biotechnology, Sher-e-Kashmir University of Agricultural Sciences & Technology of Kashmir, Srinagar, Jammu and Kashmir, India
e-mail: sayawani@gmail.com

R. Peshin, A. K. Dhawan (eds.), *Natural Resource Management: Ecological Perspectives*, Sustainability in Plant and Crop Protection,
https://doi.org/10.1007/978-3-319-99768-1_3

Keywords Environmental degradation · Economic growth · Climate uncertainties · Natural resource management

Abbreviations

WTO	World Trade Organization
SPS	Sanitary and phytosanitary
IPCC	Intergovernmental Panel on Climate Change
HKHM	Hindu Kush Himalayan Mountains
USD	United States dollar

3.1 Introduction

The Himalayas account for most of the vastly distributed mountainous areas, spreading over 12 states of India, namely, Jammu and Kashmir, Himachal Pradesh, Uttaranchal, Sikkim, Assam, Arunachal Pradesh, Manipur, Meghalaya, Mizoram, Nagaland, Tripura and West Bengal. It covers up to 2500 km in length and 250 to 400 km in breadth. Longitudinally, the Himalayas are classified as Shiwalik flat summits (600–1200 m msl altitude), middle Himalayas (65–75 km width, average height 3000 m), Greater Himalayas (average altitude 5200 m, 92 peaks over 8000 m) and Trans-Himalayas (average width 60 km, average altitude 4500 m), in addition to the mighty Himalayas. The Himalayan region is inhabited by 51 million people covering 18% of geographical area and 6% of Indian population (Wani 2011).

Himalayas have dispersed land available for cultivation with huge altitudinal variations, having diverse microclimatic conditions, with lesser potential for commercialization. However, some Himalayas niche areas possess vast potential for production of exportable commodities that are by nature organic and get accommodated under WTO (World Trade Organization) regulations, such as SPS (sanitary and phytosanitary), etc. In addition, the Himalayas provide valuable ecosystem services like fresh water, capacity to reduce disasters, huge biodiversity and a lot of space for tourism and recreation.

Considering the poor scope of industrialization in mountains and perspective of rural livelihood, agriculture remains an important source of livelihood and economic growth, despite its declining share in the economy (Table 3.1). However, in developing countries, poor accessibility to the basic benefits such as education, health, communication, road network, markets, transport and extension services remains always the major problems in the development of mountain farming. These problems are referred to as the mountain specificities. Being away from the economic, political and power centres, these populations have almost no role in decision-making, and thus people get marginalized. In addition to being insecure, these areas are affected by outmigration. Although those who leave can provide remittances, however, migration results into heavier workloads for those remaining behind such as women, children and elderly persons. Limited availability of land,

Table 3.1 The Himalayan region: demographic and agricultural indicators

State	Area (km^2)	Population (N.)	Net cropped area (000 ha)	Cropping intensity (%)	Average size of holdings per family (ha)
Northwestern hill region					
Himachal Pradesh	55,673	6,077,900	551	174	1.16
Jammu and Kashmir	222,236	10,143,700	733	147	0.67
Uttarakhand	53,483	8,489,349	788	164	1.01
Total	331,392	24,710,949	2072	160	0.97
Northeastern hill region					
Arunachal	83,743	1,097,968	166	159	3.31
Assam	78,438	26,655,528	2701	152	1.17
Manipur	22,327	2,388,634	140	142	1.22
Meghalaya	22,429	2,318,822	240	111	1.33
Mizoram	21,081	888,573	91	100	1.29
Nagaland	16,579	1,988,636	261	113	4.82
Sikkim	7096	540,851	95	127	1.65
Tripura	10,486	3,199,203	277	152	0.60
Total	262,179	39,078,215	3971	145	1.92
India	3,287,240	1,028,830,774	141,231	134	1.41

Source: http://indiastat.com/labourandworkforce/380987/employment/85/unemploymentsituation/281124/stats.aspx

which often has low productivity, coupled with poor marketability makes these fragile ecosystems get transformed into unsustainable to maintain systems, owing to over-exploitation of natural resources.

The Himalayan ecosystem is at a disadvantage for expanding livelihood opportunities. With the outmigration, while men migrate for earning livelihood, the farming land gets deprived of labour force for cultivation; therefore, family labour must be mobilized for agricultural operations. This results in poor schooling of children as they become most likely the individuals who perform such tasks. Further, the global climate change is more pronounced in the Himalayan regions having rich environments. Therefore, these regions require equally more attention for their preservation, sustenance and also fast response to changes, compared to downstream areas.

3.2 The Himalayan Environment

The Himalayan environment has three distinct features. Forest resources maintain and balance the ecology and environment, provide firewood for cooking and timber for construction and supply fresh water stored in glaciers to meet the needs of the human society and to provide national security. However, this environment has been observing multi-faceted changes affecting temperatures, weather conditions, hydrological regimes of water bodies, extended growing periods and reduction in the number and size of glaciers, resulting in the shortened snow cover. Changing the dynamics of land use system coupled with the soil health degradation and loss of biodiversity has collectively adversely impacted the socio-economic status of the mountain populations. The Himalayas are exposed to increased vulnerability and decreased resilience of the mountain sustainability, considered in the frame of the economic references. These are the transition from inaccessibility to accessibility and from poor infrastructure to improved upgraded metalled roads, high tech market infrastructure and the establishment of intra-sectoral linkages within the agricultural sector and with the overall economy. Socio-economic changes affect the livelihoods both positively and negatively, especially in the agricultural sector which is the mainstay of the Himalayan people.

Mountains and uplands in the Himalayan ranges have great potential for off-season vegetable and fruits, though these are constrained by poor market infrastructure, lack of local-specific technologies and poor extension services for transferring technology from lab to land. Additionally, mountain specificities, such as inaccessibility, fragility, poor mobility, etc., are issues contributing to either over-exploitation or nonexploitation of productive resources in these areas. At the same time, the region is bestowed with niche areas for development, especially for horticulture, including medicinal plants which provide better opportunities for a paradigm shift from traditional subsistence agriculture in favour of modern and high tech commercial agriculture (Sharma et al. 2009).

3.2.1 Climate Change Impacts and Changing Dynamics

The most visible evidence of global warming is the melting of the glaciers across the region that once possessed a cover of 9575 glaciers, with 37,466 km^2 area and a total ice volume of 2000 km^3 (Table 3.2). These values are now expected to shrink by around 40% by 2035 (IPCC, Intergovernmental Panel on Climate Change) (Raina and Srivastava 2008).

The demand for fresh water is increasing on sustained basis, partly because of urbanization and partly due to the normal population boom. On the other hand, gradual but sustained decrease in the number and volume of glaciers will put additional pressure on the scarce water resources, giving rise to disputes among rural

Table 3.2 Himalayan glacier system in India

State	Glaciers	Area (km^2)	Average size (km^2)	Glacier (%)
Jammu and Kashmir	5262	29,163	10.24	61.8
Himachal Pradesh	2735	4516	3.35	8.1
Uttarakhand	968	2857	3.87	18.1
Sikkim	449	706	1.50	8.7
Arunachal Pradesh	162	223	1.40	3.2

Source: http://niti.gov.in/content/statedata.php?type=ECONOMY &var=GSDP%20at%20constant%20(2004-05)%20 prices,%%20Percent%20Growth%20(2004-05%20to%202014-15)

and urban communities on water use and rights. This phenomenon has been observed in many states of India for the past three decades. The scientists observed that the global warming is inflicting changes in usual rainfall patterns, making farmers stop rainfed crop productions, due to the lesser reliability of rains (Haller 2012; Behnke and Kerven 2013). In addition, lessening of freezing periods and extension in warm seasons have reduced the scope for producing the traditional crops consumed mostly by local population. This has shifted the conventional agricultural paradigm to a modern and consumer-based production system. Thus, making traditional climate-resilient farming less relevant, compared to the nonfarm activities, which offer better economic revenues for the majority of the region's rural population, is non-sustainable. This transition forced many young people to abandon the traditional mountain communities and migrate to the towns in search of a better life. Increased prices for minerals in the world market led to the reopening of old mining areas and over-exploitation of the unexploited areas, by medium and small enterprise operators, led by local groups.

The Himalayan farmers are therefore searching for strategies to cope with this global change. These farmers employ traditional patterns of risk management. They are conscious of the regions geography, bestowed with diverse altitudinal variations and topography, which they use for herding and growing diverse crops. This helps them to minimize the risk of total failure. Potential of employment generation existing in urban areas, especially in the mining sector, has enabled these farm families to follow a new strategy, i.e. to pursue various economic activities at different time periods of the year by its individual members. This strategy has reduced the risks and increased the livelihood opportunities, thereby, reducing dependence on local and nonpredictable factors such as weather and climate, while increasing dependence on nonfarm economies of regional and global scope (Pratap 2015).

We know that the transformation of the Himalayan region has now settled, which is the outcome of the global warming that induced climate/environmental change processes. Climate change mitigation has been observed to be carried out by the farmers themselves by shifting the conventional system of production to commercialized, nonconventional systems of production. The economic opportunities in the Himalayas have increased and are reaching the unreached communities. The liberalized global economy has induced a dynamic change process in the Himalayan region, due to the global liberalization. There is exploitation of natural resources beyond the economic optima, due to population growth with decreasing per capita land availability. This very factor has inflicted a multidimensional change impacting the Himalayan agriculture and its sustainability. The dynamic multidimensional changes (Pratap 2015) include the virtual cycle of economic growth, owing to the convergence of policies and development. Under this change the exploitation of the niche areas of mountains will account for a sizeable income due to their exportable produce. This would further mean to redefine the poverty of states in response to the increase in per capita income and other factors contributing towards the economic development.

The rising population in the mountainous states is keeping pace with the population bloom at global level. However, a new dimension in terms of improved education and skill development has come up in these states, which will certainly push their growth beyond the targeted 4%, owing to the unprecedented economic opportunities within and outside the farm sector (Templeton and Scherr 1997). The mountainous states can experience a positive growth stimulus with a better future ahead. However, whether it can be sustained for a longer period is a question to be answered in the backdrop of receding natural resources.

The Himalayan people thus need to change from complete dependence to partial dependence on agriculture, shifting this pattern towards nonfarm sources. The past colder glaciers are rapidly melting due to longer sun exposure, and there is not as much water as before. Also when it used to rain in the past, there was more water and wetlands, enabling the pastures to remain maintained throughout the dry season. Now the situation is that the complete hydrological regime of water bodies has changed, and extended summers are observed. This has resulted in the growing of a variety of crops, increased the cropping intensity and marked the shift in favour of high tech commercialized agriculture, which leads to increased unplanned urbanization due to outmigration that has long-term negative effects on productivity and ecology. The biological degradation of support lands, farm lands, air and water, unsustainable agroecosystem, etc. could be some of the examples in this regard. The reduction in the carrying capacity of agroecosystem in the region and the natural processes going at unprecedented scale are most likely to hit the Himalayan environment, producing environmental crises bigger in magnitude, with far-reaching consequences (Wani 2011). The tendency of young people to move towards urban areas in search of white-collar jobs has been observed. These people pursue quality of higher education to get better opportunities to increase their standard of living. They lose interest in working on their fields but exploit the niche areas of production, increase their income, settle somewhere else, manage their business through remote control and enjoy their life. This way the region is put to a complete disadvantage.

3.3 Urbanization and Consequences

Urban growth has its consequences for land use and livelihoods of smallholders who live on the rural-urban fringe. The increasing demand for land and water has increased resource scarcity in the valley floor, which is the most favourable area for agricultural production. This has driven up the land prices. Many smallholders of the region own very small plots, which they mostly use for subsistence production. Thus, they depend on renting additional land for production of market-oriented crops, such as fruits, vegetables, spices, maize, etc. which provide them cash income. However, today's rising land prices have diminished smallholders' possibilities for renting such additional plots. In fact, the landowners, mostly large real estate owners, are not willing to lease their land to farmers, fearing it might restrict their ability to develop and make them fearful of selling their property as and when needed. Local smallholders perceive the urbanization as a threat to their food and income security. Many of them cope up by increasing production of home-based breeding of small animals, such as sheep and goat, and selling the meat in urban markets. They are also expanding or intensifying crop production on nearby common property resources and trying to compensate for what they have lost on the valley floor (Satterthwaite et al. 2010).

Due to socio-economic disparities in the peri-urban interface, it is crucial to consider the different stakeholders' perceptions of the urban growth impact. Peri-urban decision-making should hence be based on multiple criteria that also include the local agriculturalists' assessment, in order to prevent land use conflicts and negative effects on smallholder food and income. The peri-urban smallholders of the Himalayas understand the challenges and opportunities of urban growth and hope to profit from a growing urban market. They have developed new ways to generate income at different altitudes as an alternative to the ground lost in the valley. However, in order to create a flourishing rural-urban interface, they need the support of planners and policy-makers, especially relating to strengthening smallholder-market linkages (Mukwaya et al. 2012).

3.4 Economic Growth and Livelihoods

Economic growth is an important determinant of employment and is essential to create livelihoods and improve the standard of living of people. In the Himalayan region, the income distribution is comparatively rational, because in these regions few people are very rich and few are very poor. Though the economic activities are restricted, however, food, clothing and shelter are available to majority of the population. In order to protect the Himalayan ecology and biodiversity, an appropriate developmental strategy for creating alternative livelihoods is essential.

There are two types of economy which can normally be thought of, i.e. green economy and digital economy (Mahajan and Rahman 2015). Both these economies need to be clubbed together in order to establish a link to achieve growth. While green economy (Perlik and Kohler 2012) especially in the Himalayan states spans agriculture and its subsectors, like horticulture, animal husbandry, forestry, etc., manufacturing, like green construction, green manufacturing and green services, addresses forward and backward linkages from trade to tourism. On the other hand, digital economy could include manufacturing using light materials like electronics, information technology, the digital financial services like e-banking, etc. Therefore, while green economy generates employment through production and product differentiation, having international market, the digital economy helps in creating demand through advertisements, regulating supplies through doorstep distribution. This kind of economy would be helpful in providing employment to nearly 18–20 lakh (1.8–2 million) unemployed persons in the Himalayan states, which add up more than 2 lakh (0.2 million) every year to the workforce and make the challenge for creating livelihood even more difficult.

The wealth distribution of the Himalayan region is an important component of the growth determinants. While the Himalayan region has many potential for productive

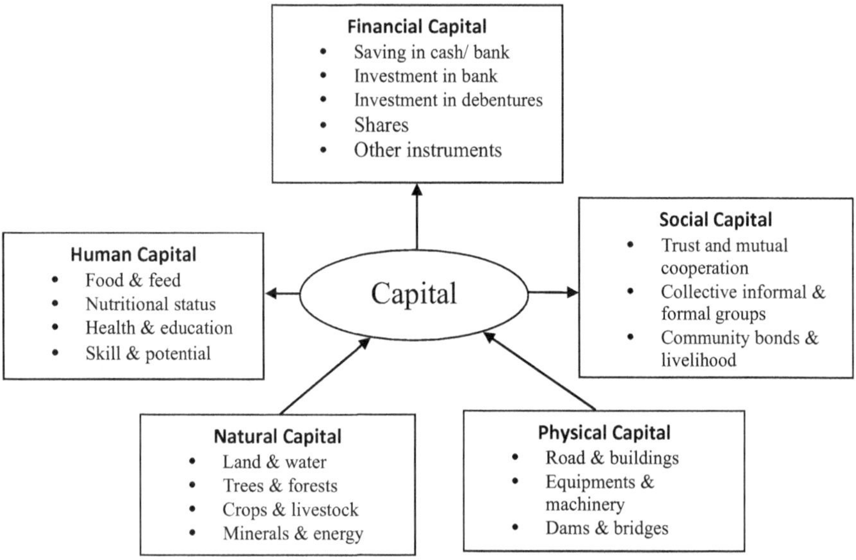

Fig. 3.1 Capital and investment requirements for the Himalayas

gains, however, it has to be followed in a rational way, taking into consideration the sustainability of resources. We will therefore necessarily need to access potential of our capital against the following forms, for its equitable distribution and sustainable consumption. For the economic advancement of Himalayas, capital requirements are needed for infrastructures, human resource development, natural resource management, community development, etc., as shown in Fig. 3.1.

3.5 Sustainability Issues and Transformation of Livelihoods

Himalayan mountains have traditionally been home to farmers accustomed to difficult conditions and the need for hard work to survive the harsh winters. When commercialization started setting in, over-exploitation of natural resources occurred, non-farm sector becoming stronger. Dividends induced many farmers to leave their villages and move to towns, to work in nonfarm sector. This scenario, however, could not be sustained for a long time, due to increased urbanization which mounted more pressure on the existing infrastructure, and available but receding resources, giving rise to major economic crisis and hampering significant investments in mountain areas/states. As nature is being destroyed nowadays, arable land is drastically reduced. Those who sell land for construction are not aware of the damage they cause and are not aware from where will the food for villages come. Many residential projects are being constructed by real estate firms, which are driving this business, and the smallholders are only the spectators in this development.

Major changes due to political transition or economic crisis can open up new opportunities if addressed in an appropriate and sustainable way. Without external support, it is difficult for small organic producers to fulfil the demanding requirements of certification schemes and to access new markets in cities or major tourist locations, which are often the first markets for organic products. Successful organic farmers could generate enough income, stay with their farming and ensure steady rural development. In turn, this will help to reduce the significant development gap between mountains and other areas.

The fact is that the Himalayas are potentially remunerative farming areas unlike mountain areas in temperate zones. Horticultural products (vegetables, flowers and spices) fetch good prices and provided quick cash, compared to other agricultural crops, like cereals and millets. Households now depend on the prices paid for these commodities at the national or world markets and at the institutions handling the products. In addition, changes in climate are reported across the Himalayas, but not much data is available to provide solid evidence of the impact. Land pressure, due to increasing rural population density, and high levels of poverty could undermine the sustainable use of these highly productive and high potential mountain agroecosystems. The outmigration of youth may help ease this pressure but represents a loss of active human capital for the Himalayan rural areas. This may negatively affect the innovative potential of these areas, including family farms.

The interplay of the above factors has transformed the traditional family farm significantly.

- Fewer social assets and weakened social cohesion within an extended family and rural community has come in, resulting in poor mobilization of community resources.
- Small-sized farms constraint agricultural productivity and hence incomes, and rural food insecurity is high.
- Less use of external inputs, such as fertilizers and pesticides, with the exception of commercial crops, due to non-enabling environments which make them inaccessible to most of small farmers. This negatively affects household and regional food security.

3.6 Future Strategies

3.6.1 Capacity Building

Farmers of the Himalayan region in India are expected to benefit from capacity building focused to develop technical and managerial skills, promoted by both the public and private sectors. Co-operative action is key to increase productivity and remove shortcomings. The collective efforts are expected to help in establishing producer's authority to negotiate in the value chain (Hoermann et al. 2010). Establishing direct links among producers, sellers and consumers will benefit farmers and reduce their vulnerability to exploitative practices of intermediaries in a market system. Moving ahead, appropriate technologies and infrastructure, such as decentralized and renewable energy supply, will be required to establish or enhance processing activities in the Himalayan areas and, in turn, provide off-farm jobs (Bradbear and Joshi 2012). Producers and processors need to maintain and communicate the quality and uniqueness and also reflect the origin of their products under a specific trade name, signifying Himalayan products to consumers, to obtain higher prices that will cover the higher labour cost required for maintaining the ecosystem services of the Himalayan areas. Certified Himalayan products guaranteeing quality through proper labels will also help in making all out efforts to market these products through information technology. Otherwise this will be beyond the capacity of marginalized groups of farmers, who always fail in being recognized in a promising market. It, therefore, needed a revisiting of the formal certification schemes, with an objective to smoothen them to be more relevant and easy to pursue. It has been observed that owing to lack of formal certification, the products of this region do not get access to international markets. This aspect needs to be considered, while revisiting the certification procedure for Himalayan produce, as marketing in national and global markets can provide a better cost-benefit ratio, improving the status of these farmers.

3.6.2 The Himalayan Product Development

The weak, Himalayan smallholder farmers integrated in commodity markets are unable to compete with downstream large-scale producers. These farmers are now able to enter into the emerging market system specialized in nutritious, healthy and organic products. Here these smallholder farmers can capitalize on such markets, owing to the differentiated organic and nutritious food products, produced by these farmers in the Himalayan region (Ramesh et al. 2010). Developing pro-poor sustainable value chains in these emerging markets with formal certification schemes that guarantee the value added of Himalayan products will help bring premium prices and can go a long way in achieving better and remunerative prices for Himalayan produce.

However, smallholder Himalayan farmers are constrained by low, dispersed and unreliable production levels, remoteness, lack of processing technology and knowledge and difficult access to market information, as well as inadequate negotiation and management skills. It remains difficult for them to raise sufficient money to meet their basic needs, invest in their farm infrastructure and fulfil personal aspirations.

The Himalayan farmers manage integrated farming systems with low input of chemical fertilizers and pesticides, all of which adds up to the potential for producing attractive, healthy and organic food for new markets. Consumers, including mountain tourists, and the private sector are rediscovering the highly nutritious and medicinal value of indigenous, underutilized and wild species (Ramesh et al. 2010).

They appreciate the qualities of organically grown or specialty products and are willing to pay heavily for these products. At the same time, some mountain areas also offer markets for locally grown products.

3.6.3 Cash Crop Production

Transitions from age old practices have taken place, giving a new dimension to the agricultural development. The horticultural crops, being cash crops, have great promise for economic upliftment of people in this region and are suited to most of the ecological zones of the Himalayan states. These have been using horticulture as a means of transforming subsistence agriculture to the commercial activity. The productivity per unit of area has increased only due to vegetables, fruits and spices, which invite attention of most of the present day society that became more healthy diet conscious. The produce, being mostly organic, have several export potential and thus provide better avenues for economic development of these areas (Sharma et al. 2009).

Cash crops, largely fruits and vegetables, suited to different agroecological conditions of Himalayan areas, have proved to be the most powerful tool that farmers across the Himalayan states have so far been able to use to transform non-viable subsistence farming into improved, high tech, food- and income-secure, commercial agriculture. For over two or more past decades, in pockets of the Himalayan region, farmers, government agencies and markets have been joining hands to develop success stories of cash cropping systems. There are now good examples of converting unsustainable subsistent agriculture into a better proposition, providing food and income security to the Himalayan farming families. Farmers, research and development agencies and markets have been improving their understanding of the potentials of developing mountain farming niches, helping harness the comparative advantage of an area/agroecological zone, valley or landscape.

Over a century ago, the 2.9 million population (1901) of the Kashmir Valley reached 8 million population in 2011, land resources remaining the same. Therefore, population growth produced a pressure due to land scarcity with no other real option but to generate enough income from cash crops in the available farm land. Even while this occurs, the 6159 million US dollars [Rs. 4000 crore, (1 crore = 10^7)] fruit economy, of which 85% comprise apples, 6% walnuts and the rest other fruits, has been sustaining the farming family's economic productivity, as much as 3.7 million people, farmers and others, of the valley. Large-scale technological inputs and wider adoption by producers are required to build a base for sustainable livelihoods and enhance growth of fruit economy, raising incomes from 61.59 US dollars to 615.81 US dollars (i.e. Rs. 4000 crores to 40,000 crores, at a conversion rate, 1 USD =64.98 INR) or more, for the farming activities for which the Kashmir Valley has potential (DES 2013).

The highland farmers of Kargil, who own dry stony sloping farmlands, are turning the old subsistent wild apricot into a good cash crop, which flourishes under such edaphic and climatic conditions. Herein, value addition and improved market access to products of wild *Prunus armeniaca* (apricots), apricot oil and dried fruits have emerged as potential products. There are more such niche fruits of Ladakh waiting to be harnessed, such as *Trichogramma* sp.-managed Kirkichoo apples of Kargil and *Hippophae rhamnoides* subsp. (Thakush) of Leh district. The widely known Leh berry "sea buckthorn" (*H. rhamnoides* L.) a wild thorny bush of Ladakh helped farmers extend benefits of farming on nonfarm government lands, i.e. forest land. Market interventions are expected to improve value of this shrub to local farmers.

Himachal Pradesh performed far better compared to other states in improving the agriculture economy of the state. Here the subsistent food grain farming is being converted to cash crops farming. The vegetable and fruit farmers have confidence of improving their livelihoods. The tribal highland farmers of Lahul are engaged mostly in cash cropping of vegetables, new fruit crops, apple and medicinal plants. The dependence of low hill areas of the state on agriculture is continuously declining and moving to the nonfarm sector. Among the Indian Himalayan farmers, higher economic returns are observed by uphill farmers compared to low hill farmers. Earlier, only Solan was known as vegetable district of Himachal, but now farmers of most districts of Himachal, Shimla, Kullu, Mandi, Sirmour, Kangra and Chamba have turned to vegetable farming (Pratap 2015). The viable farming households turned into vegetable farming for taking triple benefits: (1) there is comparative advantage of growing vegetables in the off season, (2) family labour is put to effective use, and (3) farmers are taking three crops in a year in place of one or two. In a sense, family labour is engaged on their own farms round the year, with more productivity. The positive effect of vegetable farming on small farming families is visible through the 4615 million US dollars (Rs. 3000 crore) vegetable economy. There could be further examples of success stories of cash cropping, based on improved livelihoods in villages, districts and watersheds of the northeastern states also.

The central Himalayan state of Uttarakhand is generally known for its money order economy, as the farming economy of the hill state, barring its Tarai zone, shows various indicators of unsustainability of farming-based livelihoods. Rainfed farming on the sloping landscape is marked by water scarcity, drought periods, small farms, fodder scarcity and wild animal intrusions, including monkeys. Uttarakhand, in many ways, exhibits indicators of shrinking carrying capacity and outmigration as public response, in the absence of public interventions helping to expand the carrying capacity of farming-based improved livelihoods. Weakening forestry-farming linkages leading to poor nutrient cycling and maintaining water availability is visible (Choudhary et al. 2011, 2013). There is an increasing need for chemical fertilizers to sustain crop production. Mounting difficulties of increasing costs and unsure supplies is faced by the Himalayan farmers across states, except Sikkim and few others, who opted to promote alternative systems of organic agriculture. Problems are highlighted concerning increased biological degradation of support lands across the Himalayas by unpalatable invasive species and consequent expanding threats to crop-livestock mixed farming-based livelihoods. This is a key concern of farmers in low and mid-hills (Sharma et al. 2009).

Learning from the experience of thousands of farming families inhabiting several Himalayan states, who switched over from food grain-based subsistent agriculture to fruits and vegetables-based agriculture and benefitted from marketing supply chain mechanisms and infrastructure support services of the states, it is obvious that non-viable subsistent farming by hill farmers has a chance to become viable through cash crops (Wani 2011).

3.6.4 Value Chain Development

It is necessary to develop value chains that enable the Himalayan farmers and particularly poor households to participate in and benefit from these emerging markets. Such value chains need to be developed jointly by representatives from all stakeholder groups. They need to be based on a sound analysis of mountain-specific challenges, natural resources and market potential as well as farmer's socio-economic capacities and relations among the value chain actors. Moreover, the development of a new value chain must not jeopardize the farmers' own food security and sustainable production systems (Hoermann et al. 2010).

3.6.5 Establishment of Niche Markets

Establishing niche markets under the prevailing liberal market regime in many countries requires enabling policies that acknowledge the added value of mountain products as a means to improve mountain livelihoods and regional development and, at the same time, compensate the higher labour input for maintaining critical ecosystem services. The Himalayan areas have vast potential for producing niche products, like spices, medicinal plants, fruits, vegetables, honey, mushrooms, etc., which have export potential. The efficient and proper market infrastructure will benefit the farmers to export these commodities to other countries, which will not only help them to achieve a better livelihood but will also boost tourism, thus creating additional means of earning (Pasca et al. 2010).

3.6.6 Pastoralism Nomadism

The Himalayan pastoralists graze their animals as migratory flocks. Transhumance system of pastoralism is very commonly practiced in the upper belts of Himalayan region. The pastoralists provide regulating ecosystem services such as climate regulation, flood and erosion control, food, water, genetic resources and fuel-like provisioning services, cultural services of heritage and landscaping and supporting services of nutrient cycling, habitat and primary production (Ojeda et al. 2012). A study conducted in the Naran Valley in the Pakistan Hindu Kush Himalayan

Mountains (HKHM) (Shah et al. 2012) showed that high-altitude pasture management contributes more to climate change mitigation with a superior carbon store averaging 12.2 t C per ha, as compared to cropping. The services provided by the mountain pastoralists can attract investment and benefit the society.

3.7 Policy Suggestions

- Ensured access to resources and empowering women are key requirements for promoting sustainable mountain development.
- Public investment in education, health, transport, research and extension services.
- Strengthening the availability of credit and financial inclusion of mountainous states.
- Regulation needs to be put in place for conserving biodiversity and management of degraded lands, to create resilience to global environmental changes.
- Processing units need to be created at least at *tehsil* (local administrations) level, so that grading, standardization, packaging and labelling of the Himalayan organic products is done.
- Creation of alternative livelihood through combining green economy and digital economy, to address forward and backward linkages and regulate supplies in response to market demand.
- Policies needed to be framed at national level to compensate services provided by the Himalayan people to the downstream areas in respect of eco-tourism, watershed management, biodiversity conservation, generation of hydroelectricity and supplying drinking and irrigation water.
- Specific market structure needs to be created for organic products, produced by the Himalayan region to foster trade of these commodities into the international market.
- Establishing farmer producer organizations as a strategy to lift small and marginal farmers out of poverty and enhance their competitiveness in agricultural markets.
- Ecosystem services produced by mountain farming, which are vital for downstream areas, be compensated.

References

Behnke, R. C., & Kerven, C. (2013). *Climate resilience, productivity and equity in dry lands* (Climate Change Working Paper No 4). London: IIED.

Bradbear, N., & Joshi, D. R. (2012). *Evaluation report of the project: Improving livelihoods through knowledge partnerships and value chains of bee products and services* (Unpublished report). Thimphu.

Choudhary, D., Pandit, B. H., Kinhal, G., & Kollmair, M. (2011). *Pro-poor value chain development for high value products in mountain regions: Indian Bay Leaf*. Kathmandu: ICIMOD.

Choudhary, D., Ghosh, I., Chauhan, S., Bhati, S., & Juyal, M. (2013). *Case studies on value chain approach for mountain development in Uttarakhand, India* (Working Paper 2013/6). Kathmandu: ICIMOD.

DES. (2013). *Economic survey*. Directorate of Economics and Statistics, Government of Jammu and Kashmir.

Haller, A. (2012). Vivid valleys, pallid peaks? Hypsometric variations and rural–urban land change in the Central Peruvian Andes. *Applied Geography, 35*(1–2), 439–447.

Hoermann, B., Choudhary, D., & Kollmair, M. (2010). *Integrated value chain development as a tool for poverty alleviation: An analytical and strategic framework*. Kathmandu: ICIMOD.

Mahajan, V., & Rahman, F. (2015). *Paper presented at SMDS-IV summit of IMI*, held between November 7–9, at Itanagar, Arunachal Pradesh.

Mukwaya, P., Bamutaze, Y., Mugarura, S., & Benson, T. (2012). *Rural-urban transformation in Uganda* (Working Paper). Kampala: Uganda Strategy Support Program and IFPRI.

Ojeda, G., Rueff, H., Rahim, I., & Maselli, D. (2012). Sustaining mobile pastoralists in the mountains of northern Pakistan (Evidence for Policy Series). Regional edition Central Asia, No 3. Bishkek: NCCR North-South.

Pasca, A. G., Uitton, M., & Rouby, A. (2010). *Guidelines for the development, promotion and communication of mountain food*. Brussels: Euromontana. https://www.euromontana.org/wp-content/uploads/2014/08/Policy_Recommendations_EN_doc.pdf

Perlik, M., & Kohler, T. (2012). Green economy and urbanization in mountains. In T. Kohler & others (Eds.), *Sustainable mountain development: Green economy and institutions*. From Rio 1992 to Rio 2012 and beyond. Bern.

Pratap, T. (2015). *Himalayan agriculture challenges and opportunities*. Paper presented at SMDS-IV Summit of IMI, held between November 7–9, at Itanagar, Arunachal Pradesh.

Raina, V. K, & Srivastava, D. (2008). *Glacier atlas of India Bangalore*. Published by India Geological Society of India.

Ramesh, P., Panwar, N. R., Singh, A. B., Ramana, S., Yadav, S., Shrivastava, R., & Subba Rao, A. (2010). Status of organic farming in India. *Current Science, 98*(9), 1190–1194.

Satterthwaite, D., McGranahan, G., & Tacoli, C. (2010). Urbanization and its implications for food and farming. *Philosophical Transactions of the Royal Society B, 365*(1554), 2809–2820.

Shah, I., Rahim, I., Rueff, H., & Maselli, D. (2012). *Landless mobile pastoralists: Securing their role as custodian of northern Pakistan's mountains* (Workshop Proceedings). Bern: Centre for Development and Environment, University of Bern.

Sharma, G., Liang, L., Tanaka, K., Subba, J., & Sharma, E. (2009). Sikkim Himalayan agriculture: Improving and scaling up of the traditionally managed agricultural systems of global significance. *Resources Science, 31*(9), 21–30.

Templeton, S., & Scherr, S. J. (1997). *Population pressure and the micro economy of land management in hills and mountains of developing world* (Discussion Paper 26). Washington, DC: Environment and Production Technology Division.

Wani, M. H. (2011). Hill agriculture in India: Problems and prospects of mountain agriculture. *Indian Journal of Agricultural Economics, 66*(1), 64–66.

Chapter 4
Crop Genetic Biodiversity with Special Reference to Oilseed Brassicas and Wild Allies: Conservation and Their Utilization

Surinder K. Gupta, Aditya Pratap, Vanya Bawa, and Sunil K. Rai

Abstract Crop genetic resources and their wild relatives are likely to play a significant role in securing twenty-first-century food security. This is due to their potential use in plant breeding to produce crops which withstand adverse impacts of climate change, increasing scarcity of nutrients, water and other inputs and new insect pests and diseases. A high proportion of global food production is from a small number of scientifically bred crop varieties, with narrow genetic variation. This has resulted in loss of approximately 75% of global crop genetic diversity, as these new varieties replaced a much greater range of more genetically diverse, traditional crop varieties. In cultivation, wild plants have been transformed to make them more useful to humans. The genus represents a wide range of crops including oilseed, many vegetables and fodder crops and wild species. In some species this whole range of uses is present. The wide range of crops, species and applications cause *Brassica* genetic resources to be even more scattered over collections than other crops; no single collection holds the complete genetic diversity of *Brassica*. For an integrated approach of conserving genetic resource collections, networks on species or crops are essential. By networks' collaboration, the tasks of long-term conservation of the genetic diversity and making it available to users can be more structured. Duplication of work can be avoided, and gaps in collections can be traced. For several crops there are international networks coordinated by the International Plant Genetic Resources Institute (IPGRI). For *Brassica* such a network does not exist (yet). Regeneration of *Brassica* is very costly and requires good storage conditions, in order to maintain the seed viability which is essential. It is also very important to avoid unnecessary duplicates. Consequently good and accessible information on the material in the

S. K. Gupta (✉) · V. Bawa · S. K. Rai
Division of Plant Breeding and Genetics, Sher-e-Kashmir University of Agricultural Sciences and Technology of Jammu, Chatha Jammu, India
e-mail: guptaskpbg@rediffmail.com; vanyaadarsh@gmail.com; suniliivr@gmail.com

A. Pratap
Crop Improvement Division, ICAR – Indian Institute of Pulses Research, Kalyanpur-Kanpur, Uttar Pradesh, India
e-mail: adityapratapgarg@gmail.com

R. Peshin, A. K. Dhawan (eds.), *Natural Resource Management: Ecological Perspectives*, Sustainability in Plant and Crop Protection,
https://doi.org/10.1007/978-3-319-99768-1_4

collections is needed to trace these duplicates. An international central crop database with high-quality and highly accessible data would be of great help, for efficient utilization of material. For the wild *Brassica* species, in situ conservation is of high importance; the ex situ collections help to make a small part of the diversity readily accessible for utilization. In order to guarantee availability of *Brassica* germplasm to future generations, funding agencies, such as national governments, need to commit themselves to support coherent conservation programmes.

Keywords Genetic diversity · Germplasm collection · Conservation · Gene bank · Rapeseed

Abbreviations

CBD	Convention on Biological Diversity
CIMMYT	International Maize and Wheat Improvement Centre
FAO	Food and Agriculture Organization
IGV	Institute of Plant Genetics
IPGRI	International Plant Genetic Resources Institute
UPM	Universidad Politécnica de Madrid
VIR	Vavilov Institute of Plant Genetic Resources

4.1 Introduction

Crop genetic diversity is of tremendous importance for food and nutritional security and economic development in a world of 7.6 billion people, which could even reach 9 billion by 2050. Meeting the demands of food, energy and water for this large population and supporting livestock which are of direct use to human beings are going to be a daunting task in the future. The changing average temperatures, shifting rainfall pattern, extremes of heat and cold and unpredictable droughts will further make the task far more difficult. To feed the world population, the crop yields will need to grow at a rate of 2.4% annually, even though the current growth rate is 1.3% with yield stagnating in up to 40% of the land under cereal production (Ray et al. 2013). Therefore, unparallel efforts need to be made to promote the conservation and sustainable use of plant genetic resources of different crops, for food and agriculture.

In many cases crop diversity is at the root of sustainable agriculture and provides environmental benefit for healthy, biodiverse natural environment (MEA 2005; Harlow et al. 2010). Humanity has historically used a wide variety of wild and cultivated crop plants to meet their need for food, cloth fibres, housing materials and livestock feed. All the crops originated from wild plants. However, we modified/changed them to suit our needs. The early humans used to look out for the most attractive and nutritious fruits, seeds and tubers and, therefore, inadvertently selected for these and helped their propagation, thus favouring specific species and varieties.

In modern day we started a more organized way and developed a number of high-yielding and genetically uniform varieties of crops, with consistent high yield and quality, which now dominate global food production. This signifies that more genetically diverse crop landraces and varieties are either grown to a very less extent or have become extinct.

However, an additional critical factor is the conservation and availability of genetic diversity for use in breeding, for development of new varieties. This includes not just the genetic resources of crops but also those of wild relatives (Godfray et al. 2002; Tester and Langridge 2010). In the twentieth century, the crop productivity increased. However, there were major losses of crop genetic diversity. The United Nations' Food and Agriculture Organization has estimated that 75% of crop genetic diversity was lost (FAO 1998a) due to these efforts. A survey of 75 US crop species showed that 90% of historic varieties catalogued by the United States Department of Agriculture (USDA) are now extinct (Fowler and Mooney 1990). In Germany, approximately 90% of crop genetic diversity has been lost, and in Italy 75% has disappeared (Hammer et al. 2002). Today's crop diversity ensures tomorrow's food security and the livelihood and quality of life for billions of people. In 10000 years of history of agriculture, people have discovered about 50000 varieties of edible plants. Today, human beings cultivate 7000 plants for food, and many more for fibres, medicines and other purposes (Padulosi 1999). Just three crops – *Triticum aestivum*, *Oryza sativa*, and *Zea mays* – together provide more than half of humanity's global food supply and are staple foods for 4 billion people. The problem is threefold: decreasing diversity in farmers' fields, losing crop wild relatives in nature, and the precarious state of the world's plant diversity collections. Most of the collections may have already been lost, and others are in jeopardy. At the same time, wild relatives of domesticated plants are disappearing as their natural habitats are destroyed, and the diversity sown in farmers' fields is dwindling.

Gene banks are a reservoir of traits for resistance to diseases and insect pests and to climatic and other environmental stresses and thus hold the key to improving the quality and yield of crops. Some of the plant varieties held in gene banks are now extinct in farmers' fields and in nature; others are threatened with extinction. As deforestation and habitat loss wipe out wild crop relatives, and farmers give up traditional crop varieties, gene banks become the custodians of the crops that nurtured humankind for centuries. Ideally, gene banks act as vital centres of conservation, working with governments, farmers and others to use what is conserved and to conserve what is left in nature. Gene bank collections may include seeds stored at low humidity and low temperature, plantlets cultured in test tubes (in vitro conservation) and whole plants grown in special protected fields. The majority of gene banks globally are government operated. National and regional gene banks conserve crops of greatest importance to the food security and economic stability of the nation or region concerned. The world's 15 largest national gene banks are in China, the USA, the Russian Federation, Japan, India, the Republic of Korea, Canada, Brazil, Italy, Ethiopia, Hungary, Poland, the Philippines and at two national centres in Germany. Gene banks must actively conserve their collections. Of the more than 1000 spring bread wheat (*Triticum aestivum*) varieties released in developing countries from 1966 to 1996, the number of distinct landrace ancestors per pedigree has increased significantly over time. This conscious use of diversity has improved the yield stability of wheat, decreased the amount of nitrogen fertilizer needed and increased the crop's resistance to disease and tolerance to heat and drought (Cassaday et al. 2001). There are countless examples of the use of diversity in crop improvement.

Brassica is a wide genus with some very important cultivated species (*B. carinata*, *B. juncea*, *B. napus*, *B. nigra*, *B. oleracea*, *B. rapa*). The genus represents a wide range of crops including oilseed, many vegetables and fodder crops. The genus also contains many wild species. For several crops there are international networks coordinated by the International Plant Genetic Resources Institute (IPGRI). For *Brassica* such a network does not exist (yet). Large countries like the USA and China have their own national networks. In the USA several Regional Plant Introduction Stations hold collections, amongst them is *Brassica*. In China about 20 regional crop germplasm gene banks have been established in the last 15 years.

Wild crucifers hold tremendous genetic potential for the improvement of crop brassicas either as donors of useful nuclear genes or sources of cytoplasmic and genetic male sterility. While some of them are invasive weeds in crop fields causing considerable economic losses, others such as *Crambe*, *Camelina*, *Eruca* and *Diplotaxis* (Figs. 4.1, 4.2, 4.3, 4.4 and 4.5) could offer potential industrial and domestic oil-yielding crops of the future. These could be deployed for development of specialized products and pharmaceuticals. Concerning viability, the oldest material stored in the UPM (Universidad Politécnica de Madrid) seed bank has maintained its viability practically intact (average 98.4%) after 40 years, a result by far unmatched by any other bank. The preservation procedure involves ultradrying with silica gel and keeping a small amount of this material together with the seeds inside flame-sealed glass vials. Though it was originally adapted to small-sized seeds (as those of the family Brassicaceae), the use of some larger and sufficiently hermetic containers permits the preservation of larger seed samples. In this way, present rejuvenation cycles of 25–35 years can probably be extended to one to two centuries or more.

4.2 Strategies for Conservation

After the establishment of the Nikolai Ivanovich Vavilov Research Institute (VIR) of Plant Industry in St. Petersburg, Russia, in 1921, the importance of germplasm was realized, and a number of gene banks were established across the globe. However, most gene banks were started in the last three decades, when it became increasingly clear that without an activity directed towards the conservation of genetic resources, nothing would be achieved for future generations. In the 1980s the International Board for Plant Genetic Resources (IBPGR), and several other institutions from different regions of the world, realized genetic erosion in cruciferous crops. IBPGR, therefore, constituted a team of experts and asked them to submit status reports. In order to prevent the erosion in these crops, several germplasm collection and conservation missions were initiated. In Europe, being the centre of origin and diversity for *Brassica oleracea*, landraces of cruciferous crops were collected (Meer et al. 1984). Wild species related to *Brassica oleracea* were collected in Mediterranean countries sponsored by IBPGR (Gomez-Campo and Gustafsson 1991). Systematic, planned and coordinated efforts for germplasm collection and evaluation in India resulted in augmentation of over 17,439 accessions of cultivated, wild and allied species being maintained in All India Coordinated Research Project on Oilseeds (Singh and Sharma 2007).

4.2.1 IPGRI-/IBPGR-Sponsored Collecting Missions of Wild (n = 9) **Brassica** *spp.*

The 1980s period was marked by a series of germplasm exploration-collection missions sponsored by the International Plant Genetic Resources Institute (IPGRI) (now Bioversity International) to obtain relative species of *Brassica oleracea* in the Mediterranean costs of Spain, Italy, Tunisia and Greece and of wild *B. oleracea* itself in the Atlantic coasts of North Spain, the United Kingdom (UK) and France (Gomez-Campo 2009). According to Bioversity policies, the collected material was distributed into three parts. One part was stored in the Universidad Politécnica de Madrid (UPM, Madrid, Spain) seed bank – which had previously been designed by Bioversity as the base bank for wild crucifers. A second duplicate was stored in the bank of Tohoku University (Sendai, Japan). The third duplicate was stored in a bank within the country where the material was collected, namely, the Conservatoire Botanique de Porquerolles in France, the Banca Nazionale del Germoplasma in Bari (Italy), the North Greece Gene bank of Thessaloniki (Greece) and the Kew Gardens in the UK (Gomez-Campo 2009). This series of missions was very rewarding not only because of the number of accessions collected but of the many previously undetected populations that were detected, studied and collected. Special attention was given to the demography and to the conservation status of each population. Seeds of some species such as *Brassica hilarionis* were collected for the first time (Gomez-Campo 2009).

Gomez-Campo (2009) found a very interesting picture about Sicily (Italy), in the middle of the Mediterranean Sea, because of its high diversity for four *Brassica* species, namely, *B. incana*, *B. villosa*, *B. rupestris*, and *B. macrocarpa*. Amongst these, *B. villosa* and *B. rupestris* have three subspecies each. *B. macrocarpa* inhabits in the small Egadi archipelago offshore West Sicily. Sicily might be the geographical area where the $n = 9$ *Brassica* group originated.

Gomez-Campo (2009) mentioned about the complementary expedition to Turkey, sponsored by the Tohoku University in 1983, just before the Eubea collections; as in July 1982, an expedition was held in Greece to collect *B. cretica* Lam. seeds in Eubea, Attica and Peloponissos. In July 1983 a second expedition covered the island of Crete as well as some additional areas in Eubea and Attica. The Turkey expedition included M. Ozturk from Ege University, S. Tsunoda and K. Hinata from Tohoku University and C. Gómez-Campo from the Universidad Politécnica of Madrid. They covered the east part of Anatolia and the coasts of the Black sea and the Mediterranean, where the material collected was abundant and diverse within the family Brassicaceae. Amongst other materials, several *Crambe* species of Section Crambe (syn. Sarcocrambe) were obtained. Duplicates were stored in Madrid (Spain), Sendai (Japan) and Izmir (Turkey).

4.2.2 Other Collections

The total number of known conserved *Brassica* accessions totals 74,041. From these accessions very large amounts are conserved in the East and Far East and in Europe, followed by Americas, Oceania, Australia and Africa. In Europe the collections are spread over many countries. The UK holds the maximum number of accessions, followed by Germany, France, Spain, Russia and Bulgaria. Other countries like the Czech Republic, the Netherlands, Portugal, Ukraine, Poland and Italy hold only 1%. In the East and Far East, some countries hold very extensive collections like in China (17%) and India (15%). Also considerable numbers are held by Korea, Japan, Taiwan and the Philippines.

Gomez-Campo (2009) reported that the period of 1991–1995 did not see many crucifer collections by the UPM because the time was mostly devoted to complete collections of endangered species, in general. However, within the same decade, from 1995 to 1997, the collection of accessions of the species *Eruca vesicaria* is very fruitful. This species has a maximum diversity in the Iberian Peninsula with two subspecies – subsp. *vesicaria* and subsp. *sativa* – which grow on different types of soils. At least 100 populations were first collected and then morphologically characterized (Gómez-Campo 2005). Collections of wild *Eruca* have been also carried out by D. Pignone of the former CNR Institute of Plant Genetics (actually Institute of Biosciences and Bioresources, IBBR) in Bari, Italy.

In Europe, where many national collections of different size exist, IPGRI, as of December 2006, operates under the name Bioversity International, co-ordinating the European Cooperative Programme for Crop Genetic Resources Networks (ECP/GR), which was founded in 1980. In this programme, crop working groups were established. The *Brassica* Working Group was established in 1989, and the first meeting took place in 1991 (IBPGR 1993). In the Working Group meeting, activities are discussed regarding collection, regeneration, safe duplication, descriptor lists, in situ conservation, and research and international collaborations (Gass et al. 1995; Maggioni et al. 1997; Maggioni 1998). An important activity of the working groups is the establishment of European databases. The European *Brassica* database (Bras-EDB) was created by the CGN after a decision in the 1991 Working Group meeting and is continuously being updated (Hintum and Boukema 1993; Boukema and Hintum 1998). Although no worldwide *Brassica* network exists, for overviews of the world collections of *Brassica*, the inventory made by FAO of all world collections can be used (FAO 1998).

4.3 Seed-Drying Procedures

Several seed-drying procedures at the International Crops Research Institute for the Semi-Arid Tropics (Rao and Bramel 2000) methods are available for drying seeds available. The most common and safe methods used for drying are dehumidified drying and silica gel drying.

4.3.1 Dehumidified Drying

The FAO/IPGRI Genebank Standards recommend the use of 15 ± 5% RH and 15 ± 5 °C temperature for drying seeds. For smaller gene banks, seed-drying cabinets are designed to provide these environmental conditions. Larger gene banks, however, need modular walk-in seed-drying rooms. The seeds packed in muslin cloth bags are placed on the open racks of the drying room or seed-drying cabinet or in the drying room or cabinet, until the moisture content is likely to be in the range required for storage.

4.3.2 Silica Gel Drying

Small samples can be dried using silica gel. Dried silica gel (deep blue in colour) is placed in desiccators or glass jars with an airtight seal. The weight of the silica gel used should be equal to the seeds for efficient drying. Place the seeds in muslin cloth bags, and keep them in close proximity to the silica gel, and keep the desiccators at 15–20°C. Change the silica gel daily or when the colour changes from beep blue to pink or pale blue.

4.3.3 Seed-Drying and Seed Storage Procedures in the UPM Seed Bank

The procedure used to preserve crucifer seeds in the UPM seed bank was designed with complete independence from the procedures used by the other eight to nine banks existing since 1966. Instead of placing the emphasis on low temperatures, the emphasis was rather placed on low moisture content. The low moisture content was obtained by placing some dehydrated silica gel together with the seeds in sealed grass tubes. The silica gel absorbs moisture from the seeds and levels down to approximately 2% moisture content, a value which falls within what is practically considered ultra-desiccation. The use of silica gel has several additional advantages. It is manufactured with an indicator which warns – by changing its colour – if moisture accidentally enters the container. It also absorbs some toxic gases produced during seed ageing, which eventually constitute the ultimate cause of seed death.

Thus, silica gel protects the seeds by means of two independent mechanisms, reducing water vapours and reducing toxic gases. Of particular practical interest is the possibility of a visual control from outside of possible entrances of water vapour.

To keep the seeds ultradry, it is necessary to use reliable waterproof containers. Flame-sealed glass provides the most reliable solution. Standard laboratory glass vials are stretched and closed using the flame of a Bunsen burner (helped or not by a blower, depending upon the size of the tubes). Some 100 of such ampoules are usually made for each accession.

To prevent the breakage of the sealed glass tip, an external label is rolled around, and a melted mixture of wax (2/3) and resin (1/3) is poured inside (Gómez-Campo 2006).

The ampoules are then stored in a cold room at a moderately low temperature usually between 5°C and −5°C.

4.4 Conservation Approaches

4.4.1 In Situ Approaches

The conservation method used depends on the biology of the plant. Many food crops, including the most common – wheat, rice and maize – produce seed that if dried, and kept dry and cold, can remain viable for many years. Typically, this means placing the dried seed in sealed jars or packets and storing them for 20–30 years at temperatures of −5°C. Seeds in long-term storage for up to 100 years are kept at even lower temperatures. Some plant species have short-lived seeds that are sensitive to drying and cooling and are more difficult and expensive to maintain. To a large extent, conservation of these species must be individually tailored to their needs. A number of crops are propagated vegetatively – by tubers, roots and cuttings. These are stored either as whole plants grown in field gene banks or in vitro – as plantlets in test tubes. They include many important species such as potato, sweet potato, banana, cassava and yam. Finally, it is becoming more common to store cells for long and even indefinite periods by 'cryopreservation': freezing in liquid nitrogen at −196°C. To ensure safety of the collections, gene banks must create duplicate backup collections for safe storage in a separate location. In addition, it is important for gene bank workers to maintain the health of collections in a separate location, with periodic inspection of plants and cleaning the collections of diseases and pests. Wheat provides a particularly good example. Sonalika is one of the best known wheat varieties ever released in developing countries. To develop this variety, CIMMYT breeders used landraces from about 17 countries and breeding lines from 14 countries. All were sourced from the in-trust collection. Six continents contributed to just one small section of Sonalika's pedigree. Two East Asian landraces provided the dwarfing genes that made wheat plants shorter, helping to create the semidwarf wheat varieties that have dramatically improved wheat yields over the past 30 years.

In situ conservation is a strategy complementary to the ex situ approach of gene banks, in which germplasm is conserved in its normal habitat, i.e. in nature reserves or 'on farm'. Though in situ methods are generally unsuitable for conserving cultivated material, this approach is particularly recommended for conserving wild species. One of the reasons for this is that most wild species are very difficult to regenerate ex situ. Wild species are mainly included in ex situ collections to make them readily available to the users. A strategy for in situ conservation of wild species related to *Brassica oleracea* was proposed by Gustafsson (1995), which was elaborated later (Maggioni et al. 1997; Maggioni 1998).

Availability of genetic resources held in gene banks and public research institutes in industrialized and developing countries has, traditionally, always been freely accessible, especially for public-oriented research, unlike the material in collections held by private industries. However, this free access will no longer be guaranteed in the future due to a number of developments. There has been a trend in recent years to privatize breeding and research institutes. As a result, germplasm is considered an asset, and, therefore, access to third parties is no longer obvious in all cases. Furthermore, there is an ongoing discussion on an international level concerning access to plant genetic resources. The Convention on Biological Diversity (CBD) gave nations the jurisdiction over their biological resources.

4.5 Species and Crops

Gomez-Campo (1999) reported that *B. oleracea* had the highest number of accessions (27%) followed by *B. rapa* (25%). *B. napus* and *B. juncea* are each represented with 18% of the collections, while *B. carinata* and *B. nigra* are much less represented, with only 2% and 1% of the total collections, respectively. The wild species are represented with less than 1% of the total diversity collected. In about 7% the collected proper taxonomical classification is not known or is difficult to interpret owing to various reasons including morphological similarity. Amongst cultivated brassicas, *B. oleracea* var. *capitata* (cabbage) had the highest number of accessions (37%), followed by *B. oleracea* var. *botrytis* (20%), comprising cauliflower (the main part) and broccoli and *B. oleracea gemmifera* (Brussels sprouts). In 28% of collections, no intraspecific name is provided.

A large number of *Brassica* accessions are conserved in collections all over the world. The high number of *B. oleracea* in European collections, *B. juncea* in India and China, *B. carinata* in Ethiopia, and *B. rapa* in Europe and the East and Far East reflects the centres of origin and diversity of these species (IBPGR 1981). Regeneration of *Brassica* is very costly. Therefore good storage conditions in order to maintain the seed viability are essential. It is also very important to avoid unnecessary duplicates, to avoid huge investments in maintaining them. An international central crop database with high-quality and highly accessible data would be of great help. This would also encourage a more efficient use of the material. For the wild *Brassica* species, in situ conservation is of high importance; the ex situ collections help to make a small part of the diversity readily accessible for utilization. In order

Fig. 4.1 Morpho-physiological variation in *Diplotaxis*: (**a**) pre-flowering plants of *D. cretacea*. (**b**) Siliqua arrangement in *D. cretacea*. (**c**) Flowers of *D. cretacea* and (**d**) siliqua of *D. cretacea*. (**e**) A young plant of D. *berthautii*. (**f**) A young plant of *D. harra* ssp. crassifolia. (**h**) Siliqua structure of *D. harra* ssp. crassifolia. (**i**) Flowering plants of *D. muralis*. (**j**) Siliqua arrangement of *D. muralis*. (**k**) Leaf structure of *D. muralis*. (**l**) Siliqua structure of *D. muralis*. (**m**) A flowering plant of *D. siettiana*. (**n**) Siliqua arrangement in *D. siettiana*. (**o**) Siliqua structure in *D. siettiana*. (**p**) Flowers of D. siifolia. (**q**) Flowering plants of *D. siifolia*. (**r**) Leaf structure of *D. siifolia*. (**s**) Flower of *D. siifolia*. (**t**) Siliqua arrangement in *D. siifolia*. (Source: Pratap and Gupta 2009)

Fig. 4.2 Morpho-physiological variation in genus *Brassica*: (**a**) pre-flowering plants of *B. fruticulosa*. (**b**) Leaf structure in *B. fruticulosa*. (**c**) Fruiting plants of *B. tournefortii*. (**d**) Flowers of *B. Fruticulosa* ssp. cosoniana. (**e**) Constricted siliquae of *B. fruticulosa* ssp. cosoniana. (**f**) A young plant of *B. oxyrrhina*. (**g**) Siliquae of *B. fruticulosa* ssp. cosoniana. (**f**) A young plant of *B. oxyrrhina*. (**g**) Siliquae pattern in *B. desnottessi*. (**h**) Flowers of *B. desnottessi*. (**i**) Leaf structure in *B. desnottessi.* (Source: Pratap and Gupta 2009)

to free availability of *Brassica* germplasm to future generations, funding agencies, such as national and international governments, need to commit themselves to support coherent conservation programmes. The present author's observations on the morpho-physiological variation in different wild allies have already been discussed along with the recent literature in biology and ecology of crucifers (Pratap and Gupta 2009) and also presented in Figs. 4.1, 4.2, 4.3, 4.4 and 4.5. Some of the species have great importance in breeding programme of crop *Brassicas*, either as donor of useful genes or to supply andro-sterility (Gomez-Campo 1980; Prakash and Hinata 1980.

Fig. 4.3 Morpho-physiological variation in *Erucastrum*: (**a**) flowers of *E. abyssinicum*. (**b**) *Siliquae of E. abyssinicum*. (**c**) A young plant of *E. arabicum*. (**d**) Pre-flowering plant of *E. virgatum*. (**e**) *Flowers of E. gallicum*. (**f**) Pod arrangement in *E. gallicum*. (**h**) *Flowers of E. gallicum*. (**h**) Flowers of E. laevigatum. (**i**) Pre-flowering plant of *E. laevigatum*. (**j**) Leaf structure in *E. laevigatum*. (**k**) Pre-flowering plant of *E. laevigatum* ssp. *elongatum*. (**l**) Pre-flowering plant of *E. nasturtifolium*. (**m**) Flowers of *E. nasturtifolium*. (**n**) A young plant of *E. littoreum*. (Source: Pratap and Gupta 2009)

Fig. 4.4 Morpho-physiological variation in *Trachystoma*, *Moricandia*, *Crambe*, *Hirschfeldia and Sinapis*: (**a**) Flowers of *T. ballii*. (**b**) Pre-flowering plants of *T. ballii*. (**c**) Leaf structure of *T. ballii*. (**d**) A flowering plant of *M. arvensis*. (**e**) Plants of *Crambe abyssinica* at reproductive phase. (**f**) Inflorescence of *C. abyssinica*. (**g**) Spherical fruits of *C. Abyssinica*. (**h**) Leaf structure of *C. abyssinica*. (**i**) Pre-flowering plants of *H. incana*. (**j**) Flowers *of H. incana*. (**k**) Leaf structure of *C. incana*. (**l**) Pre-flowering plants of *H. incana* ssp. *cosobrina*. (**m**) A young plant of *S. alba*. (**n**) Pre-flowering plants of *S. pubescens*. (**o**) Flowers of *S. pubescens*. (**p**) Leaf structure of *S. pubescens*. (Source: Pratap and Gupta 2009)

Fig. 4.5 Morpho-physiological variation in *Sisymbrium*: (**a**) Pre-flowering plant of *S. irio*. (**b**) Pre-flowering of plants of *S. officinale*. (**c**) Leaf structure of *S. officinale*. (**d**) Mature pods of *S. orientale*. (**e**) Siliqua pattern in *T. arvense*. (**f**) Mature siliquae of *T. arvense*. (**g**) *Leaf structure of T. arvense*. (Source: Pratap and Gupta 2009)

References

Boukema, I. W., van Hintum, Th. J. L. (1998). The European Brassica database. In G. Thomas & A. Monteiro (Eds.). Proceedings of an international symposium on *Brassicas*, ISHS. *Acta Horticultura,* 459:249–254.

Cassaday, K., Smale, M., Fowler, C., & Heisey, P. W. (2001). Benefits from giving and receiving genetic resources: The case of wheat. *Plant Genetic Resources Newsletter*, No. 127.

FAO. (1998). *The state of the World's plant genetic resources for food and agriculture*. Rome: Food and Agriculture Organisation.

Fowler, C., & Mooney, P. (1990). *The threatened gene – Food, politics, and the loss of genetic diversity*. Cambridge: The Lutworth Press.

Gass, T., Gustafsson, M., Astley, D., & Frison, E.A., compilers. (1995). *Report of a working group on Brassica*. 2nd meeting, 13–15 November 1994, Lisbon.

Godfray, H. C. J., Beddington, J. R., Crute, I. R., Haddad, L., Lawrence, D., Muir, J., Hammer, F., Gladis, K. T., & Diederichsen, A. (2002). *In situ* and on-farm management of plant genetic resources. *European Journal of Agronomy, 19*, 509–517.

Gómez-Campo, C. (1980). Studies on Cruciferae: VI. Geographical distribution and conservation status of Boleum Desv., Guiraoa Coss. and Euzomodendron Coss. *Anales del Instituto Botánico A.J. Cavanilles, 35*, 165–176.

Gómez-Campo, C. (1999). *Biology of Brassica coenospecies*. Developments in plant genetics and breeding. *Elsevier Science B.V.*, 490.

Gómez-Campo, C. (2009). Collection, preservation, and distribution of wild genetic resources. In S. K. Gupta (Ed.), *Biology and breeding of crucifers* (pp. 361–374pp). Boca Raton: CRC Press/Taylor & Francis Group.

Gomez-Campo, C., & Gustafsson, M. (1991). Germplasm of wild n=9 Mediterranean species of Brassica. *Botanika Chronika, 10*, 429–434.

Gómez-Campo, C., Aguinagalde, I., Ceresuela, J. L., Lázaro, A., Martinez-Laborde, J. B., ParraQuijano, M., Simonetti, E., Torres, E., Tortosa, M. E. (2005). An exploration of wild *Brassica oleracea* L. germplasm in Northern Spain. *Genetic Resources and Crop Evolution, 52*, 7–13.

Gómez-Campo, C., Aguinagalde, I., Ceresuela, J., Lázaro, A., & Martınez-Laborde, J. (2006). Erosion of genetic resources within seed gene banks: The role of seed containers. *Seed Science Research, 16*, 291–294.

Gustafsson (1995). A strategy for in situ conservation of wild Brassica, Pp. 75–76. In: Gass, T.,Gustaffson, M., Astley, D. & Frison, E. (Compilers), Report of a working group on Brassica, second meeting, 13–15 November 1994, Lisbon. ECP/GR. International Plant Genetic Resources Institute, Rome.

Hammer, K., Gladis, T. H., & Diederichsen, A. (2002). In situ and on-farm management of plant genetic resources. *The European Journal of Agronomy, 19*, 509–517.

Harlow, J., Hopkins, J., Stone, D., Waters, R., Lovett, J., & Dorling, D. (2010). No charge? Valuing the natural environment: Technical report. *Natural England research reports,* Number032. file:///C:/Users/INDIAN/Downloads/NERR032%20edition%201.pdf.

IBPGR. (1981). *Genetic resources of cruciferous crops*. The *International Board for Plant Genetic Resources* (IBPGR), Secretariat consultation on the genetic resources of cruciferous crops, 17–19 November 1980.

IBPGR (1993). Report of working group on Brsicas (First meeting 1991). Europn cooperative programme for the conservation nd exchange of crop genetic resources (ECP/GR), IBPGR, Rome.

Maggioini, L. (1998). Activities and achievements of the ECP/GR Brassica working group. In: Thomas, G. and A. Monteiro (Eds.), *Proceedings of an international symposium on Brassicas*, ISHS. *Acta Horticultura*, 459:243–248

Maggioni, L., Astley, D., Gustafsson, M., Gass, T., & Lipman, E., compilers. (1997). *Report of a working group on Brassica*. Third meeting, 27–29 November 1996, International Plant Genetic Resources Institute, Rome, Italy, 92pp.

MEA (MILLENNIUM ECOSYSTEM ASSESSMENT). (2005). Ecosystems and human Well-being: Synthesis. Washington DC: Island Press.

Padulosi, S. (1999). *Final report: Conservation and use of underutilized Mediterranean species*. International Plant Genetic Resources Institute, Regional Office for Central and West Asia and North Africa (CWANA), Aleppo, Syria Panayotou.

Prakash, S., & Hinata, K. (1980). Taxonomy, cytogenetics and origin of crop *Brassicas*, a review. *Opera Botanica, 55*, 1–57.

Pratap, A., & Gupta, S. K. (2009). Biology and ecology of crucifers. In S. K. Gupta (Ed.), *Biology and breeding of crucifers*. Boca Raton: CRC Press/Taylor & Francis Group.

Rao, N. K., & Bramel, P. J. (2000). *Manual of gene bank operations and procedures*. Technical manual no. 6. Patancheru, A. P. 502 324, India: International Crops Research Institute for the Semi-Arid Tropics. 190 pp.

Ray, D. K., Mueller, N. D., West, P. C., & Foley, J. A. (2013). Yield trends are insufficient to double global crop production by 2050. *PLoS One, 8*(6), e66428. https://doi.org/10.1371/journal.pone.0066428.

Singh, R., Sharma, S.K., (2007). Evaluation, maintenance and conservation of germplasm. In S. K. Gupta (Ed.), *Advances in botanical research–rapeseed breeding* (Vol. 45, pp. 465–481). Londo: Academic Press.

Tester, M., & Langridge, P. (2010). Breeding technologies to increase crop production in a changing world. *Science, 327*, 818–822. https://doi.org/10.1126/science.1183700.

van der Meer, Q. P., Toxopeus, H. C., Roelofsen, H., & Astley, D. (1984). The collection of landraces of cruciferous crops in the EC countries. *Final report of the EC research programme* 0890. IVT, Wageningen. 209pp.

van Hintum, T. J. L., & Boukema, I. W. (1993). The establishment of the European database for Brassica. *Plant Genetic Resources Newsletter, 94/95*, 11–13.

Chapter 5
Integrated Farming System: Enhancing Income Source for Marginal and Small Farmers

Sohan S. Walia, Vikrant Dhawan, Ashok K. Dhawan, and N. Ravisankar

Abstract Integrated farming is defined as a biologically integrated system, which integrates natural resources in a regulated mechanism into farming activities to achieve maximum replacement of off-farm inputs and sustain farm income. The productivity of a diversified cropping system always tends to increase when it is integrated with dairy, poultry or fishery components. An integrated farming system (IFS) helps farmers, especially small and marginal, to achieve maximum returns and income from different integrated components, thereby improving their standard of living. The IFS also acts as a means for providing nutritional security to a farmer's family as the farmer is able to provide various IFS components such as vegetables, fruits, egg, milk, fish, etc. to his family and get the income from the surplus amount of these components. The higher returns with the farming system were not only due to higher productivity of the system but also due to lower cost of production and recycling of by-products of crop components. There is an increase in the value for labour absorption in IFS farms due to additional components brought into integration within the farm. The IFS is feasible with respect to socio-economic imperatives, but actual adoption rates of integrated farming are limited and unevenly spread among farmers. Thus, in order to develop a nation, farmers should be properly made aware of the use and management of IFS.

S. S. Walia (✉)
Punjab Agricultural University, Ludhiana, Punjab, India
e-mail: sohanwalia72@yahoo.co.in

V. Dhawan
Agricultural Technology Management Agency (ATMA), Department of Agriculture and Farmer Welfare, Punjab, Roopnagar, Punjab, India
e-mail: vikrantdhawann@yahoo.com

A. K. Dhawan
The Indian Ecological Society, Ludhiana, Punjab, India
e-mail: indianecologicalsociety@gmail.com

N. Ravisankar
ICAR-Indian Institute of Farming System Research, Meerut, India
e-mail: agrosankar2002@yahoo.co.in

R. Peshin, A. K. Dhawan (eds.), *Natural Resource Management: Ecological Perspectives*, Sustainability in Plant and Crop Protection,
https://doi.org/10.1007/978-3-319-99768-1_5

Keywords Integrated farming · Farmers · Productivity · Income · Economics

Abbreviations

IFS	Integrated farming system
NEPZ	North-eastern plains zone
RGEY	Rice grain equivalent yield
CFS	Conventional farming system
NMR	Net monetary return
MSC	Multiple species cropping
FFB	Fresh fruit bunches
MRS	Most remunerative sequence
SBM	Sorghum-berseem-maize

5.1 Introduction

The most important achievement of the Green Revolution in India is the dramatic increase in crop productivity, particularly of cereals, namely rice, wheat and maize. As a result, the total food grain production witnessed manyfold increase in a short span of 50 years. During the 1970s and mid-1980s, the agricultural growth rate increased rapidly and thereafter started steady declining and disturbing the environment, as a whole. Nowadays, agriculture is faced with the challenge of producing food for a rapidly growing world population. To meet this demand, a sound "farming system" is crucial.

The household, its resources and the resource flows and interactions at the individual farm levels are together referred to as a farming system. It is a set of agro-economic activities that are interrelated and interact among themselves, in a particular agrarian setting. Furthermore, it is a complex interrelated matrix of soil, plants, animals, implements, power, labour, capital and other inputs controlled in parts by farming families and influenced to varying degrees by political, economic, institutional and socials forces. These operate at many levels and therefore designate a set of agricultural activities that are organized while preserving land productivity.

A farming system can be described and understood by its structure and functioning. The structure in its wider sense includes land use pattern, production relations, land tenures, size of holdings and their distribution, irrigation, marketing including transport and storage, credit institutions, financial markets, research and education. To achieve this goal, the individual farmer allocates certain quantities and qualities of four production factors – land, labour, capital and management – which are necessary processes in crop, livestock and off-farm enterprises. In describing farming systems and their characteristics, it is noteworthy why farming in a specific case is carried out in one way rather than another.

A chain of interactions exists among the components within farming systems, and it becomes difficult to deal with such complex interlinked systems. This is one of the reasons for slow and inadequate progress in the field of farming system research in the country. This problem can be overcome by construction and application of suitable whole farm models (Dent 1990). However, it should be mentioned that inadequacy of available data from the whole farm perspective currently constrains the development of whole farm models.

5.2 Types of Farming Systems

The classification of farming systems has been based on a number of key factors, including (i) the available natural resource base; (ii) the dominant pattern of farm activities and household livelihoods, including relationship to markets; and (iii) the intensity of production activities. These criteria were applied to each of the six main regions of the developing world. The 72 farming systems identified in the 6 developing regions can be grouped into 8 major categories. These are (i) smallholder irrigated farming systems; (ii) wetland rice-based farming systems; (iii) rainfed farming systems in humid areas; (iv) rainfed farming systems in steep and highland areas; (v) rainfed farming systems in dry or cold areas; (vi) dualistic farming systems with both large-scale commercial and smallholder farms; (vii) coastal artisanal fishing mixed farming systems; and (viii) urban-based farming systems. Except in the case of the dualistic category, smallholder producers dominate these system types (Dixon et al. 2001). However, Gill et al. (2009) reported that the study conducted by the All India Coordinated Research Project (AICRP) on cropping system, based on survey through resource characterization in different state agricultural universities, elucidates that there are 14 predominant farming systems in the country.

5.3 Landholding Scenario in India

The Indian economy is predominantly rural and agricultural, and the declining trend in the size of landholding poses a serious challenge to the sustainability and profitability of farming. It may be noted that Indian agriculture is home for small and marginal farmers (80%). Therefore, the future of sustainable agriculture growth and food security in India depends on the performance of such farmers. According to the Agricultural Census (2010–2011), there were an estimated 98 million small and marginal holdings, out of around 120 million total land households in the country. The average size of landholdings has declined from 2.3 ha in 1970–1971 to 1.37 ha in 2000–2001, which further declined to 1.23 ha in 2005–2006 and 1.15 ha in 2010–2011. If this trend continues, the average size of holding in India would be a mere 0.68 ha in 2020 and would be further reduced to a low of 0.32 ha in 2030 (DES 2009).

Small and marginal farmers (<2 ha) account for more than 80% of total farm households, but their share in operated area is around 44%. Thus, there are significant land inequalities in India. As shown in Table 5.1, the share of marginal and small

Table 5.1 Changes in percentage distribution of farm holdings and operated area

	Distribution of farm holdings (%)				Distribution of operated area (%)			
Land class[a]	1960–1961	1981–1982	1991–1992	2002–2003	1960–1961	1981–1982	1991–1992	2002–2003
i. Marginal	39.1	45.8	56.0	62.8	6.9	11.5	15.6	22.6
ii. Small	22.6	22.4	19.3	17.8	12.3	16.6	18.7	20.9
Sub-total [(i+ii) Small and marginal]	61.7	68.2	75.3	80.6	19.2	28.1	34.3	43.5
iii. Semi medium	19.8	17.7	14.2	12.0	20.7	23.6	24.1	22.5
iv. Medium	14.0	11.1	8.6	6.1	31.2	30.1	26.4	22.2
v. Large	4.5	3.1	1.9	1.3	29.0	18.2	15.2	11.8
Total (i+ii+iii+iv+v)	100	100	100	100	100	100	100	100

Source: NCEUS (2008). National Sample Survey Land Holdings 8th, 17th, 26th, 37th, 48th, 55th rounds, Central Statistical Organization, Government of India

[a]Marginal = 0.01 to 1.00 ha; small = 1.01 to 2.00 ha; semi-medium = 2.01 to 4.00 ha; medium = 4.01 to 10.00 ha; large >10 ha

farmers accounted for around 81% of operational holdings in 2002–2003 as compared to about 62% in 1960–1961. Similarly, the area operated by small and marginal farmers has increased from about 19% to 44% during the same period. Recent data for 2010–2011 shows that the share of small and marginal farmers in landholdings was 85% (MoA 2015).

5.3.1 Farm Size and Productivity

In terms of production, small and marginal farmers provide a larger contribution to the production of high-value crops. They contribute around 70% to the total production of vegetables, 55% to fruits against their share of 44% in land area (Birthal et al. 2011). Their share in cereal production is 52% and 69% in milk production. Thus, small farmers contribute to both diversification and food security.

The results of The National Sample Survey Office (NSSO) 2003 farmers' survey has empirically established that small farms continue to be productive more in value terms per ha than medium and large farms. It shows that, from an efficiency point of view, smallholdings are equal to or better than large holdings.

5.3.2 Smallholders and Livelihoods

It is true that smallholdings have higher productivity than medium and large farms, but this is not enough to compensate for the disadvantage of the small area of holdings. The cost of cultivation per hectare is also high on small and marginal farmers than medium and large farms (Fig. 5.1).

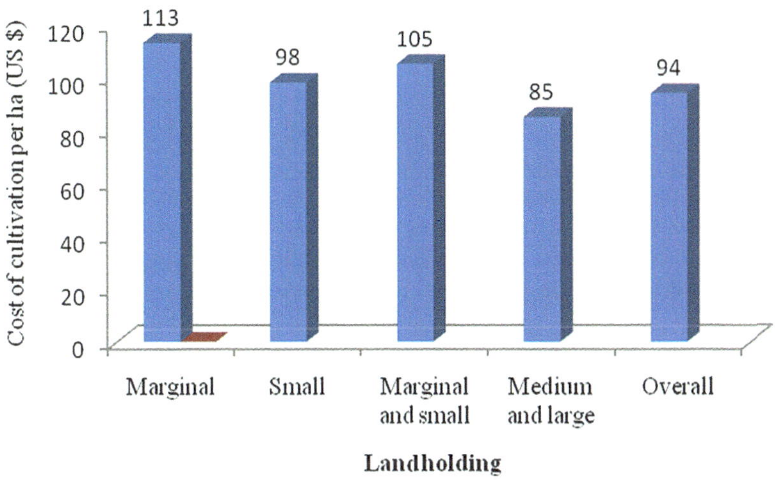

Fig. 5.1 Cost of cultivation per hectare. Marginal landholding ≤1 ha, small landholding = 1–2 ha, medium landholding = 4–10 ha, large landholding ≥10 ha. Source: Assessment Survey of Farmers 2003

Chand et al. (2011) observed that small farms in India are superior in terms of production performance but weak in terms of generating adequate income and sustaining livelihoods. The study shows that smallholdings below 0.8 ha do not generate enough income to keep farm family out of poverty, despite high productivity.

The burden on marginal farmers (≤ 1ha) is becoming unbearable. Under the gradual shrinking of landholding, it is necessary to integrate land-based enterprises like fishery, poultry, duckery, apiary, field and horticultural crops, etc. within the biophysical and socio-economic environment of the farmers to make farming more profitable and dependable (Behera et al. 2004). No single farm enterprise is likely to be able to sustain small and marginal farmers without resorting to integrated farming systems (IFS) for the generation of adequate income and gainful employment year round (Mahapatra 1994). Hence, it is necessary for such vulnerable sections of the farming community to adopt a "farming system approach".

Farming system approach to agricultural research and development efforts would accelerate agricultural growth in the country and thereby provide leverage for transforming poverty-prone rural India into a prosperous country, by strengthening its rural economy. Certainly this will play a key role and enable the academicians and policymakers to formulate and implement appropriate policies for a balanced, integrated and, overall, an agricultural revolution in the twenty-first century, which is very much important to make India a developed nation.

Farming system approach, therefore, is a valuable approach to address the problems of sustainable economic growth in farming communities in India.

5.4 Integrated Farming System (IFS)

Integrated farming is defined as biologically integrated system which, in a regulated mechanism, integrates natural resources into farming activities to achieve maximum replacement of off-farm inputs and sustain farm income (Titi et al. 1993). It emphasizes on and requires a holistic approach. Such an approach is essential because agriculture has a vital role to play, a role much wider than just production of crops, including providing diverse, attractive landscapes and encouraging biodiversity and conserving wildlife. The basic aim of IFS is to derive a set of resource development and utilization practices, which lead to substantial and sustained increase in agricultural production (Kumar and Jain 2005). Integrated farming seeks to reinforce the positive influences of agricultural production while meeting its negative impacts. It is a means for achieving a sustainable agriculture and is an indispensable part of sustainable development.

The future agricultural system should be reoriented from the single commodity system to food diversification approach to sustain food production and income. Integrated farming systems, therefore, assume greater importance for sound management of farm resources to enhance farm productivity, which will reduce environment degradation and improve the quality of life of resource-poor farmers, maintaining agricultural sustainability as well.

5.4.1 Advantages of Integrated Farming Systems

1. Security against complete failure of a system
2. Minimization of dependence for external inputs
3. Optimum utilization of farm resource
4. Efficient use of natural resources, such as sunlight, water, land, etc.

5.4.2 Productivity Enhancement by Integrated Farming Systems

The productivity of a diversified cropping system always tends to increase when it is integrated with dairy, poultry or fishery components. Agriculture in north-eastern plains zone (NPEZ) of Uttar Pradesh is subsistent in nature due to which farmers with a lowland base are trapped in a vicious circle of poverty→low income→low savings→low investment→low productivity→poverty. Singh et al. (2006a, b) compared seven crop sequences, namely rice (*Oryza sativa* L.)-mustard (*Brassica* spp.)-sorghum (*Sorghum bicolor*, C_1); rice-pea (*Pisum sativum*)-onion (*Allium cepa*)-(C_2); rice-pea-okra (*Abelmoschus esculentus*, C_3); rice-wheat-fallow (C_4); rice-wheat-moong (*Vigna radiata*, C_5); rice-berseem (*Trifolium alexandrinum*)-sorghum (C_6); and sorghum-berseem-maize (*Zea mays*, C_7), in order to diversify the rice-wheat

Table 5.2 Productivity (rice grain equivalent yield) of an integrated farming system (0.25 ha) (mean data of 2 years)

	Component productivity (kg)					System productivity (kg/ha)
Farming system	Crop	Dairy	Poultry	Fish	Total	Total
Rice-mustard-sorghum	2734.43				2734.43	10,937.72
Rice-pea-onion	3062.85				3062.85	12,251.38
Rice-pea-okra	4567.68				4567.68	18,270.72
Rice-wheat-moong bean	2455.24				2455.24	9820.90
Rice wheat fallow	2080.21				2080.21	8320.80
Rice-berseem-sorghum	2898.29				2898.29	11,593.10
Sorghum-berseem-maize	2590.82				2590.82	10,363.20
MRS + sorghum-berseem-maize + dairy	3527.43	26,588.11			30,115.54	1,20,462.10
MRS + sorghum-berseem-maize + poultry	3527.43		3131.165		6658.595	26,634.30
MRS + sorghum-berseem-maize + fish	3061.09			1066.98	4128.065	16,512.20
MRS + sorghum-berseem-maize + dairy + poultry	3475.62	26,588.11	3131.165		33,194.89	1,32,779.50
MRS + sorghum-berseem-maize + dairy + fish	3009.27	26,588.11		1066.98	30.664.35	1,22,657.00
MRS + sorghum-berseem-maize + fish + poultry	3009.27		3131.165	1066.98	7207.405	28,829.60
MRS + sorghum-berseem-maize + dairy + fish + poultry	2957.00	26,588.11	3131.165	1066.98	33,743.705	1,34,974.80

MRS (most remunerative sequence) = rice-pea-okra
Source: Singh et al. (2006a, b)

system. They observed that rice-pea-okra sequence was the most productive in terms of rice grain equivalent yield (RGEY). Further, when C_3 and C_7 were integrated with dairy, poultry and fishery, it resulted in a markedly higher system productivity (Table 5.2).

Singh et al. (2007) also reported an experiment carried out at the Agricultural Research Farm, Institute of Agricultural Sciences, Banaras Hindu University, Varanasi, during 2002–2003 and 2003–2004 to identify an efficient farming system for irrigated agroecosystem of eastern Uttar Pradesh. These findings also identified that the farming system comprising crop components (rice-pea-onion and rice-pea-okra), dairy, poultry and fishery was the most suitable and efficient farming system model, giving the highest system productivity, i.e. rice grain equivalent yield of 12.16 and 17.09 tons/ha, respectively, under irrigated agroecosystem of eastern Uttar Pradesh (Table 5.3).

Table 5.3 Productivity (rice grain equivalent yield) of integrated farming system (mean data of 2002–2003 and 2003–2004)

Farming system	Component productivity (kg)					System productivity (kg/ha)
	Crop	Dairy	Poultry	Fish	Total	Total
Rice-mustard-sorghum	2916				2916	11,664
Rice-pea-onion	3541				3541	12,164
Rice-pea-okra	4272				4272	17,088
Rice-wheat-green gram	3006				3006	12,024
Rice-wheat	2161				2161	8644
Rice-berseem-sorghum	2695				2695	10,780
Sorghum-berseem-maize	2405				2405	9619
MRS + sorghum-berseem-maize + dairy	3270	23,633			26,903	107,612
MRS + sorghum-berseem-maize + poultry	3270		3037		6307	25,228
MRS + sorghum-berseem-maize + fish	2857			1506	4363	17,452
MRS + sorghum-berseem-maize + dairy + poultry	3242	23,633	3037		29,912	119,648
MRS + sorghum-berseem-maize + dairy + fish	2809	23,633		1506	27,948	111,792
MRS + sorghum-berseem-maize + fish + poultry	2809		3037	1506	7352	29,408
MRS + sorghum-berseem-maize + dairy + fish + poultry	2761	23,633	3037	1506	30,937	123,748

Source: Singh et al. (2007)

The Tungabhadra project area of Karnataka is said to be the rice bowl, where 85% of farms are small and medium. Agriculture in the Tungabhadra project area of Karnataka is dominated by the rice-rice monocropping system. Out of 3.49 lakh (0.349 million) hectares, rice occupies an area more than 2 lakh (0.2 million) hectares. The natural resource is fatigued. The need for diversification in some of this area is clear, since the income of farmers who depend solely on the produce of their traditional monocrops of rice pattern is decreasing, due to narrow margins of profitability and changed food consumption habits. This demanded urgent development of a profitable IFS model equivalent or superior to the rice-rice system.

Channabasavanna et al. (2002) reported that a field experiment was conducted during the wet and dry seasons of 1996–1998 at the Agricultural Research Station, Siruguppa (Karnataka), the pooled data of which showed that the maximum production is in rice (fish)-rice (fish) with poultry (11,133 kg/ha), which was significantly superior to others (Table 5.4).

Channabasavanna and Biradar (2007) developed a rice-fish-poultry system model for rice-growing farmers of Karnataka, and out of the designed models, the

Table 5.4 Total production potential, cost of cultivation, net return and benefit/cost ratio, as influenced by rice-based farming systems

Treatments	Total productivity (kg/ha/year)				Cost of cultivation ($/ha/year)			Net returns ($/ha/year)			Benefit/cost ratio		
	1996–1997	1997–1998	1998–1999	Pooled	1996–1997	1997–1998	1998–1999	1996–1997	1997–1998	1998–1999	1996–1997	1997–1998	1998–1999
Rice-rice	8223	6970	741[illegible]	7535	338	371	364	419	397	470	2.24	2.07	2.29
Rice-sesame (*Sesamum indicum*)	6814	6140	6795	6583	269	300	303	350	373	465	2.30	2.24	2.54
Rice-chilli (*Capsicum frutescens*)	5766	5557	6314	5879	325	365	341	208	244	384	1.64	1.67	2.13
Rice-wheat (*Triticum aestivum*)-GM	7086	6912	7314	7104	326	345	341	314	410	488	1.96	1.92	2.43
Rice + fish (Ri)-rice + fish (dung)	8783	10,030	10,030	9614	558	537	449	285	556	680	1.51	2.03	2.51
Rice + fish (Ri) + rice + fish (poultry)-poultry	8670	12,457	12,273	11,133	556	1132	451	324	216	930	1.58	1.19	2.73
Rice + fish (Ri)-rice + fish (fish feed)	8476	9754	10,205	9478	689	655	508	154	411	640	1.22	1.63	2.26
Rice-rice-banana on bunds	8169	9156	10,130	9152	419	449	435	327	560	695	1.78	2.21	2.60
Rice-rice-drumstick (*Moringa oleifera*) and curry leaf on bunds	8307	7553	7761	7874	440	456	439	331	373	431	1.75	1.82	1.98
Rice-rice-para grass (*Brachiaria mutica*) and cowpea (*Vigna unguiculata*) on bunds + dairy	8101	10,270	10,127	9499	426	453	455	332	644	681	1.78	2.42	2.49
Rice (*Oryza sativa*) GLM-rice	9458	9408	9498	9455	367	401	395	494	624	674	2.35	2.55	2.70
Rice-rice-GM	8407	8454	8816	8559	354	389	382	450	539	614	2.18	2.38	2.41

Source: Channabasavanna et al. 2002

one with rice-fish (pit at the centre) and poultry (reared separately) showed highest productivity (17,502 kg/ha/year) (Table 5.5).

Channabasavanna et al. (2009) reported that IFS comprising components such as cropping, vegetables, fishery, poultry and goat rearing was studied at the Agricultural Research Station, Siruguppa, Karnataka, India, during the wet and dry seasons of 2003–2004 and 2005–2006. The aim was to find the productivity of IFS over conventional rice-rice system in the Tungabhadra project area of Karnataka. The integrated farming system approach recorded 26.3% higher productivity over the conventional rice-rice system (Table 5.6).

Murugan and Kathiresan (2005) conducted studies at Annamalai University, Annamalai Nagar, Tamil Nadu, in order to trace the impact of integrating farming enterprises, namely fish culture, poultry rearing and rabbit rearing along with lowland transplanted rice as crop yield. They found that the highest grain

Table 5.5 Productivity (rice grain equivalent), economics (pooled over 2 years) and employment generation of rice-fish poultry system

Treatment	Productivity (kg/ha/year)	Net returns ($/ha/year)	B/C ratio	[a]Employment generation (man-days)
Integrated farming system				
Rice-fish (pit at one side) and poultry (shed on fish pit)	15,291	795	1.73	600
Rice-fish (pit at one side connected by trenches) and poultry (shed on fish pit)	15,152	770	1.14	612
Rice-fish (pit at the centre) and poultry (reared separately)	17,502	1016	1.91	625
Rice-fish (pit at one centre connected by trenches) and poultry (reared separately)	14,605	609	1.57	625
Rice-fish (pit at four corners connected by trenches) and poultry (reared separately)	15,227	729	1.63	628
Mean	15,555	784	1.60	618
Conventional cropping system				
Control (rice-rice system)	6667	348	1.90	6667

Source: Channabasavanna and Biradar (2007)
[a]Mean over 2 years

Table 5.6 Productivity (rice equivalent yield), profitability and employment generation under integrated farming system and conventional cropping system (pooled data of 3 years)

	Productivity (kg/ha/year)	Net returns ($)	Employment generation (man-days/ha/year)
Integrated farming system	7088	369	275
Conventional cropping system	5611	279	459

Source: Channabasavanna et al. (2009)

Table 5.7 Productivity of integrated farming systems (mean data over 2 years)

Farming system	Component productivity (kg/ha)[a] Crop	Poultry	Mushroom	Rice grain equivalent yield (kg/ha)
Rice cropping alone	4311	–	–	4311
Rice-groundnut + mushroom + poultry	6557 (39)	6060 (36)	4305 (25)	16,922
Rice-cowpea + mushroom + poultry	7662 (43)	6060 (34)	4305 (23)	18,027
Rice-brinjal + mushroom + poultry	11,122 (52)	6060 (28)	4305 (20)	21,487
Rice-sunn hemp + mushroom + poultry	4933 (33)	6060 (39)	4305 (28)	15,358

Source: Manjunath and Itnal (2003)
[a]Figures in parentheses indicate percent contribution to total system productivity

yield of rice (5.67 tonnes/ha and 5.25 tonnes/ha) was obtained with integrated rice-fish-poultry farming systems, during the first and second seasons, respectively.

Goa, a small state in the west coast of India, has a geographical area of 370,100 ha with 34,112 ha landholdings, more than 80% of which is below 1 ha (NABARD 2001). The average landholding of a farmer is below 0.5 ha. Manjunath and Itnal (2003) reported that a field experiment was conducted at the ICAR research complex for Goa, Old Goa, during 1999–2000 and 2000–2001 to assess the productivity of rice-based cropping sequences, namely rice-brinjal, rice-cowpea, rice-groundnut (*Arachis hypogaea*), rice-sunn hemp (*Crotalaria juncea*) and sole rice integrated with poultry and mushroom production.

The highest system productivity (21,487/kg/ha/year) of rice grain equivalent yield was recorded with rice-brinjal (*Solanum melongena*) system integrated with mushroom and poultry, followed by rice-cowpea (18,027 kg/ha/year) and rice-groundnut system (16,922 kg/ha/year) (Table 5.7). The contribution of crops towards the system productivity ranged from 33% to 52%, while the share of poultry and mushroom production was 28–39% and 20–28%, respectively.

Mohanty et al. (2010a, b) reported on-farm experiment conducted at Khentalo village of Cuttack district of Odisha to know the impact of rice-fish-prawn culture productivity. A rice grain yield of 3.04 tonnes/ha in the rice-fish-prawn system was obtained, which was 16.9% higher than rice monocrop.

5.4.3 *Economic Benefits of Integrated Farming Systems*

IFS helps farmers, especially those small and marginal, to achieve maximum returns and income from different integrated components and thus improves their standard of living. The higher returns were not only due to higher productivity of the farming system but also due to lower costs of production and recycling of by-products from crop components.

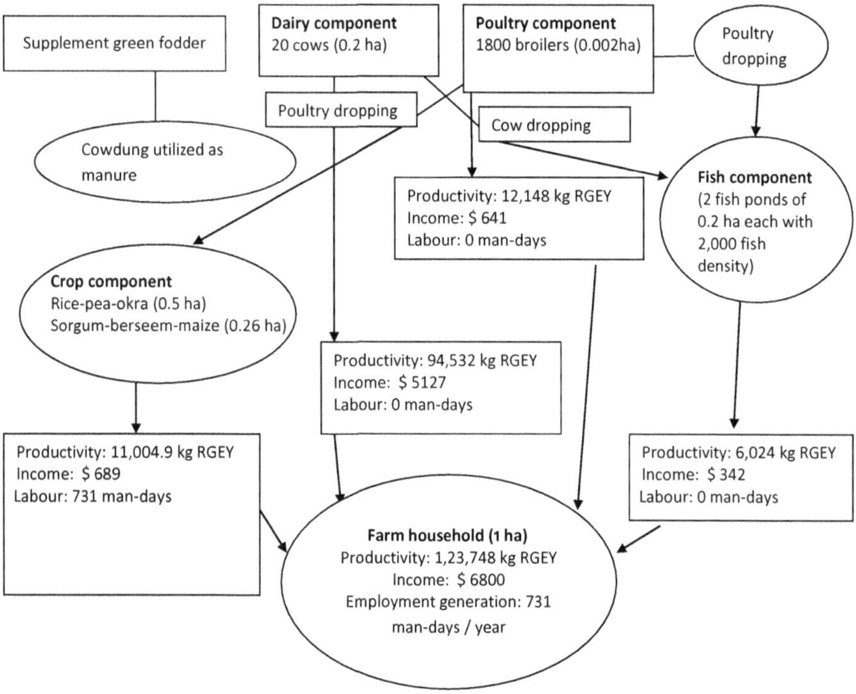

Fig. 5.2 Sustainable farming system model for irrigated agroecosystem of Varanasi and Chandauli region (NEPZ) of Uttar Pradesh. (Singh et al. 2006a, b)

Based on the research of Singh et al. (2006a, b), three models were developed for marginal and small farmers of NPEZ of Uttar Pradesh with 1 ha land. In all the three models, 0.5 ha area was allocated to rice-pea-okra sequence, which was the most remunerative sequence (Fig. 5.1). In a small area of the farm, the fodder sequence involving sorghum-berseem-maize was taken in all the three models to meet the green fodder requirement of the dairy component round the year (Fig. 5.2).

The feed requirement of fingerlings in the fish pond was met from the cow unit in models 1 and 3 and also through poultry droppings in model 1, while the surplus was utilized as manure. The recycling of waste of one component into the other reduced the input requirements and generated more income. Therefore, this will increase the profitability of small and marginal farmers. As regards the economics of the three models, the gross as well as net returns were maximum in model 1 involving crop, dairy, poultry and fishery components, closely followed by model 2 (Table 5.8).

The study by Channabasavanna et al. (2002) revealed that, at the end of the third year, the rice (fish)-rice (fish)-poultry system gave the highest net return and benefit/cost ratio ($ 930/ha/year and 2.73, respectively, see Table 5.5).

Table 5.8 Economics of different farming system models for the north-eastern plains zone of Uttar Pradesh

Model	Cost of cultivation/ production ($/ha)	Gross returns ($/ha)	Net returns ($/ha)	B/C ratio
1. MRS (0.5 ha) + SBM (0.26 ha) + dairy (0.02 ha) + poultry (0.02 ha) + fishery (0.2 ha)	4962	11,538	6576	1.33
2. MRS (0.5 ha) + SBM (0.46 ha) + dairy (0.02 ha) + poultry (0.02 ha)	4869	11,351	6481	1.33
3. MRS (0.5 ha) + SBM (0.28 ha) + dairy (0.02 ha) + poultry + fishery (0.2 ha)	4456	10,485	6029	1.35

MRS, most remunerative sequence; SBM, sorghum-berseem-maize
Source: Singh et al. (2006a, b)

The rice-fish-poultry system model, with rice-fish (pit at the centre) and poultry (reared separately), developed by Channabasavanna and Biradar (2007) for rice-growing farmers of Karnataka also revealed the profitability ($ 1016/ha/year) and B/C ratio (1.91) (Table 5.6).

The integrated farming system (IFS), comprising components like cropping, vegetables, fishery, poultry and goat rearing studied by Channabasavanna et al. (2009), reported 32.3% higher profitability over the conventional rice-rice system (Table 5.7).

The studies conducted at Annamalai University, Annamalai Nagar, Tamil Nadu, by Murugan and Kathiresan (2005) showed the highest net return of $ 2515/ha and $ 3679/ha during the first and second season, respectively, obtained with integrated rice + fish + poultry farming systems (Table 5.9).

It was followed by model C, with a net income of $ 229/ha/year. Based on the sustainability index derived for different models, model E was found superior with maximum sustainability for net returns (65.3%).

Nageswaran et al. (2009) took into account various parameters, such as input use pattern, cost, gross and net income and benefit/cost ratio, to compare IFS and conventional farming system (CFS) at Keela Manakudi village, Parangipettai block, Chidambaram (Tamil Nadu), during 1997–2002. Data on IFS components were collected from demonstration plot while that of conventional farming practices were collected from 32 randomly selected households. Twenty-three components were initially identified for integration. The IFS plot was taken on lease from a farmer for a duration of 3 years and later extended for 2 years. It was observed that the IFS approach realized higher returns in terms of net revenue. The average annual net revenue per acre of IFS was more than 2.5 times that of CFS (Table 5.11).

Govindan et al. (1990) studied the financial budgets of farms in Tamil Nadu. The study was designed to assess the financial viability of a poultry and fish culture system. They concluded that, under the prevailing conditions, higher incomes and on-farm labour use can be achieved by integrating different enterprises on the farm. Similarly, Rangasamy et al. (1996) studied the integration of poultry, fish and mushroom with rice cultivation over a 5-year period.

Table 5.9 Economics of integrated farming systems during two seasons

Components	Cost of cultivation		Grain yield		Meat yield		Fish yield		Net income	
	\$/40 m^2	\$/ha	\$/40 m^2	\$/ha	\$/40 m^2	\$/ha	\$/40 m^2	\$/ha	\$/40 m^2	\$/ha
First season (Samba)										
Rice	0.7	161.6	0.3	66.6					0.8	194.8
Rice + fish	0.9	226.6	0.3	65.3			0.0	5.6	1.2	287.9
Rice+ poultry	6.9	1721.9	0.4	92.9	0.2	60.3			9.6	2387.6
Rice + rabbit	2.7	674.1	0.3	85.2	0.1	3.6			2.1	455.7
Rice + fish + poultry	7.2	1795.2	0.4	91.5	0.2	60.3	0.0	6.9	10.1	2514.8
Rice + fish + rabbit	3.0	747.4	0.3	83.4	0.1	3.7	0.0	5.9	2.5	374.0
Second season (Navarai)										
Rice	0.7	165.6	0.3	67.1					0.7	181.9
Rice + fish	0.9	225.6	0.3	63.2			0.0	5.3	1.0	262.0
Rice + poultry	10.1	2537.3	0.3	86.1	0.4	93.6			14.1	3474.5
Rice + rabbit	3.2	790.4	0.3	76.5	0.1	16.4			2.4	588.3
Rice + fish + poultry	10.4	2609.0	0.3	84.7	0.4	93.6	0.0	7.8	14.7	3678.9
Rice + fish + rabbit	3.4	8620.6	0.3	75.3	0.1	0.2	0.0	6.6	2.8	708.6

Source: Murugan and Kathiresan (2005). Solaiappan et al. (2007) reported that five IFS models were examined at Kovilpatti in Tamil Nadu during 2003, 2004 and 2005 to identify a superior model for attaining maximum net returns and benefit/cost ratio. The different models assessed were (A) conventional cropping, (B) crop + poultry (20) + goat (4), (C) crop + poultry (20) + goat (4) + dairy (1), (D) crop + poultry (20) + goat (4) + sheep (6) and (E) crop + poultry (20) + goat (4) + sheep (6) + dairy (1). Among the models examined, model (E) recorded a maximum net income of \$ 284/ha/year (Table 5.10)

Manjunath and Itnal (2003) conducted a field experiment during 1998–2001 at the ICAR Research Complex for Goa, Old Goa, to evaluate suitable integrated farming systems in different topographies. Integration of dairy with coconut through forage intercropping enhanced the net returns nearly five times (\$ 522/ha) in comparison to monocropped situation (\$ 100/ha), although the cost of maintenance was enhanced by five times. Similarly, when rice-based cropping system was integrated with poultry and mushroom, the gross and net returns improved by four times (\$ 2396/ha and \$ 1216/ha, respectively) over monocropping, while the cost of production was enhanced by five times (\$ 1181/ha), leading to a reduction in benefit/cost ratio (from 1.27 to 1.03) (Table 5.12 and Fig. 5.3).

IFS, comprising components such as cropping, vegetables, floriculture, fishery, poultry, duckery and cattle/bullock, cow and calf rearing, was undertaken by Ravisankar et al. (2007) in Calicut village of South Andaman during 2004–2006, to study the productivity and profitability for slopy upland areas of Bay Islands. Among the components evaluated, the highest net returns were obtained from crop

Table 5.10 Economics, sustainability index and employment generation of IFS

IFS model	Income from crop ($/ha)	Income from animals ($/ha)	Gross income ($/ha)	Expenditure for crops + animals ($)	Net income ($/ha)	Net income ($/ha/year)	Benefit/cost ratio	Sustainability index	Employment generation (man-days/ha/year)
A. Crop alone	454.8	0.0	454.8	0.4	99.5	33.2	1.28	23.0	185
B. Crop + goat + poultry	437.7	397.3	835.0	434.7	400.3	133.5	1.92	12.3	297
C. Crop + goat + poultry + dairy	394.8	1203.3		878.4	687.5	229.2	1.72	46.1	343
D. Crop + goat + poultry + sheep	455.0	638.1	1093.2	741.6	351.5	117.2	1.47	6.6	343
E. Crop + goat + poultry + sheep + dairy	472.3	1510.3		1151.1	851.5	289.6	1.75	65.3	389
Mean	452.6	937.3	1202.4	724.3	397.4	159.4	1.63		311
SD	13.3	510.0	614.8	332.0	295.2	98.4	0.25		78
CV (%)		0.9	0.8	0.7	1.0	1.0	15.5		25

Table 5.11 Year-wise overall returns of IFS and CFS (1997–2002)[a]

Year	IFS Total cost ($)	IFS Total revenue ($)	IFS Revenue net ($)	IFS Benefit/cost ratio	IFS Labour days/ha	CFS Total cost[b] ($)	CFS Total revenue ($)	CFS Revenue net ($)	CFS Benefit/cost ratio	CFS Labour days/ha[c]
1997–1998	183	173	−10	0.94	430	90	140	50	1.55	219
1998–1999	303	457	153	1.51	510	100	136	36	1.36	199
1999–2000	345	543	213	1.57	408	106	160	62	1.51	186
2000–2001	376	646	270	1.72	414	114	281	167	2.46	280
2001–2002	628	937	314	1.49	560	116	157	53	1.36	236
Mean	367	551	188	1.50	464	105	175	73	1.66	224
% between 1997–1998 and 2001–2002	243.19	442.23	2976.93		30.28	28.35	12.07	7.35		7.88

Source: Nageswaran et al. (2009)
[a]All figures calculated for an acre of land
[b]Includes inputted value of family labour
[c]Based on the average of 32 CFS farms and compared with IFS demo plot

Table 5.12 Economic analysis of different farming systems (mean data of 1999–2000 and 2000–2001)

Production system	Gross returns ($/ha)	Cost of cultivation ($/ha)	Net returns ($/ha)	Benefit/ cost ratio
Monocropping				
Cashew (*Anacardium occidentale*)	507	103	404	3.91
Coconut (*Cocos nucifera*)	296	196	100	0.51
Rice	558	246	312	1.27
Total	1361	545	816	1.50
Improved cropping system				
Cashew	507	103	404	3.91
Coconut + forage	880	468	413	0.88
Rice-brinjal (0.5 ha) + rice-cowpea (0.5 ha)	1226	508	718	1.41
Total	2613	1079	1534	1.42
Integrated farming system				
Cashew	849	263	586	2.23
Coconut + forage + dairy	1530	1009	522	0.52
Rice-brinjal (0.5 ha) + rice-cowpea (0.5 ha) + mushroom + poultry	2396	1181	1215	1.03
Total	4775	2452	2323	0.95

Source: Manjunath and Itnal (2003)

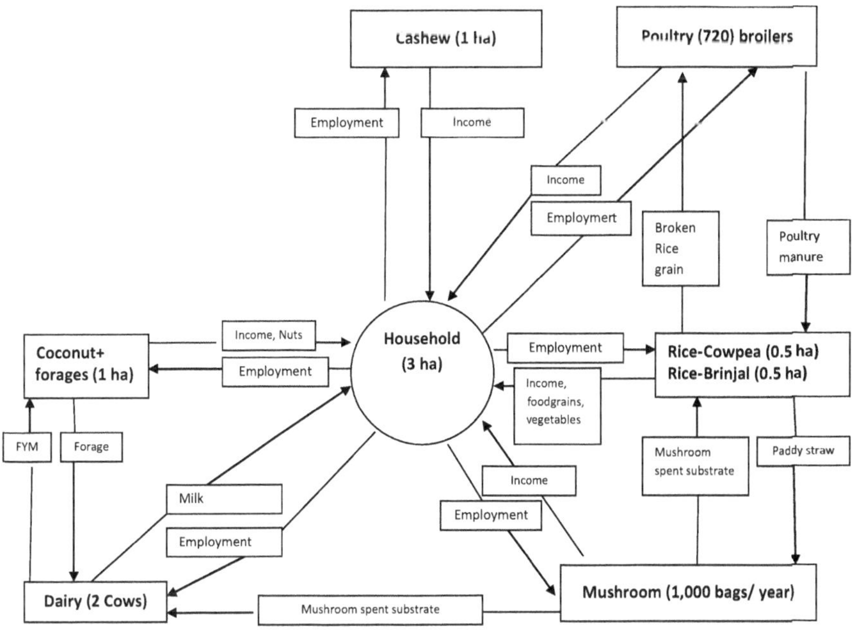

Fig. 5.3 A sustainable schematic model showing synergistic relationships among components of IFS at Goa. (Manjunath and Itnal 2003)

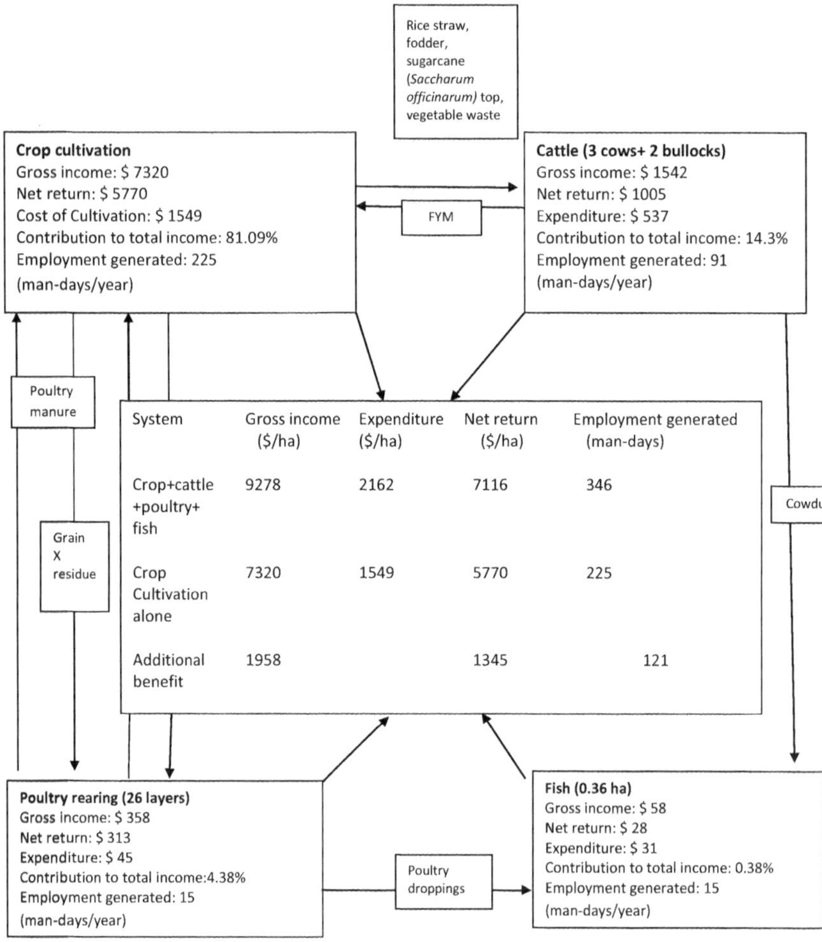

Fig. 5.4 Productivity, profitability and employment generation in integrated farming system in hilly upland areas of Bay Islands. (Ravisankar et al. 2007)

(81.09%), followed by livestock (14.3%), poultry (4.38%) and fish (0.38%). Employment generation was 346 man-days/ha/year under IFS. The net return obtained from all components was also higher under this system with a benefit/cost ratio of 3.30. To improve productivity, economic returns and employment generation for family labour, integration of all these components could be adopted instead of cultivating the crop alone in the hilly upland situations of the Andaman and Nicobar Islands (Fig. 5.4).

It is paradoxical to note that Odisha, with rich natural resource, is one of the poorest states of the country because of poor resource base of farming community. Most farmers are small and marginal (84%) having too small a farm size to employ the family labour force round the year, due to monocropping. Under these

Table 5.13 Economics of integrated farming system

Component	Area (sqm)	Gross return ($)	Investment ($)	Net return ($)	Benefit/cost ratio	Labour needed (man-days)
Rice	1600	34	18	16	1.91	14
Vegetables	1420	567	149	418	3.79	96
Fruits	540	161	33	129	4.90	39
Floriculture	60	5	1	4	6.00	1
Agroforestry	210	34	6	28	5.60	7
Fishery	120	37	11	26	3.28	8
Dairy	50	235	8	155	2.92	83
Total	4000	1073	299	775	3.59	248

Source: Sahoo et al. (2012)

circumstances, integration of more than one production component is desirable for better utilization of available resources at the command of farmers, i.e. IFS approach.

Sahoo et al. (2012) reported that Ranjan Kumar Bhuyan of Balasore district of Odisha is a small and marginal farmer who followed a pond-based farming system. It was characterized by small farm houses (240 m^2), a general playing are for children (120 m^2) and 4000 m^2 land under pond-based farming system. The economic analysis of pond-based farming system revealed that the gross income of $ 1073 was obtained by investing $ 299 with a benefit/cost ratio of 3.59. The net income from the pond-based farming system was $ 775. The risk of economic loss in this system was minimum (Table 5.13 and Fig. 5.5).

Earlier, Behera et al. (2006) reported that a study was undertaken in a farmer's field (0.4 ha) at Bashathi village of Bhadrak district of coastal Odisha, during 2003–2005 to generate adequate income round the year, including enterprises like field and horticultural crops, fishery, dairy and agroforestry. Net returns of $ 771 were obtained with an investment of $ 299. Adoption of integrated farming involving these enterprises helped the farmer earn his livelihood more comfortably than in the previous year (Table 5.14).

The farmer earned a monthly income of $ 64 excluding his labour utilized in the farm. The production of fish, milk, foodgrains, vegetables, fruits, etc. within the farm helped in improving the standard of living and provided better nourishment to the family members.

Mohanty et al. (2010a, b) reported that a poor tribal (Talajang village in Tarangada GP of Gumma block in Gajapati district of Odisha) farmer usually grows paddy and miner millets in his land of 2.5 acres, struggling to get one square meal every day. Farmers developed an integrated farming system which fetches high income and food security. By adopting this farming system module, the farmer earned seven times higher net monetary return (NMR) as compared to the traditional method in 2008–2009, as depicted in Table 5.15.

The benefit/cost ratio of the IFS model is 2.70, whereas in the traditional system it was 2.08 only. Furthermore, sustainability was found in the integrated farming system model.

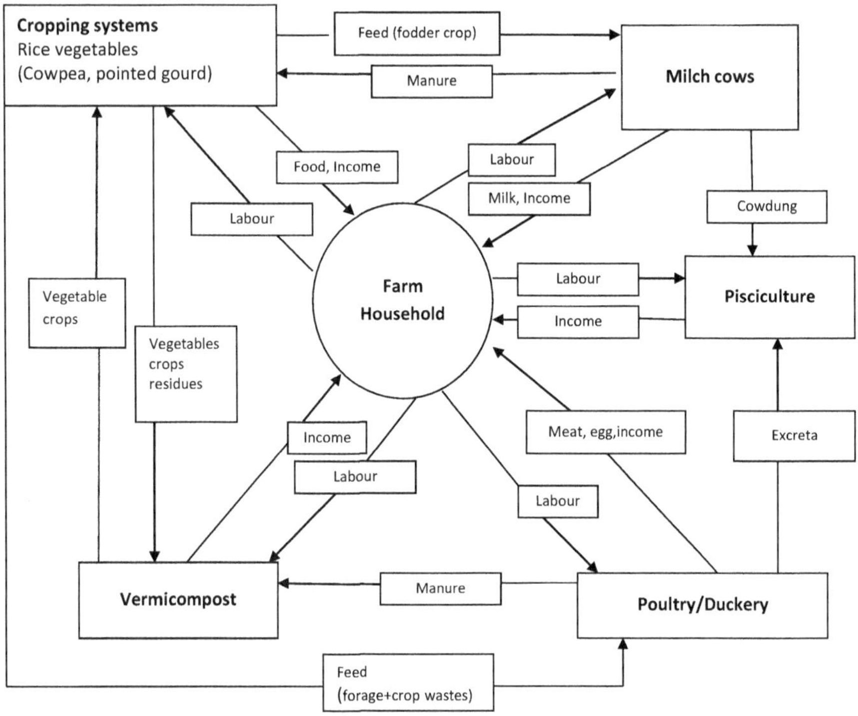

Fig. 5.5 Resource flow model diagram in pond-based farming system. (Sahoo et al. 2012)

Table 5.14 Income and employment generation from different components of farming system during 2004–2005

Enterprise	Area (m^2)	Gross returns ($)	Investment ($)	Net returns ($)	Labour requirement (man-days)
Rice	1600	34	18	16	14
Vegetables	1420	567	149	418	96
Fruits	540	161	33	129	39
Floriculture	60	0	1	0	1
Agroforestry	210	34	6	28	7
Fishery	120	37	11	26	8
Dairy	50	235	81	155	83
Total	4000	1068	299	771	248
Preceding year	2760	610	197	413	136

Source: Behera et al. (2006)

Table 5.15 Farming systems followed in Talajang village

Components		Area (0.4 ha)	Expenditure ($)	Return ($)	NMR ($)	B/C ratio
Integrated farming system						
Crop components	Rice (SRI)-fallow	1.2	74	232	158	3.13
	Ragi (*Eleusine coracana*)-horse gram (*Macrotyloma uniflorum)*	0.5	23	52	29	2.28
	Ragi + arhar (*Cajanus cajan)*	0.3	12	34	22	2.83
	Maize-horse gram	0.2	17	29	12	1.71
	Groundnut	0.1	9	25	16	2.70
Horticulture	Tissue culture banana	0.10	36	90	54	1.67
	Yam (*Dioscorea alata)*	0.1	25	55	30	2.20
	Tapioca (*Manihot esculenta)*	0.2	30	66	36	2.18
	Runner beans (*Phaseolus coccineus)*-brinjal	0.2	52	173	121	3.34
Poultry	Vanaraja	30nos	29	118	89	4.05
Vermicompost	Compost	1no	48	84	35	1.33
	Total		355	957	602	2.70
Conventional cropping system						
Crop Components	Rice + fallow	1	42	69	27	1.65
	Ragi + jowar + minor millets + arhar + suan + cowpea	1.5	19	53	34	2.75
	Runner bean	0.4	19	45	26	2.33
	Total		81	168	87	2.08

Source: Mohanty et al. (2010a, b)

Table 5.16 Details of year-wise expenditure and returns from the integrated farming unit at Khentalo

Items	1989	1990	1991	1992	1993	1994	1995	1996	1997	1998
Total expenditure ($)	634	663	695	777	816	871	944	1056	1202	1282
Total returns ($)	2806	3089	3532	3194	3694	3919	4024	4473	4887	5444
Net profit ($)	2172	2426	2837	2416	2877	3061	3210	4223	3685	4161

Source: Singh et al. (2006a, b)

Singh et al. (2006a, b) reported that the model involving fish, prawn, horticulture (papaya, banana, coconut, teak and vegetables), poultry and mushroom was developed in Khentalo village of Cuttack, Odisha, under the active involvement of farmers and scientists. An area of 2 ha ricefield was converted into this model which was otherwise yielding poor returns ($ 242/ha/year). But the net profit from this farming system varied from $ 2172 to 4161/ha/year, and the net returns increased year after year (Table 5.16).

Coconut is a smallholder crop in India, and more than 90% of the five million coconut holdings in the country are less than 1 ha in size. These smallholding coconut farms often do not provide adequate income to the dependent families (Das

Table 5.17 Returns from coconut-based mixed farming systems

Year	Gross return ($/ha)	Net return ($/ha)
1989–1990	2976	868
1990–1991	3037	942
1991–1992	3074	573
1992–1993	3095	540
1993–1994	3737	897
1994–1995	4102	1276
1995–1996	4592	1435
1996–1997	4679	1442
1997–1998	4740	2047

Source: Maheswarrapa et al. (2001)

1991), not providing gainful employment opportunities for the family labour throughout the year. Maheswarrapa et al. (2001) reported that a model was studied at Central Plantation Crops Research Institute, Kasaragod (Kerala), integrating grasses, dairy, poultry, rabbitry and fish culture during 1989–1990 to 1997–1998, in order to perform an economic analysis of integrated mixed farming systems in coconut gardens. Realized gross returns increased from $ 2982 to $ 4740 (by 59.29%), and the net returns increased from $ 868 to $ 2047 (by 151.33%) (Table 5.17).

Reddy and Sang-Arun (2011) reported that in order to create multiple sources of food, income and employment from each holding under coconut, the Peekay Tree Crops Development Foundation successfully demonstrated an integrated ecological farming system in small and marginal landholdings, covering 395 ha of Kerala which benefited a total of 2150 households (PTCDF 2009). The survey conducted in four villages, namely Aroor, Vayalar, Pattanakkad and Kodamthuruth of Pattanakkad block, Kerala, presents the findings on integrated farming practices, such as growing of tree species of medicinal importance and arable crops, livestock maintenance and waste biomass utilization at both household and community levels. Figure 5.6 shows an integrated coconut farm model.

All farmers in the study area possess smallholdings ranging in area from 0.01 to 0.4 ha. The farmers adopted various combinations of IFS (Table 5.18). Examples of plant and animal species found in the project area are listed in Table 5.19. Apart from the ecological advantages, it was also noticed that introduction of a variety of crops, animals, poultry and aquaculture into the coconut farm increased the net household income from US$ 400 to US$ 600–700/ha/year, besides freeing coconut farmers from the risk involved in monoculture (Table 5.20).

Promoting the cultivation of medicinally important tree species along with cash crops on a coconut farm helps in creating a dependable supply source of raw material for the Ayurvedic medical system and generating organic waste and biomass for fuel, thereby helping to increase the availability of per capita energy consumption.

Ramrao et al. (2006) reported that investigations were carried out in Durg district of Chhattisgarh to find out a sustainable mixed farming model, which is economically viable, integrating different components like crop, livestock, poultry and duck

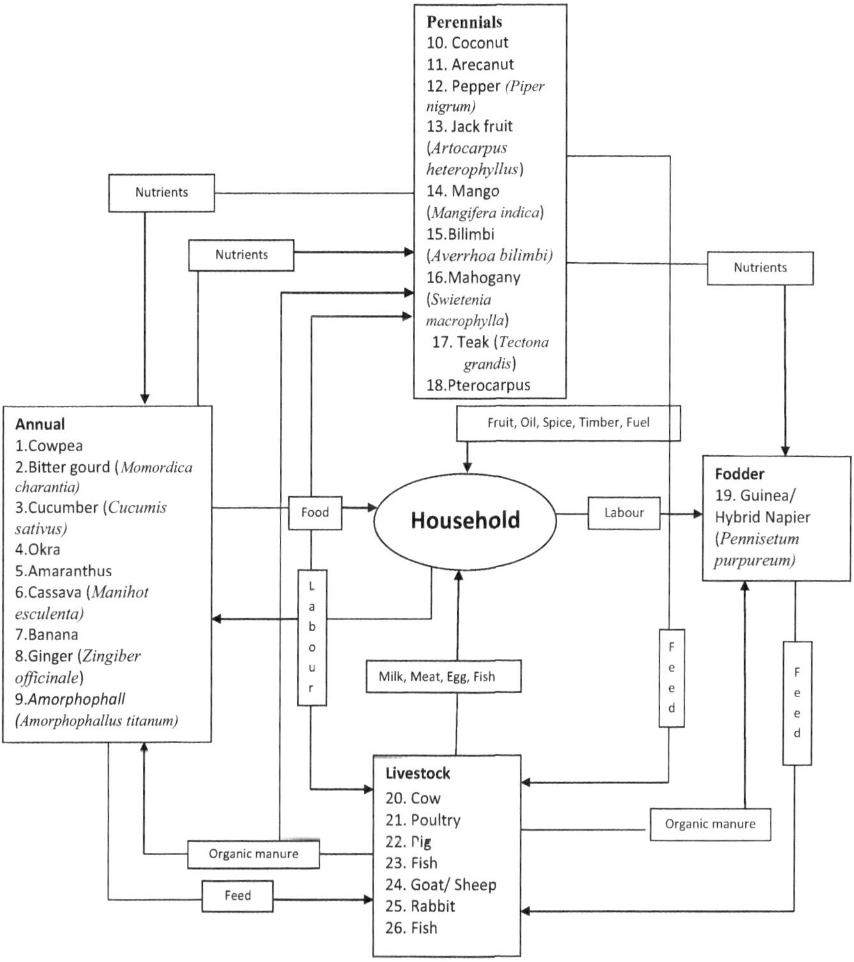

Fig. 5.6 A model of integrated coconut-based farming system followed in Kerala. (Thampan and Vasu 2007)

on a 1.5 acre marginal farmer landholding. Different viable modules were developed to find out the best package on the landholding of 1.5 acre suitable for the tribal region: (T_1) arable, (T_2) crop + 2 bullocks + 1 cow, (T_3) crop + 2 bullocks + 1 buffalo, (T_4) crop + 2 bullocks + 1 cow + 1 buffalo, (T_5) crop + 2 bullocks + 1 cow + 1 buffalo + 10 goats and (T_6) crop + 2 bullocks + 1 cow + 1 buffalo + 10 goats + 10 poultry + 10 ducks, were developed. A model having 2 bullocks + 1 cow + 1 buffalo + 10 goats + 10 poultry + 10 ducks along with crop cultivation was the best, with a net income of $ 534 per year against arable farming (crop farming) alone ($ 127per year), with cost returns of 1:2.238 (Table 5.20 and Fig. 5.7).

Table 5.18 Overview of ecological integrated farming system in Pattanakkad block[a]

Items	Progress report PIA, 2009		Field observation from sample survey, 2010
	Before project	After project	
Average landholding (ha)	0.01 to 0.4	0.01 to 0.4	0.01 to 0.2
Number of households benefited	None	1500 + 650	100% family benefited from integrated farming
Type of integration	Only coconut	Coconut with multiple tree species, vegetables, ornamental plants, animal, bird, fishes	MSC-rabbits-fish-bird farming, MSC-cow farming, MSC-duck farming, MSC-duck-chicken farming, MSC-duck-chicken-goat
Average household income (USD/ha/yr)	<400	658	600–700
Other outcome	Scarcity in food, medicine, biodiversity loss	Increased biodiversity and food security	Increased in species diversity, composition, increased food security, increased soil moisture content and nutrients through mulching and biomass degradation, increased family nutritional status, revitalization of local Ayurvedic medical system, increased family income and employment opportunity, increased energy, security
	Soil has low water holding capacity	2329 tonnes carbon stock through plant biomass	
	Low nutrient availability	About 173 tonne carbon sink in the soil	
	Energy crises		

[a]MSC, multiple species cropping. Source: Reddy and Sang-Arun (2011)

Table 5.19 Examples of diversification in integrated coconut farm in Kerala

Items	Species name/common name
Multiple species cropping (MSC)	*Acacia catechu*, *Aegle marmelos*, *Caesalpinia sappan*, *Garcinia cambogia*, *Gmelina arborea*, *Melia dubia*, *Myristica fragrans*, *Phyllanthus emblica*, *Pongamia pinnata*, *Pterocarpus marsupium*, *Punica granatum*, *Strychnos nux-vomica*, *Tamarindus indica*, *Terminalia arjuna*, *Terminalia bellirica*, *Terminalia chebula*, banana, coco, turmeric, ginger, *Anethum graveolens*
Animal	Goat, cow, rabbit
Poultry	Chicken, duck, other birds
Aquaculture	*Tilapia nilotica*, *Poecilia reticulata*, *Carassius auratus*, *Pterophyllum scalare*, etc.

An on-farm experiment conducted at Khentalo village in Cuttack district of Odisha by Mohanty et al. (2010a, b) also showed that net returns enhanced by 23-fold in rice-fish-prawn system in comparison to rice monocrop. Significantly higher net returns of \$ 1284/ha and higher ratio of the output value to cost of cultivation (1.6) in the rice-fish-prawn system inferred that rice-fish-prawn culture was more beneficial when adopted in lowland/waterlogged areas.

Table 5.20 Income and expenditure of different mixed farming modules for marginal holders

Treatment		Expenditure[1] (Rs)	Gross income (Rs)	Net income (Rs)	Cost/return ratio	Employment days
T_1	Crop (1.5 acre)	$200^a \pm 1.34$	$326^a \pm 1.02$	$127^a \pm 1.50$	$1.63^a \pm 0.01$	$165^a \pm 1.55$
T_2	Crop + 2 bullocks + 1 cow	$305^b \pm 1.32$	$534^b \pm 0.85$	$229^b \pm 1.50$	$1.75^b \pm 0.007$	$273^b \pm 1.15$
T_3	Crop + 2 bullocks + 1 buffalo	$309^c \pm 1.38$	$604^c \pm 0.88$	$295^c \pm 1.50$	$1.95^c \pm 0.008$	$273^b \pm 1.02$
T_4	Crop + 2 bullocks + 1 cow + 1 buffalo	$344^d \pm 1.33$	$690^d \pm 0.83$	$346^d \pm 1.48$	$2.00^d \pm 0.007$	$291^c \pm 0.99$
T_5	Crop + 2 bullocks + 1 cow + 1 buffalo + 10 goats	$376^e \pm 1.30$	$850^e \pm 0.83$	$474^e \pm 1.45$	$2.26^e \pm 0.007$	$308^d \pm 0.98$
T_6	Crop + 2 bullocks + 1 cow + 1 buffalo + 10 goats + 10 poultry + 10 ducks	$402^f \pm 1.28$	$935^f \pm 0.88$	$533^f \pm 1.47$	$2.23^f \pm 0.007$	$316^e \pm 0.90$

Values with different superscript in the same column differ from each other significantly ($P < 0.01$)
[1](Mean ± SE)
Source: Ramrao et al. 2006. Data in columns flanked by the same letters do not differ by Duncan's multiple range test, at $P < 0.05$

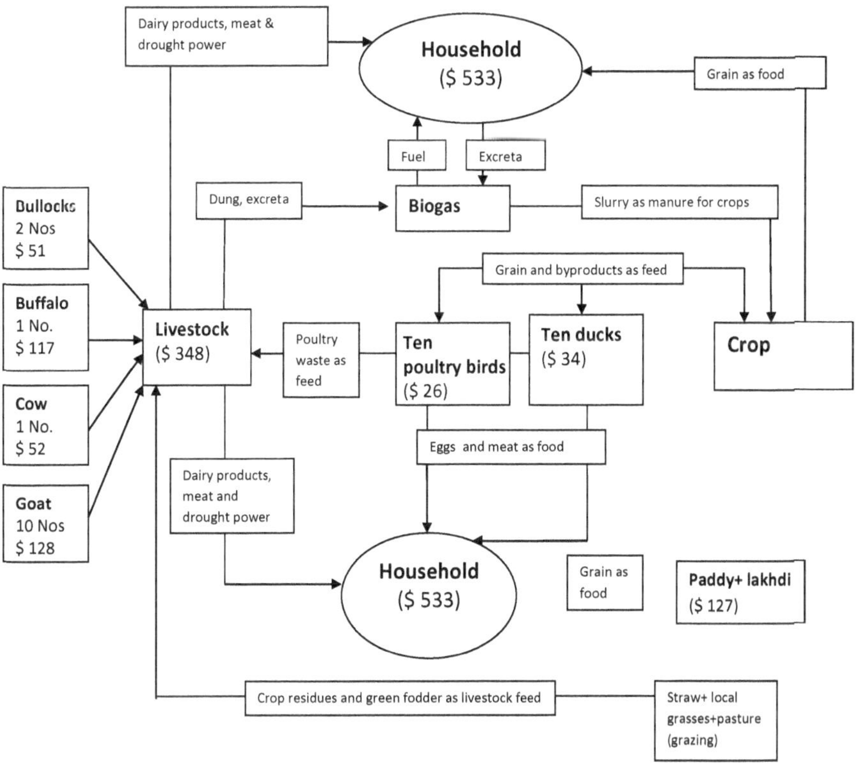

Fig. 5.7 An ideal model of mixed farming for marginal farmer (1.5 acre) as practiced in Durg district of Chhattisgarh. (Ramrao et al. 2006)

Table 5.21 Economics (X 10^3/$/ha) of various farming systems in Punjab, India

Systems (area)	REY (t)	Total cost	Net return	Employment generated (man-days)	Employment generation over rice-wheat (man-days)	B/C ratio
Surinder Kaur farm						
Rice-wheat (2.0)	18.0	0.6	1.1	–	–	1.67
Rice-wheat (1.78) + (0.22) dairy	29.7	1.6	1.2	–	138.2	1.74
Rice-wheat (1.22) + (0.22) dairy + fishery (0.56)	30.4	1.6	1.3	–	136.6	1.80
Rice-wheat (1.14) + dairy (0.22) + fishery (0.56) + piggery (0.24)	32.2	1.7	1.4	–	171.9	1.83
Shri Mahinder Pal Singh, village Anihar						
Rice-wheat (2.0)	17.8	0.7	1.0	34.2	–	2.45
Rice + wheat (1.4) + dairy (0.6)	27.2	1.3	1.1	155.4	69.3	1.86
Rice-wheat (0.4) + dairy (0.6) + fishery (1.0)	29.9	1.3	1.3	154.3	70.4	2.00
Shri Bhupinder Singh, village Birk						
Rice-wheat (1.0)	12.5	0.4	0.8	12.8	–	2.82
Rice + wheat (0.6) + dairy (0.4)	25.0	1.5	0.9	171.5	158.8	1.65
Rice-wheat (0.6) + dairy (0.4) + poultry (100)	25.5	1.5	1.0	175.1	162.3	1.64

Source: Gill et al. (2009)
REY, rice equivalent yield

The results of a study on IFS conducted in Punjab in a farmer participatory mode showed the beneficial effect of this approach (Gill et al. 2009). The rice-wheat system gave net returns of $ 1072/ha while rice-wheat + dairy with an advantage of $ 141/ha. Rice-wheat + dairy + fishery further supplemented the net return up to $ 223, and when it was strengthened with piggery, the net advantage/ha was around $ 324. In addition, the dairy enterprises generated 138 man-days and piggery generated 28 man-days employment over the rice-wheat system. The cost/benefit ratio was also in the viable range of 1.74–1.83 (Table 5.21). The average of 2 years at the village Anihar gave for the dairy enterprises $ 119 over the rice-wheat system with a net profit margin ($/ha) enhanced to 273 when coupled with fishery enterprise. The integrated approach generated 70 man-days/year over the rice-wheat system and the cost/benefit ratio varied from 1.86 to 2.00 (Table 5.21).

Similarly at another village farm, the dairy enterprise gave $ 174/ha, and poultry also enhanced the margin to $ 186/ha. These enterprises could generate additional manpower of 173 man-days/year (Table 5.21). The profit margin increase was directly linked to the management at the farmer level. After 3–4 years, the progeny of milch animals also added to the profit margin.

The integrated and ecological farming system projects carried out in different countries of Europe revealed that the inputs of fertilizer nitrogen, herbicides, fungicides, insecticides and plant growth regulators were substantially reduced for

Table 5.22 Reduction of inputs at Saint-Hilaire (1991–1995) and reduction of wheat yields at Boigneville (1991–1994) for IFS compared with CFS

Saint-Hilaire (1991–1995)	Boigneville (1991–1994)
Input reduction (%)	Yield reduction (%)
Seed (+2.6)	1991 (−34.6)
Herbicide (−10.1)	1992 (−19.6)
Insecticide (−28.3)	1993 (−15.9)
Fertilizer (−41.3)	1994 (−14.8)
Fungicide (−89.8)	Mean (−21.2)

the integrated-ecological regimes compared with conventional farming systems. Overall, the reduction in pesticide inputs for the integrated-ecological systems varied depending on the cropping system and agroclimatic factors.

Network trials were conducted at various sites in France during 1991–1995 to determine the performance of IFS with conventional farming systems, based on the levels of production inputs and crop yields. The results at Saint-Hilaire and at two other sites show that it is possible to reduce the total production inputs from 25% to 37%. The amount of herbicides, insecticides, chemical fertilizers and fungicides used in IFS compared to CFS decreased by 10.1, 28.3, 41.3 and 89.8%, respectively. The results suggest that IFS poses a much lower risk to environment than CFS. The results obtained at Boigneville (Table 5.22) show the extent to which winter wheat yields were reduced in IFS, as the level of inputs decreased.

Wheat yield was reduced by 35% in 1991 and by 1994 had narrowed to about 15%, compared with CFS. This would suggest that management practices for the IFS had improved substantially over time. An economic analysis of the results obtained at three trial sites, including Montgaillard, Boigneville and Saint-Hilaire, was conducted using 1995 prices and accounting for any compensatory aid, i.e. subsidies. It is most interesting that despite the lower yields associated with IFS, the net returns per hectare were greater for IFS compared with CFS at all three sites. The probable reason for this is that IFS are more economical for both operating costs and mechanization efficiency.

A significant number of studies have been undertaken in Thailand into the financial viability of IFS. In a study by Tokrishna (1992), integration of duck raising and fish enterprises resulted in farmers being able to earn a net profit of US$ 1850/ha of which 87% came from fish with yields of 3.5 t/ha. In the Nakhon Ratchasima and Khon Kaen provinces, Kaewsong et al. (2001) evaluated the socio-economic status of 30% of the members of a farmer network that promoted IFS in 2001. The study revealed that the average total income of the members was higher than in other areas in the northeast region.

Integrated tree crops-ruminant systems are potentially very important but are underestimated in Southeast Asia. The inclusion of animals provides the entry point for development and has the twin advantages of increasing the supplies of animal proteins and also value addition in the oil palm. The benefits of integration are considerable and are mediated through positive crop-animal-soil interactions and merit expanded development. The reasons for low adoption of the systems are poor awareness of the potential of integrated systems, resistance by the crop-oriented plantation sector and inadequate technology application. Integrated oil palm-based systems are an important pathway for integration with ruminants (buffaloes, cattle, goats and sheep) and provides an entry point for development. The advances in research and development in Southeast Asia highlight demonstrable increased productivity from animals and meat offtakes, value addition to the oil palm crop, sustainable development and distinct economic impacts. The results from 12 out of 24 case studies concerning oil palm over the past three decades showed increased yield of 0.49–3.52 mt of fresh fruit bunches (FFB)/ha/yr; increased income by about 30%; savings in weeding costs by 47–60% equivalent to 21–62 RM/ha/yr ($ 5-15/ha/year at 2018 rates); and an internal rate of return of 19% based on actual field data. The results provide important socio-economic benefits for resource-poor small farmers. Potential increased offtakes and additional income exist with the integration of goats. Devendra (2011) in a synthesis of 24 case studies from 7 countries over the period 1984–2007, which includes 12 from Malaysia, observed:

- Distinct economic benefits were reported, e.g. crop yields and savings on weeding costs.
- Most of the reports deal with individual benefits and not the system as a whole.
- Very few studies were concerned with quantitative animal productivity.
- Issues of sustainability were not addressed.

5.4.4 *Employment Generation*

There is an increase in the value for labour absorption in IFS farms due to additional components brought into integration within them. Channabasavanna and Biradar (2007) also concluded that rice-fish-poultry system model IFS showed 4.7% higher man-days over CFS (Table 5.5). Solaiappan et al. (2007) reported maximum employment generation (389 man-days/ha/year) with model E followed by model C, with employment generation of 343 man-days/ha/year (Table 5.10). Comparative analysis was also carried for determining the days of labour requirement by the MS Swaminathan Research Foundation (Nageswaran et al. 2009) in IFS and CFS (Table 5.11). Ravisankar et al. (2007) revealed an employment generation of 346 man-days/ha/year under the integrated farming system at Calicut village of South Andaman. Sahoo et al. (2012) reported the system followed by Ranjan Kumar Bhuyan, and Behera et al. (2006) reported that a study in Miniati Behera's farm generates an employment opportunity of 248 man-days (Table 5.14 and Table 5.15 respectively). Ramrao et al. (2006) also reported that the model having 2 bullocks + 1 cow + 1 buffalo + 10 goats + 10 poultry + 10 ducks has an employment generation of 316 man-days (Table 5.20).

5.4.5 *Constraints of Integrated Farming Systems*

The IFS is feasible with respect to socio-economic imperatives; actual adoption rates of integrated farming are limited and unevenly spread among farmers. The study by Ngambeki et al. (1992) in Cameroon revealed that the major production constraints are animal feed shortages throughout the year, labour bottlenecks and soil degradation. Csavas (1992) reported that in most farms studied in China there was a dependency on imported feed rather than internal recycled inputs. This was concluded to be due to resource-poor farmers in general not having feed lot-type systems in which to undertake livestock production. Lightfoot (1997) suggested that the four main constraints to adoption of integrated farming systems in the Philippines and Ghana were:

(1) The long transition period that often occurs when implementing an integrated production system. This lead-in time can vary between 3 and 10 years. Farmers could not afford declines in food production and income generation over this period.
(2) Labour shortages, especially where the family size is small, which effectively prevented them from adopting integrated farming techniques.
(3) Lack of secure land rights.
(4) Disincentives for adopting integrated farming, resulting from government subsides, credits for fertilizers and herbicides.

Banerjee et al. (1990) assessed the impact on allocating farm area to different types of crops and livestock. The study revealed that there are few opportunities for increasing farm net returns with the limited amount of capital available. This conclusion is supported by the study of Tipraqsa (2006), who alludes to the fact that high start-up costs may constrain farmers from switching to integrated farming and from exploiting the benefits of resource integration. In the case of northeast Thailand, the study by Thamrongwarangkul (2001) reported that resource-poor farmers often cannot go beyond the transition period due to their need for food and for immediate economic returns to meet cash needs, such as schooling, medical treatment and loan repayment. Contrasting this, Tokrishna (1992) pointed out that a farmer who becomes successful and wants to expand the area of his integrated farm in Thailand would be limited by access to adequate water supply, animal feed and market outlets.

5.5 Conclusion

It can be concluded that the philosophy of IFS revolves around better utilization of the available resources. It enhances production and helps in achieving higher economic returns. Moreover, farm families get a scope for gainful employment round the year. Thus, IFSs are the need of the hour in India to help our small and marginal farmers, which ensure good income and better standard of living, playing a role in the development of the nation as a whole.

References

Banerjee, B. N., Sarker, S. C., & Maity, A. K. (1990). Impact of resource optimization on cropping pattern and income on crop-dairy mixed farm. *Indian Journal of Dairy Science, 43*(3), 295–301.

Behera, U. K., Jha, K. P., & Mahapatra, I. C. (2004). Integrated management of available resources of the small and marginal farmers for generation of income and employment in eastern India. *Crop Research, 27*(1), 83–89.

Behera, U. K., Behera, B. C., & Mohapatra, P. C. (2006). Diversification of farming system for the marginal farmers under flood prone coastal ecosystem. *Indian Farming* (August): 9–13.

Birthal, P. S., Joshi, P. K., & Narayanan, A. V. (2011). *Agricultural diversification in India: Trends, contribution to growth and small farmer participation*. Patancheru: The International Crops Research Institute for the Semi-Arid Tropics (ICRISAT), mimeo.

Chand, R., Prasanna, P. A., & Singh, A. (2011). Farm size and productivity: Understanding the strengths of smallholders and improving their livelihoods. *Economic and Political Weekly, 46*, 26–27.

Channabasavanna, A. S., & Biradar, D. P. (2007). Rice-fish poultry system model for rice growing farmers of Karnataka. *Indian Farming, 57*(5), 20–22.

Channabasavanna, A. S., Itnal, C. J., & Patil, S. G. (2002). Productivity, economic analysis and changes in physic-chemical properties of soil as influenced by integrated rice (*Oyrza sativa*) based farming systems. *Indian Journal of Agronomy, 47*(1), 1–5.

Channabasavanna, A. S., Biradar, D. P., Prabhudev, K. N., & Hegde, M. (2009). Development of profitable integrated farming system model for small and medium farmers of Tungabhadra project area of Karnataka. *Karnataka Journal of Agricultural Science, 22*(1), 25–27.

Csavas, I. (1992). Regional review on livestock-fish production systems in Asia. In *Proceedings of the FAO/IPT workshop on integrated livestock-Fish production systems*, 16–20 December 1991, Institute of Advance Studies, University of Malaya, Kuala Lumpur, Malaysia.

Das, T. K. (1991). Economic viability of coconut based farming systems in India. *Journal of Plantation Crops, 19*, 191–201.

Dent, J. B. (1990). *Systems theory applied to agriculture and the food chain* (Vol. 365). Amsterdam: Elsevier.

DES. (2009). *Agricultural statistics at a glance*. New Delhi: Directorate of Economics and Statistics, Department of Agriculture & Cooperation, Ministry of Agriculture and Farmers Welfare, Government of India.

Devendra, C. (2011). Invited review – Integrated tree crops-ruminants systems in South East Asia: Advances in productivity enhancement and environmental sustainability. *Asian-Australasian Journal of Animal Sciences, 24*, 587–602.

Dixon, J., Gulliver, A., & Gibbon, D. (2001). Introduction. In M. Hall (Ed.), *Summary, farming systems and poverty, improving farmers' livelihoods in a changing world* (pp. 1–7). Rome/Washington DC: FAO/World Bank.

Gill, M. S., Singh, J. P., & Gangwar, K. S. (2009). Integrated farming system and agriculture sustainability. *Indian Journal of Agronomy, 54*(2), 128–139.

Govindan, R., Chinnaswami, K. N., Chandrasekaran, B., Budhar, M. N., & Prince, J. (1990). Poultry-cum-fish culture in rice farming in Cauvery delta region of Tamil Nadu. *Indian Journal of Agronomy, 35*(1/2), 23–29.

Kaewsong, B., Suwannimitara, A., & Suraseang, S. (2001). *Sustainable farming development for small scale farmers in South Khon Kaen – North Korat: Assessment report*. Sustainable Agriculture Charity, Thailand.

Kumar, S., & Jain, D. K. (2005). Are linkages between crops and livestock important for the sustainability of the farming system? *Asian Economic Review, 47*(1), 90–101.

Lightfoot, C. (1997). Promoting low external inputs agriculture (LEIA) from LEIA to intensive integrated farming systems. In F. Dolberg & P. H. Petersen (Eds.), *Proceedings of a workshop, maximizing the influence of the user:* Alternatives *to the* training *and visit system.* April 7–11, 1997, Tune Landboskole, Denmark.

Mahapatra, I. C. (1994). Farming system research – A key to sustainable agriculture. *Fertilizer News, 39*(11), 13–25.

Maheswarrapa, H. P., Hegde, M. R., Dhanapal, R., Sairam, C. V., & Singh, T. V. (2001). Impact of integrated mixed farming system in coconut (*Cocos nucifera*) garden on coconut yield and economic analysis. *Indian Journal of Agronomy, 46*(1), 56–63.

Manjunath, B. L., & Itnal, C. J. (2003). Farming system options for small and marginal holdings in different topographies of Goa. *Indian Journal of Agronomy, 48*(1), 4–7.

MoA. (2015). *Agriculture census 2010–11*. New Delhi: Agriculture Census Division, Department of Agriculture & Cooperation, Ministry of Agriculture.

Mohanty, D., Patnaik, S. C., Das, P. J., Parida, N. K., & Nedunchezhiyan, M. (2010a). Integrated farming system for sustainable livelihood: A success story of a tribal farmer. *Indian Farming, 60*(6), 41–43.

Mohanty, R. K., Thakur, A. K., Ghosh, S., & Patil, D. U. (2010b). Impact of rice fish prawn culture on rice field ecology and productivity. *Indian Journal of Agricultural Sciences, 80*, 597–602.

Murugan, J., & Kathiresan, R. M. (2005). Integrated rice farming systems. *Indian Farming, 55*(5), 4–6.

NABARD. (2001) *A report on the development of plantation and horticulture sector in Goa*. National Bank for Agriculture and Rural Development, Goa.

Nageswaran, M., Selvaganapathy, E., Subbiah, V. R., & Nair, S. (2009). *Demonstration and replication of integrated farming systems at Chidambaram* (pp. 16–53). Chennai: M S Swaminathan Research Foundation.

NCEUS. (2008). *A special programme for marginal and small farmers*, A report prepared by the national commission for enterprises in the unorganized sector. New Delhi: NCEUS.

Ngambeki, D. S., Deuson, R. R., & Preckel, P. V. (1992). Integrating livestock into farming systems in northern Cameroon. *Agricultural Systems, 38*, 319–338.

PTCDF (Peekay Tree Crops Development Foundation). (2009). *Promoting coconut- based agro-ecosystem and efficient product utilization for augmenting on-farm income, Improving quality of environment and conserving natural resources* (unpublished).

Ramrao, W. Y., Tiwari, S. P., & Singh, P. (2006). Crop-livestock integrated farming system for the Marginal farmers in rain fed regions of Chhattisgarh in Central India. *Livestock Research for Rural Development, 18, Article No.102.*

Rangasamy, A., Venkitaswamy, R., Purushothaman, S., & Palaniappan, S. (1996). Rice poultry-fish-mushroom integrated farming systems for lowlands of Tamil Nadu. *Indian Journal of Agronomy, 41*(3), 344–348.

Ravisankar, N., Pramanik, S. C., Rai, R. B., Nawaz, S., Biswas, K. R., & Bibi, N. (2007). Study on integrated farming system in hilly upland areas of Bay Islands. *Indian Journal of Agronomy, 52*, 7–10.

Reddy, V., & Sang-Arun, J. (2011). *Promoting coconut-based agro-ecosystem and efficient product utilisation for augmenting on-farm income, improving quality of environment and conserving natural resources*. Occasional Paper, IGES. p. 25.

Sahoo, B., Lenka, A., & Nedunchezhiyan, M. (2012). Sustainable livelihood support through enterprise convergence in pond based farming system. *Odisha Review*, April: 46–51.

Singh, K., Bohra, J. S., Singh, Y., & Singh, J. P. (2006a). Development of farming system model for north-eastern plains zone of Uttar Pradesh. *Indian farming, 56*(2), 5–11.

Singh, Y. V., Jha, K. P., & Sahoo, R. (2006b). Integrated farming system model for a two hectare lowland farmer of alluvial zone. *Indian Farming, 56*(2), 12–14.

Singh, K., Bohra, J. S., Singh, T. K., Singh, J. P., Singh, Y., & Singh, C. S. (2007). Productivity and economics of integrated farming system in irrigated agroecosystem of Uttar Pradesh. *Indian Journal of Agronomy, 52*, 7–10.

Solaiappan, U., Subramanian, V., & Maruthi, G. R. (2007). Selection of suitable integrated farming system model for rain-fed semi-arid vertic inceptisols in Tamil Nadu. *Indian Journal of Agronomy, 52*(3), 194–197.

Thampan, P. K., & Vasu, K. I. (2007). Coconut for rural welfare. In *Proceedings of international coconut summit*, Kochi, India.

Thamrongwarangkul, A. (2001). *For our Thailand.* Annual report on sustainable community development for good livelihood and environment project, Khon Kaen, Thailand.

Tipraqsa, P. (2006). Opportunities and constraints of integrated farming system in Northeast Thailand. A case study of the Huai Nong Ian catchment, Khon Kaen province. In M. Denich, C. Martius, & C. Rodgers (Eds.), Ecology and Development Series No. 35 University of Bonn, Cuvillier Verlag Gottingen, Germany.

Titi, A. L, Boller, E. F, & Gendrier, J. P (Eds.). (1993). Integrated production principles and technical guidelines. International Organization for Biological and Integrated Control (IOBC). IP-Guidelines and Endorsement. *Technical Bulletin, 16* (1).

Tokrishna, R. (1992). Integrated livestock-fish farming systems in Thailand. In T. K. Mukherjee, P. S. Moi, J. M. Panandam, & Y. S. Yang (Eds.), *Proceedings of the FAO/IPT workshop on integrated livestock-fish production systems*. 16–20 December 1991. Institute of Advanced Studies, University of Malaya, Kuala Lumpur, Malaysia.

Chapter 6
Health Effects of Changing Environment

Randeep Guleria, Vartika Mathur, and Ashutosh Dhanuka

Abstract Environment plays a crucial role in our economic, social and cultural behaviour as well as on health. However, since the beginning of industrialization era, focus on economic development has caused detrimental effects on the environment. Last two centuries have witnessed changes in global environmental factors such as rise in temperature leading to global warming, depletion of stratospheric ozone layer, loss of biodiversity and marked degradation in air and water quality due to atmospheric pollution, thereby causing upsurge in infectious and non-infectious diseases. Environmental health has emerged as an important part of medicine. The World Health Organization (WHO) estimates that 24% of global disease burden and 23% of all deaths can be attributed to environmental factors. Deaths from heart disease, cancer, respiratory disorders and many vector-borne diseases such as malaria, dengue, chikungunya and cholera have increased due to changes in climate, especially in developing countries. Besides limited attention to sanitation, hygiene, as well as quality of food and drinking water, factors such as deforestation, increasing vehicular traffic, migration from rural to urban areas, decreasing water resources and inadequate drainage systems contribute to increase incidence of diseases. The need of the hour is to sensitize ourselves about the way our ecology is being degraded and the health effects it is causing. A holistic view is needed to address the problem of environmental health where agriculture, animal husbandry, public health, water safety and air pollution need to be looked at in a combined manner for education, planning and resource allocation. Therefore, a close

R. Guleria
Department of Pulmonary Medicine and Sleep Disorders, All India Institute of Medical Sciences, New Delhi, India
e-mail: randeepguleria2002@yahoo.com

V. Mathur (✉)
Department of Zoology, Sri Venkateswara College, University of Delhi, New Delhi, India
e-mail: vmathur@svc.ac.in

A. Dhanuka
Geetanjali Medical College and Hospital, Udaipur, India
e-mail: ashutruth@gmail.com

R. Peshin, A. K. Dhawan (eds.), *Natural Resource Management: Ecological Perspectives*, Sustainability in Plant and Crop Protection,
https://doi.org/10.1007/978-3-319-99768-1_6

association between scientists, public health professionals and administrators is needed for integrated design and development of framework to attain harmony between man and nature.

Keywords Pollution · Infectious diseases · Indoor pollution · Changing climate · Air pollution · Water pollution

Abbreviations

AIIMS	All India Institute of Medical Sciences
COPD	Chronic obstructive pulmonary disease
ENSO	El Niño Southern Oscillation
IPCC	Intergovernmental Panel on Climate Change
WHO	World Health Organization

6.1 Introduction

An ecosystem is defined as a community of living beings surviving and interacting in mutual and interdependent relationship with their physical environment. For thousands of years, man has lived in harmony with their natural surroundings. Environment has played a crucial role in his economic, social and cultural behaviour as well as on his health. The role of environment in various diseases has been well documented, both in communicable and non-communicable diseases. Since the dawn of industrialization era in Europe 400 years ago and its subsequent spread to the rest of world, economic development and physical comfort for mankind have increased at a tremendous pace. This increase is perhaps the most rapid over the last three decades. Often this has been done, knowingly or unknowingly, at the cost of our environment. The last 200 years have witnessed a sharp increase in the global temperature from its levels around 4000 years ago, owing to industrialization (Mann et al. 2008; Marcott et al. 2013). Moreover, other concerns such as depletion of carbon fuels at alarming rates, damage to the ozone layer and rise in seawater levels, combined with global warming, have damaged our environment extensively, leading to changes in the aquatic biodiversity and to the extinction of many species of plants and animals (Thomas et al. 2004). Increase in urbanization has led to loss of dense forests. Air pollution has risen to the extent that many big cities in the world have a highly toxic air quality. However, very little has been done by various governmental and non-governmental agencies with almost no visible results. The need of the hour is to sensitize the scientific community, as well as the common man, about the way our ecology is being degraded and the health effects it is causing and to suggest ways to get remedies for this situation.

Environmental health has emerged as an important part of medicine, due to the rapid environmental changes linked to industrialization and urbanization. It is being increasingly recognized that environmental factors play a key role in human health and are linked to many chronic and infectious diseases. Deaths from heart disease and respiratory illness are increasing, and many diseases such as malaria, dengue, chikungunya and cholera are sensitive to changes in the climate (McMichael et al. 2006; Patz et al. 2005).

According to the World Health Organization (WHO), 'In its broadest sense, environmental health comprises those aspects of human health, disease and injuries that are determined or influenced by factors in the environment. This includes the study of both direct pathological effects of various chemical, physical and biological agents, as well as effects on health of the broad physical and social environment, which include housing, urban development, land use and transportation, industry and agriculture'. The WHO estimates that 24% of the global disease burden and 23% of all deaths can be attributed to environmental factors. Moreover, environmental factors have a much bigger impact in developing countries than developed ones, and this effect is seen much more in the vulnerable population such as children and elderly.

In a developing country like India, the burden of various diseases is increasing due to environmental factors and the changes in our environment. It is estimated that 94% diarrhoeal disease burden may be attributed to environmental factors such as unsafe food and drinking water, as well as poor sanitation and hygiene. Similarly, in India there is strong evidence linking lower respiratory tract infection to indoor air pollution caused by the use of solid fuels in household. Almost 42% of acute lower respiratory tract infections in developing countries are attributable to environmental factors. Besides this, a close association of vector-borne diseases and environmental conditions has been established. Furthermore, factors such as deforestation, increasing vehicular traffic, migration from rural to urban areas, decreasing water resources and inadequate drainage system are important environmental and ecological factors that contribute to infectious diseases.

6.2 Temperature Changes

The temperature of the earth has increased by about 0.6 °C over the past 100 years (Griggs and Noguer 2002; McCarthy 2001). Winters are shortening and average temperature is rising. Intergovernmental Panel on Climate Change (IPCC) of United Nations predicts that the global temperature will rise by 1.8–5.8 °C by the turn of this century, if no remediable actions are taken (Houghton et al. 2001). This will lead to rise in sea level by 9–88 cm and drowning of coastal cities, which comprise 50% of world's major cities (Crutzen 2006; FitzGerald et al. 2008; Nicholls and Cazenave 2010). Higher temperatures will lead to melting of polar ice, melting of glaciers, floods and droughts (Patz et al. 2001). Average temperature shall rise during both summers and winters. Heat waves will increase, and average annual precipitation will also increase correspondingly. Heat waves, floods and droughts lead to natural calamities, shortage of food supplies, increased risk of infectious diseases and increased human mortality (Haines et al. 2006).

6.3 Water Extremes and Pollution

Clean water is essential for the survival of humans. Water pollution due to environmental changes therefore constitutes another serious risk to the health of our planet. Water pollution occurs when energy and substances are released and degrades the quality of water for other users. Anything that is added to water, which is more than its capacity to break it down, constitutes water pollution. Anthropogenic activities such as industrial waste effluents, sewage disposal and agricultural activities are some of the major causes of water pollution (Manivasakam 2005; Tilman et al. 2001). Chemical pollution of surface water causes major health problems as it can be used directly for drinking or it may contaminate shallow wells, used for drinking. Ground water, which is much deeper, has very few pathogens as it gets filtered when it passes through many underground layers. It can be polluted by toxic chemicals such as fluoride and arsenic which may be present in the soil or the rock layers. Similarly, pollution of coastal water can cause contamination of sea food (Guleria 2013). Changing environment has a serious effect on safe water, affecting not only human health but also changing the ecology of plants. The global effect of water pollution has not been studied in detail and is limited to mainly outbreaks of waterborne infections or certain chemical toxins in limited areas, such as arsenic in drinking water in Bangladesh, 'Minamata' disease in Japan, etc. (Argos et al. 2010; Harada 1995). The burden of waterborne diseases is grossly underreported in India due to lack of data, poor surveillance and reporting. According to a report from the Ministry of Health and Family Welfare, nearly 40 million people are affected by waterborne diseases such as diarrhoea, enteric fever, amoebiasis and helminthic infestations, every year. WHO estimates >330,000 deaths annually, in India alone, due to contaminated water consumption. Moreover, floods and droughts also affect human health. Floods lead to physical injury as well as spread of waterborne diseases such as diarrhoea, enteric fever and viral hepatitis. Overcrowding occurs and sanitation is affected, leading to respiratory infections. Diseases such as malaria and dengue may turn into epidemics. On the other hand, drought leads to lack of sanitation, decreased food production and ultimately malnutrition.

Another aspect of waterborne diseases is chemical contamination leading to diseases such as fluorosis and methemoglobinemia, due to contamination of soil water owing to fluoride and fertilizers. Chronic exposure to contaminated water can cause significant health effects and can lead to liver and kidney damages. This occurs due to chronic exposure to copper, cadmium, arsenic, mercury, chromium and chlorobenzene. Endocrine effects have been reported, and problems relating to reproduction, development and behaviour have also been observed.

6.4 El Niño Southern Oscillations (ENSO)

ENSO is a cycle of seawater temperature and pressure changes occurring over the southern Pacific Ocean at an interval of 2–7 years and lasting for 6–18 months. This leads to episodes of floods in the southwest United States, Mexico and Western coast of Latin America and droughts in Southeast Asia and the Pacific islands (Kovats et al. 2003). This may be followed by cold waves called La Niña. Higher global temperatures are predicted to lead to more frequent and severe ENSOs, and this will lead to significant effects on human health, in the coming years (Bouma et al. 1997).

6.5 Infectious Diseases and Environment

Change in the global climate has led to higher temperatures, humidity and floods, which has made the environment more conducive for parasites such as mosquitoes and fleas (Patz et al. 2000). Malaria, dengue and other vector-borne diseases are expected to increase both in magnitude and their geographical reach (Haines et al. 2006). People living at higher altitudes may also likely experience resurgence in vector-borne diseases, due to a rise of average temperatures in these regions. Moreover, these diseases can spread to any part of the world in a very short time (at times during the incubation period) and cause an outbreak in a community where these diseases do not usually occur, resulting in diagnostic difficulties. This has recently been seen during the Ebola and the MERS coronavirus outbreaks. Other factors such as breakdown of public health infrastructure, shortage of medical supplies and changes in land use also contribute to adversities in health, due to water pollution.

6.6 Air Pollution

Air pollutants affect the human body through the inhalational route. Environmental changes due to industrialization have drastically altered the quality of the air we breathe. There are hundred substances that pollute the air and may harm human health. Pollutants are generally classified as primary or secondary pollutants. Chemicals that are directly emitted from a source are known as primary pollutants. These include sulphur dioxide, nitrogen oxides, carbon monoxide, volatile organic compounds, etc. Moreover, particulate matter emitted due to combustion from automobile exhaust, heating, cooking and industrial sources are also primary pollutants. Secondary pollutants, such as formaldehyde, nitric acid and different aldehydes, on the other hand, are formed from chemical or photochemical reaction in the atmosphere. On exposure to sunlight, volatile organic compounds and nitrogen oxides react photochemically, producing pollutants such as ozone.

Air pollution and occupational exposure may cause a variety of negative health outcomes, including reduced lung function in children as well as increased susceptibility to infections, airway inflammation and cardiovascular diseases. Respiratory disorders due to air pollution are emerging to be a major contributor to mortality, according to recent epidemiologic studies. Moreover, low-level air pollution is recently being recognized as a risk factor for lung diseases and death from COPD (Bosson and Blomberg 2013). With newer insights into the immunopathogenesis of asthma, the contribution of air pollution to allergen sensitization and airway hyperresponsiveness are being established. For example, increased exposure to nitrogen dioxide during infancy correlates with increased risk for asthma in later childhood. Ozone can produce significant adverse effects on human health (Gryparis et al. 2004; Teague and Bayer 2001; Uysal and Schapira 2003). Moreover, recent research is now linking air pollution to increased risk of respiratory symptoms and duration of respiratory tract symptoms. International Agency for Research on Cancer recently designated diesel exhaust as a human carcinogen.

Sulphur oxides are produced mainly from industrial activities processing materials that contain sulphur, such as generation of electricity from coal, oil or gas, as well as by combustion of fossil fuels. Sulphur dioxide is also present in motor vehicle emission. Together with ozone, it is known to cause foliar injury and reduction in plant growth (Smith 2012; Tingey and Reinert 1975). It is mainly absorbed in the upper airways as it is water soluble. Its exposure is known to cause symptoms such as nose and throat irritation. It may travel to the lower airways and cause bronchoconstriction and dyspnoea, especially in asthmatic individuals, thus worsening their condition (Balmes et al. 1987; Ierodiakonou et al. 2015).

Nitrogen oxides are emitted primarily from motor vehicle exhausts, as well as from stationary sources such as electric utilities and industrial boilers. Compounds such as sulphur and nitrogen oxides cause chemical reactions in air and acid rains. Although acid rains do not affect humans radically, they may indirectly cause health problems, particularly difficulty in breathing and, in extreme cases, lung problems such as asthma or chronic bronchitis. Moreover, nitrogen oxides are the main precursors in the formation of tropospheric ozone. They also form nitrate particles and acid aerosols. Exposure to nitrogen dioxide for a short term leads to changes in airway responsiveness and deterioration in pulmonary function in individuals with underlying lung disease. Long-term exposure may lead to increased chances of recurrent respiratory tract infections and alter lung mechanics (Berglund et al. 1993).

Carbon monoxide is produced mainly due to motor vehicle emission. In urban areas more than 80% of the carbon monoxide emission may be due to motor vehicles. Besides this, the combustion of coal, oil and gas also leads to carbon monoxide production. Moreover, tobacco smoke is one of the main sources of indoor pollution of carbon monoxide. High levels of carbon monoxide are extremely dangerous to humans, more so because it is colourless, tasteless and odourless and therefore cannot be detected by humans. Early symptoms of carbon monoxide include weakness, headache, nausea, dizziness, confusion, disorientation and visual instability. Carbon monoxide quickly enters the blood stream and forms carboxyhaemoglobin which causes more systemic effects. It reduces oxygen delivery to the tissues and may have a serious health threat to those with underlying heart disease (Badman and Jaffé 1996). Prolonged or severe exposure may result in lethal arrhythmias, electrocardiographic changes, pulmonary oedema, various neurological symptoms as well as death, most likely due to cardiac failure. Carbon monoxide is known to cause foetal development disorders, brain lesions and, in extreme cases, even mortality (Raub et al. 2000).

The atmospheric levels of lead have decreased due to the use of unleaded fuel. However, lead toxicity continues to be a problem, due to the exposure occurring in drinking water. Lead exposure leads to adverse effects on the central nervous system, causing neurological symptoms such as sleep disorders, fatigue, trembles in limbs, blurred vision and slurred speech, as well as kidney and liver disorders (Kampa and Castanas 2008). Lead toxicity can lead to lower intelligence, learning deficits and behavioural disturbances.

Ozone is an important secondary pollutant and is a component of photochemical smog. It is a pulmonary irritant and an oxidant. It may produce significant adverse effects on human health. Exposure to ozone causes airway inflammation, airway hyperreactivity and a decline in lung functions. Ozone exposure causes cough, chest tightness and wheezing. The increase in the levels of tropospheric ozone is associated with reduced baseline lung functional as well as structural abnormalities, exacerbation of asthma and premature mortality. Recent studies have shown increased admissions for chest complaints and worsening of asthma on exposure to even low levels of ozone. Studies looking at long-term exposure to ozone suggest that a cumulative long-term exposure in childhood may affect lung function, especially that of the small airways of the lung, in adult life (Künzli et al. 1997).

Ozone also affects mucous membrane and causes pulmonary inflammation and has both a local and systemic effect on the immune system. Patients with underlying respiratory illness such as asthma and chronic obstructive airway disease are more prone to the harmful effect of ozone. High ozone concentrations have been linked with increased hospital admissions for pneumonia, COPD and asthma (Gryparis et al. 2004; Teague and Bayer 2001; Uysal and Schapira 2003).

Particulate matter consists of liquid or solid mass contained in an aerosol. It is a mixture of numerous different chemicals, with varying properties. Major sources of particulate matter are factories, power plants, incinerators, motor vehicles, construction activities, fire and dust. Broadly particulate matters from 2.5 to 10 μm in diameter are coarse particulate matter. Coarse particulate matter consists mainly of airborne soil dust and elements such as silicon and aluminium. Fine particles of less than 2.5 μm are composed mainly of sulphate and organic material. Particulate matter in air is associated with allergic rhinitis, lung inflammation, pulmonary disorders, cardiac arrhythmia, ischemic cardiovascular events, higher incidences of cancer and shortening of life (Carlsten and Georas 2014; Dockery et al. 1993; Kampa and Castanas 2008; Pope III et al. 1995; Raaschou-Nielsen et al. 2013).

Only recently we began to understand the cardiovascular effects of air pollution. High levels of air pollution worsen underlying heart disease. But now it is becoming clear that persistent exposure to high levels of air pollution may also lead to heart disease. This is especially true for particulate matter. Inflammation in lungs also

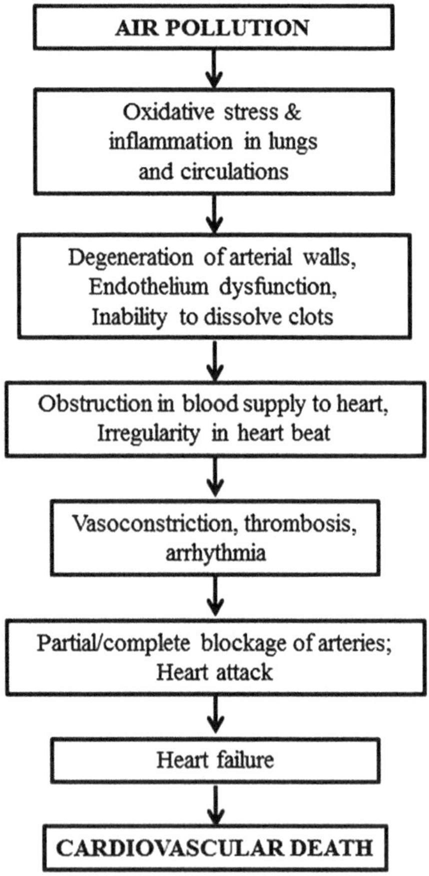

Fig. 6.1 Adverse cardiovascular effects of air pollution

causes inflammation in the blood, leading to atherosclerosis and an increase incidence of coronary artery disease that may be fatal (Fig. 6.1). Many well-conducted studies have demonstrated a 12–25% higher risk of coronary artery disease in individuals exposed to high levels of air pollution, for many years (Cesaroni et al. 2014; Miller et al. 2007). This increased risk has been linked to higher levels of PM2.5 in the ambient air. Studies have also looked at subclinical atherosclerosis, which is the pathological process associated with coronary artery disease. A positive association between subclinical atherosclerosis in the carotid and the coronary arteries has been observed with long-term exposure to high levels of air pollution (Künzli et al. 2010; Künzli et al. 2005). There is therefore now a significant body of evidence linking air pollution to cardiovascular diseases and increased mortality. Many investigators argue that air pollution should now be considered as a preventable risk factor like smoking and dyslipidemia for the development of coronary artery disease, and steps should be taken to bring down the exposure to air pollution.

Indoor air pollution and its effect on human health are important as individuals spend more than 50% of their time indoors. Cooking is an integral part of indoor human activity. The WHO has estimated that about 50% of the world's population, or about 3 billion people, still uses solid fuel for their household energy needs. Of these, about 2.4 billion people use biological material (wood, charcoal, crop waste and dung), and the remaining use coal. In India, about 58% of the population has been estimated to depend upon wood, and about 11% depend upon dung for energy. Although this number is slowly decreasing and moving towards the use of other fuels such as liquefied petroleum gas (LPG) and kerosene, it is still very significant. In India in 2010, of the 1.2 billion people, about 700 million still used solid fuel for cooking or heating. Many studies over the last three decades have documented the link between solid fuel exposure and different respiratory diseases. Lim et al. (2012) estimated more than 100 million premature deaths per year due to indoor air pollution, because of solid fuel used for cooking purposes. Exposure to high concentrations of harmful substances in smoke during use of biomass fuel causes significant illness amongst homemakers and young children. It has been shown that biomass fuel is a less efficient means of energy production and a number of carcinogenic constituents are released during biomass combustion (Chafe et al. 2014; Smith and Sagar 2014). Inhalation of these particles in high concentration leads to 'lung overloading' and sustained inflammation. This results in the release of reactive oxygen that causes Deoxyribonucleic acid (DNA) damage. Indoor smoke produced due to burning of solid fuel contains many pollutants. Particulate matter, nitrogen oxides, carbon monoxide, benzene, 1.3 butadiene, polycyclic aromatic hydrocarbons, free radicals and volatile organic compounds are many of the toxic substances that have been found in smoke produced by burning solid fuels. Chronic exposure to these harmful substances leads to lung fibrosis and subsequently the development of lung cancer. The evidence for the development of lung cancer due to biomass exposure has been shown in experimental animals, but the evidence in humans is not that strong.

Indoor air pollution thus accounts for a significant proportion of the global burden of disease in developing countries. The link between solid fuel exposure and chronic obstructive lung disease in women and acute respiratory tract infection in children is strong. The commendable initiative by the government of India called 'give it up' is a step in trying to decrease the effects of indoor air pollution on human health. Also steps to improve ventilation in kitchens or use smokeless stoves *chulla* may also help in reducing the exposure to indoor air pollution (Reddy et al. 2004).

6.7 Electronic Waste (E-Waste)

Waste generated from used electronic devices and household appliances constitutes e-waste. It comprises of a wide range of equipments and devices falling under 'hazardous' and 'non-hazardous' categories such as computers, mobile phones, refrigerators, washing machines, air conditioners, personal stereos, consumer electronics, etc., that are discarded by users (Puckett et al. 2013). Pollution due to electronic and electrical waste has rapidly grown over the last decade due to progressive increase in production of electronics, lack of proper disposal facilities in India and dumping of e-waste from developed countries. In 2010 alone, India generated about 0.4 million tons of e-waste (double the amount as compared to 2006), which is progressing rapidly. E-waste may contain many toxic substances which may be harmful to the environment and human health. This can have a significant economic and social impact on society. Iron and steel constitute about 50% of the e-waste followed by plastics (21%), nonferrous metals (13%) and other constituents (10%). Others include nonferrous metals like copper, aluminium, silver, gold, platinum, palladium, etc. The presence of elements such as lead, mercury, arsenic, cadmium, selenium and hexavalent chromium, with flame retardants beyond threshold quantities of e-waste, classifies them as hazardous waste. Manual recycling of e-waste is done predominantly via the unorganized sector, and the work force involved consists predominantly of individuals with low literacy and hardly any training to protect themselves from ill effects and to identify warning signals of toxicity. Accordingly, a significant percentage of health problems due to e-waste results from direct contact with harmful materials and inhalation of toxic fumes. Moreover, these materials may get accumulated in the food and water and are consumed. Heavy metals such as lead can cause kidney failure, neurologic manifestations and hypertension. Mercury toxicity can lead to central and peripheral nervous system damage and hepatic and renal toxicity (Guleria 2013).

Furthermore, uncontrolled burning, disposal and dismantling of e-waste can cause a number of problems including air pollution and water pollution. There is a lack of an environmentally effective recycling infrastructure for e-waste, and this leads to pollution of the environment. This is gradually changing our ecology. There is, therefore, a need to increase public awareness about the harmful effects of e-waste and develop an effective recycling and disposal plan, to prevent or minimize air and water pollution.

6.8 Conclusion

There should be general awareness of how changes in climate and environment lead to significant acute and chronic effects on human health. These effects can be both for infectious and non-infectious illnesses. A holistic view is needed to address the problem of environmental health where agriculture, animal husbandry, public health, water safety and air pollution need to be looked at in a combined manner for education, planning and resource allocation. General population should also be made aware about the ways to reduce harm to our environment. Intergovernmental efforts should be made to check climate change, avoid deforestation and use alternative sources of energy like solar energy instead of petroleum products. Ultimately, as embedded in its definition, ecosystem is a community, and unless all people in community put efforts to conserve it, no amount of individual effort can suffice. Therefore, a close teamwork between scientists, public health professionals and administrators is needed for integrated vertical and horizontal planning.

References

Argos, M., Kalra, T., Rathouz, P. J., Chen, Y., Pierce, B., Parvez, F., Islam, T., Ahmed, A., Rakibuz-Zaman, M., & Hasan, R. (2010). Arsenic exposure from drinking water, and all-cause and chronic-disease mortalities in Bangladesh (HEALS): A prospective cohort study. *The Lancet, 376*(9737), 252–258.

Badman, D. G., & Jaffé, E. R. (1996). Blood and air pollution; state of knowledge and research needs. *Otolaryngology--Head and Neck Surgery, 114*(2), 205–208.

Balmes, J. R., Fine, J. M., & Sheppard, D. (1987). Symptomatic bronchoconstriction after short-term inhalation of sulfur dioxide. *American Review of Respiratory Disease, 136*, 1117–1121.

Berglund, M., Boström, C.-E., Bylin, G., Ewetz, L., Gustafsson, L., Moldéus, P., Norberg, S., Pershagen, G., & Victorin, K. (1993). Health risk evaluation of nitrogen oxides. *Scandinavian Journal of Work, Environment & Health*, 1–72.

Bosson, J. A., & Blomberg, A. (2013). Update in environmental and occupational medicine 2012. *American Journal of Respiratory and Critical Care Medicine, 188*(1), 18–22.

Bouma, M., Kovats, R., Goubet, S., Cox, J. S. H., & Haines, A. (1997). Global assessment of El Niño's disaster burden. *The Lancet, 350*(9089), 1435–1438.

Carlsten, C., & Georas, S. N. (2014). Update in environmental and occupational lung diseases 2013. *American Journal of Respiratory and Critical Care Medicine, 189*(9), 1037–1043.

Cesaroni, G., Forastiere, F., Stafoggia, M., Andersen, Z. J., Badaloni, C., Beelen, R., Caracciolo, B., de Faire, U., Erbel, R., & Eriksen, K. T. (2014). Long term exposure to ambient air pollution and incidence of acute coronary events: Prospective cohort study and meta-analysis in 11 European cohorts from the ESCAPE project. *BMJ, 348*, f7412.

Chafe, Z., Brauer, M., Klimont, Z., Van Dingenen, R., Mehta, S., Rao, S., Riahi, K., Dentener, F., & Smith, K. R. (2014). Household cooking with solid fuels contributes to ambient PM2.5air pollution and the burden of disease. *Environmental Health Perspectives, 122*(12), 1314.

Crutzen, P. J. (2006). *The "anthropocene"* (pp. 13–18). Berlin: Springer.

Dockery, D. W., Pope, C. A., Xu, X., Spengler, J. D., Ware, J. H., Fay, M. E., Ferris, B. G., Jr., & Speizer, F. E. (1993). An association between air pollution and mortality in six US cities. *New England Journal of Medicine, 329*(24), 1753–1759.

FitzGerald, D. M., Fenster, M. S., Argow, B. A., & Buynevich, I. V. (2008). Coastal impacts due to sea-level rise. *Annual Review of Earth and Planetary Sciences, 36*, 601–647.

Griggs, D. J., & Noguer, M. (2002). Climate change 2001: The scientific basis. Contribution of working group I to the third assessment report of the intergovernmental panel on climate change. *Weather, 57*(8), 267–269.

Gryparis, A., Forsberg, B., Katsouyanni, K., Analitis, A., Touloumi, G., Schwartz, J., Samoli, E., Medina, S., Anderson, H. R., & Niciu, E. M. (2004). Acute effects of ozone on mortality from the "air pollution and health: A European approach" project. *American Journal of Respiratory and Critical Care Medicine, 170*(10), 1080–1087.

Guleria, R. (2013). Basic considerations of environmental and occupational diseases. In Y. P. Munjal & S. K. Sharma (Eds.), *API textbook of medicine* (pp. 1990–1992). Mumbai: The Association of Physicians of India.

Haines, A., Kovats, R. S., Campbell-Lendrum, D., & Corvalán, C. (2006). Climate change and human health: Impacts, vulnerability and public health. *Public Health, 120*(7), 585–596.

Harada, M. (1995). Minamata disease: Methylmercury poisoning in Japan caused by environmental pollution. *Critical Reviews in Toxicology, 25*(1), 1–24.

Houghton, J. T., Ding, Y., Griggs, D. J., Noguer, M., van der Linden, P. J., Dai, X., Maskell, K., & Johnson, C. (2001). Climate change 2001: The scientific basis.

Ierodiakonou, D., Zanobetti, A., Coull, B. A., Melly, S., Postma, D. S., Boezen, H. M., Vonk, J. M., Williams, P. V., Shapiro, G. G., & McKone, E. F. (2015). Ambient air pollution, lung function, and airway responsiveness in asthmatic children. *Journal of Allergy and Clinical Immunology, 137*(2), 390–399.

Kampa, M., & Castanas, E. (2008). Human health effects of air pollution. *Environmental Pollution, 151*(2), 362–367.
Kovats, R. S., Bouma, M. J., Hajat, S., Worrall, E., & Haines, A. (2003). El Niño and health. *The Lancet, 362*(9394), 1481–1489.
Künzli, N., Lurmann, F., Segal, M., Ngo, L., Balmes, J., & Tager, I. B. (1997). Association between lifetime ambient ozone exposure and pulmonary function in college freshmen—Results of a pilot study. *Environmental Research, 72*(1), 8–23.
Künzli, N., Jerrett, M., Mack, W. J., Beckerman, B., LaBree, L., Gilliland, F., Thomas, D., Peters, J., & Hodis, H. N. (2005). Ambient air pollution and atherosclerosis in Los Angeles. *Environmental Health Perspectives*, 201–206.
Künzli, N., Jerrett, M., Garcia-Esteban, R., Basagaña, X., Beckermann, B., Gilliland, F., Medina, M., Peters, J., Hodis, H. N., & Mack, W. J. (2010). Ambient air pollution and the progression of atherosclerosis in adults. *PLoS One, 5*(2), e9096.
Lim, S. S., Vos, T., Flaxman, A. D., Danaei, G., Shibuya, K., Adair-Rohani, H., AlMazroa, M. A., Amann, M., Anderson, H. R., Andrews, K. G., & Aryee, M. (2012). A comparative risk assessment of burden of disease and injury attributable to 67 risk factors in 21 regions 1990–2010: A systematic analysis for the Global Burden of Disease Study 2010. *The Lancet, 380*(9859), 2224–2260.
Manivasakam, N. (2005). Physico-chemical examination of water sewage and industrial effluents. Physico-chemical examination of water sewage and industrial effluents (Ed. 5).
Mann, M. E., Zhang, Z., Hughes, M. K., Bradley, R. S., Miller, S. K., Rutherford, S., & Ni, F. (2008). Proxy-based reconstructions of hemispheric and global surface temperature variations over the past two millennia. *Proceedings of the National Academy of Sciences, 105*(36), 13252–13257.
Marcott, S. A., Shakun, J. D., Clark, P. U., & Mix, A. C. (2013). A reconstruction of regional and global temperature for the past 11,300 years. *Science, 339*(6124), 1198–1201.
McCarthy, J. J. (2001). *Climate change 2001: Impacts, adaptation, and vulnerability: Contribution of working group II to the third assessment report of the intergovernmental panel on climate change*. Cambridge: Cambridge University Press.
McMichael, A. J., Woodruff, R. E., & Hales, S. (2006). Climate change and human health: Present and future risks. *The Lancet, 367*(9513), 859–869.
Miller, K. A., Siscovick, D. S., Sheppard, L., Shepherd, K., Sullivan, J. H., Anderson, G. L., & Kaufman, J. D. (2007). Long-term exposure to air pollution and incidence of cardiovascular events in women. *New England Journal of Medicine, 356*(5), 447–458.
Nicholls, R. J., & Cazenave, A. (2010). Sea-level rise and its impact on coastal zones. *Science, 328*(5985), 1517–1520.
Patz, J. A., Graczyk, T. K., Geller, N., & Vittor, A. Y. (2000). Effects of environmental change on emerging parasitic diseases. *International Journal for Parasitology, 30*(12), 1395–1405.
Patz, J. A., McGeehin, M. A., Bernard, S. M., & Ebi, K. L. (2001). The potential health impacts of climate variability and change for the United States: Executive summary of the report of the health sector of the US National assessment. *Journal of Environmental Health, 64*(2), 20.
Patz, J. A., Campbell-Lendrum, D., Holloway, T., & Foley, J. A. (2005). Impact of regional climate change on human health. *Nature, 438*(7066), 310–317.
Pope, C. A., III, Thun, M. J., Namboodiri, M. M., Dockery, D. W., Evans, J. S., Speizer, F. E., & Heath, C. W., Jr. (1995). Particulate air pollution as a predictor of mortality in a prospective study of US adults. *American Journal of Respiratory and Critical Care Medicine, 151*(3_pt_1), 669–674.
Puckett, J., Byster, L., Westervelt, S., Gutierrez, R., Davis, S., & Hussain, A. (2013). Exporting harm: The high-tech trashing of Asia. Basel action network. E-waste/technotrashfinalcom.pdf.2000. Accessed 17/05.

Raaschou-Nielsen, O., Andersen, Z. J., Beelen, R., Samoli, E., Stafoggia, M., Weinmayr, G., Hoffmann, B., Fischer, P., Nieuwenhuijsen, M. J., & Brunekreef, B. (2013). Air pollution and lung cancer incidence in 17 European cohorts: Prospective analyses from the European Study of Cohorts for Air Pollution Effects (ESCAPE). *The Lancet Oncology, 14*(9), 813–822.

Raub, J. A., Mathieu-Nolf, M., Hampson, N. B., & Thom, S. R. (2000). Carbon monoxide poisoning—A public health perspective. *Toxicology, 145*(1), 1–14.

Reddy, T., Guleria, R., Sinha, S., Sharma, S., & Pande, J. (2004). Domestic cooking fuel and lung functions in healthy non-smoking women. *Indian Journal of Chest Diseases &Allied Sciences, 46*(2), 85–90.

Smith, W. H. (2012). *Air pollution and forests: Interactions between air contaminants and forest ecosystems*. New York: Springer.

Smith, K. R., & Sagar, A. (2014). Making the clean available: Escaping India's Chulha trap. *Energy Policy, 75*, 410–414.

Teague, W. G., & Bayer, C. W. (2001). Outdoor air pollution: Asthma and other concerns. *Pediatric Clinics of North America, 48*(5), 1167–1183.

Thomas, C. D., Cameron, A., Green, R. E., Bakkenes, M., Beaumont, L. J., Collingham, Y. C., Erasmus, B. F., De Siqueira, M. F., Grainger, A., & Hannah, L. (2004). Extinction risk from climate change. *Nature, 427*(6970), 145–148.

Tilman, D., Fargione, J., Wolff, B., D'Antonio, C., Dobson, A., Howarth, R., Schindler, D., Schlesinger, W. H., Simberloff, D., & Swackhamer, D. (2001). Forecasting agriculturally driven global environmental change. *Science, 292*(5515), 281–284.

Tingey, D. T., & Reinert, R. A. (1975). The effect of ozone and Sulphur dioxide singly and in combination on plant growth. *Environmental Pollution (1970), 9*(2), 117–125.

Uysal, N., & Schapira, R. M. (2003). Effects of ozone on lung function and lung diseases. *Current Opinion in Pulmonary Medicine, 9*(2), 144–150.

Chapter 7
An Approach to Cancer Risk Assessment and Carcinogenic Potential for Three Classes of Agricultural Pesticides

Chanda Siddoo-Atwal

Abstract Sometimes, it is not for decades that the devastating effects of pesticides are recognized scientifically. Even then, specific substances can become clouded in controversy, while their victims continue to suffer, due to differing scientific opinions. Thus, a comprehensive approach to cancer risk assessment is required, based on the latest models of carcinogenesis. Pesticides are a broad category of chemicals including insecticides, fungicides, and herbicides which help to eliminate the various agricultural pests encountered in farming. There are a number of parameters that are relevant to investigations into the deleterious effects of chemical pesticides. The main ones include toxicity, mutagenicity, tumorigenicity, reproductive disruption, endocrine disruption, DNA damage, necrosis, apoptosis, and teratogenicity. Generally, the more of these categories a chemical profile fits, the greater the likelihood that it is potentially dangerous and carcinogenic. The greater number of species a chemical affects adversely, the more likely it is to have a negative impact on living organisms as a whole. There are basically three major methods available to cancer researchers conducting experiments into specific pesticides. Firstly, there are the epidemiology and statistics from farming communities or other populations at risk. However, it can be difficult and challenging to identify exposure to specific chemicals, due to the diverse nature of farming activities. Secondly, animal studies can provide useful information, but sometimes these are limited in scope due to a completely different biochemical detoxification pathway in humans. Thirdly, there are human cell culture studies. These can be highly informative in healthy cells but are of limited value in transformed or cancer cells since these also often display different biochemical pathways. An approach to carcinogenic potential based on all these criteria is adopted for three classes of pesticides including organochlorines,

C. Siddoo-Atwal (✉)
Moondust Cosmetics Ltd., West Vancouver, BC, Canada
e-mail: moondustcosmetics@gmail.com

R. Peshin, A. K. Dhawan (eds.), *Natural Resource Management: Ecological Perspectives*, Sustainability in Plant and Crop Protection,
https://doi.org/10.1007/978-3-319-99768-1_7

organophosphates, and dithiocarbamates. One example is used from each category: DDT and some of its metabolites are the chosen example for organochlorines; chlorpyrifos is the example investigated for organophosphates; and mancozeb is the example explored in the category of dithiocarbamates.

Keywords Pesticides · Toxicity · Mutagenicity · Tumorigenicity · Teratogenicity · Apoptosis · Organochlorines · Organophosphates · Dithiocarbamates · Glyphosate

Abbreviations

AT	Ataxia-telangiectasia
Bcl-2	B-cell lymphoma 2
DDD	1,1-dichloro-2,2-bis(4-chlorophenyl)ethane
DDE	dichlorodiphenyldichloroethylene
DDT	dichlorodiphenyltrichloroethane
IFN	Interferon
NF-κB	Nuclear factor kappa B
PCB	Polychlorinated biphenyl
PI3-K	Phosphatidylinositol 3-kinase
UV	Ultraviolet

7.1 Introduction

Pesticides are a broad category of chemicals including herbicides, insecticides, and fungicides which help to eliminate the various agricultural pests encountered during farming and other human activities. With an ever-increasing world population, there is a continued demand for increased food production. There is also an ongoing battle with environmental pathogens when a diligent program of vector control is required, in many countries. Over the past century, there have been several "chemical revolutions" hailing new classes of pesticides fulfilling a wide variety of roles. So, the chemical industry has served a real purpose in this regard. It is only when the negative effects of specific chemicals become apparent that it is necessary to take some positive action and limit their use, or ban them altogether, based on the extent of the available scientific findings. In this case, the chemical industry must also lead the way in finding ecologically viable alternatives.

Organochlorines

$$Cl-CH_3$$

The distinguishing feature of this category of chemicals is that each contains at least one covalently bonded atom of chlorine as a functional group. In nature, chlorine is relatively harmless, but, when produced industrially, it can be deadly. Chlorine is made by the transformation of salt into chlorine gas which is then used in the manufacture of organochlorines. Mercury is also employed in the chlor-alkali process, through which they are produced. Some of the major uses of organochlorines include pesticides, solvents, PVC plastic, and pulp and paper bleaching. These also represent the main sources of pollution (Thornton 2002).

DDT, the most famous of the organochlorine chemicals, was discovered in 1874 by H. Müller, but its insecticidal properties were not recognized until 1939. Large-scale industrial production began in 1943, and it helped to control malaria and typhus during and after World War II. Its application as an agricultural pesticide started after 1945 and reached a peak in the 1960s, worldwide (Turusov et al. 2002).

The detrimental aspects of the organochemicals started to become apparent only in the 1970s. As a class, they are generally resistant to degradation and tend to persist in air, water, and soil. Organochlorines, including DDT, polychlorinated biphenyls (PCBs), and the dioxin family, accumulate in human adipose tissue. Pregnant women can pass them on to their fetuses and, in some cases, to lactating infants in breast milk. The developing nervous system is particularly vulnerable to their effects, resulting in neurobehavioral deficits and short-term memory loss. Contaminated foods such as meat, fish, dairy, and fruit are responsible for 80% of human organochlorine exposure (Hall 1992).

Organophosphates

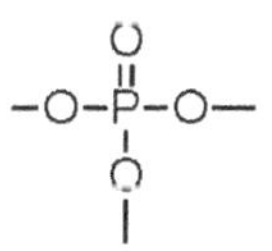

This is the general name for the esters of phosphoric acid. Organic phosphorous chemistry started in 1820 when M. Lassaigne reacted ethanol with phosphoric acid to produce triethyl phosphate. However, it was not until the 1900s that the insecticidal properties of organophosphates were discovered by G. Schrader, who is recognized as the "father" of this family of pesticides. Although his main interest was insecticides, he also went on to synthesize two nerve gases, tabun and sarin, as a result of this research into organophosphates.

Currently, organophosphates are employed for their herbicidal and fungicidal properties as well. They are also used as acaricides, nematocides, and helminthicides. This group of pesticides inhibits the enzyme, acetylcholinesterase, which is vital to normal nerve function. It is this action that is largely responsible for the acute lethality of this class of chemicals. In fact, acute mammalian toxicity was a common feature of all the early organophosphate insecticides. By the 1970s, organophosphates had become the dominant class of pesticides worldwide due to the discovery of lower toxicity compounds and the growing pest resistance to organochlorine chemicals (Chambers 1992).

Dithiocarbamates

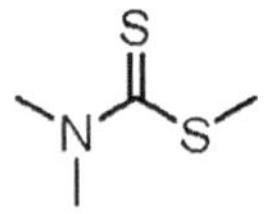

In dithiocarbamates, both oxygen atoms are replaced by sulfur in the carbamate group. There are more than 15 known dithiocarbamates, and most of these were developed after World War II. Their applications include use as insecticides, herbicides, and fungicides in agriculture. Industrial and commercial applications include use as biocides. Some are employed in vector control. Certain dithiocarbamates with hydrophilic groups form water-soluble, heavy metal complexes resulting in high biological activity. Generally, these are readily absorbed through the mucous membranes and the skin. However, other dithiocarbamate metal complexes used as fungicides are insoluble in water and are soluble in nonpolar solvents (WHO 1988).

7.2 Cancer Methodology

There are a number of parameters of study that are relevant for testing the deleterious effects of chemical pesticides. The main ones include toxicity, mutagenicity, tumorigenicity and/or carcinogenicity, reproductive disruption, endocrine disruption, DNA damage, necrosis, apoptosis (cell death), and teratogenicity (birth defects). Generally, the more of these categories a chemical profile fits, the greater the likelihood that it is potentially dangerous and carcinogenic. The greater number of species a chemical affects adversely, the more likely it is to have a negative impact on living organisms, as a whole.

Toxicity can be further subdivided into major groups such as immunotoxicity, neurotoxicity, genotoxicity, and reproductive toxicity. Immunotoxicity refers to detrimental effects upon the immune system such as self-reactive lymphocytes. Neurotoxicity indicates negative effects upon specific nerve cells and the whole nervous system resulting in conditions such as altered cognition or motor control. Genotoxicity refers to different kinds of DNA damage (potentially lethal) caused to single cells, including those involved in reproduction. Genotoxicity to germ cells or damage to the reproductive system, in general, is known as reproductive toxicity.

There are basically three major methods available to cancer researchers conducting investigations into specific pesticides. All of these can be subjected to statistical analysis.

Firstly, there are the epidemiology and statistics derived from farming communities or other populations deemed to be at risk. It can be difficult and challenging to identify exposure to specific chemicals, due to the diverse nature of farming activities. However, some excellent studies exist in which specific farming areas are limited to growing specific crops. This implies the use of very specific pesticides, unique to the pests attacking those particular plant species.

Secondly, in vivo and in vitro animal studies can provide useful information, but sometimes these are limited in scope due to a completely different biochemical detoxification pathway in humans. Moreover, ethical considerations are causing more and more researchers to avoid these. Nevertheless, from an ecological view, these studies can provide valuable data about the effects of various chemicals on marine life like fish, which are an important food source to other species.

Thirdly, there are human cell culture studies. These can be highly informative in healthy human cells but are of limited value in transformed or cancer cells, since these often display altered biochemical pathways. Techniques include various DNA damage assays (some measure apoptotic changes in cell populations like the DNA diffusion assay), apoptosis (cell death) assays involving methods like flow cytometry, and Raman microspectroscopy of single-cell changes in the nucleus and various cellular membranes. There are a number of experimental assays available to the investigator to measure genetic damage, including chromosomal aberrations, DNA single- and double-strand breaks, DNA fragmentation, and the appearance of micronuclei.

7.3 Current Models of Carcinogenesis

Traditionally, experimental carcinogenesis is a complex, multistage process including initiation, promotion, and malignant progression in which the failure of DNA repair mechanisms and the subsequent clonal expansion of damaged cells play a pivotal role (El-Abaseri et al. 2006). However, more recently, it has become apparent that the pathogenesis of cancer is closely connected with aberrantly regulated apoptotic cell death and the resulting deregulation of cell proliferation (Denmeade and Isaacs 1996; James et al. 1998). "Programmed cell death" is a general term for a biochemical pathway of cell death which is essential to metazoans in maintaining tissue homeostasis. "Apoptosis" (*falling leaves* in Greek) specifically refers to one particular mode of programmed cell death, which is responsible for the elimination of potentially deleterious, mutated cells.

Typically, the targets include single cells that are aged, are dysfunctional, or have been damaged beyond repair by external stimuli such as ultraviolet (UV) irradiation, as in the case of apoptotic sunburn experienced at the beach (Siddoo-Atwal 2009, 2011). Specifically, double-strand breaks in DNA caused by radiation or other carcinogens can trigger apoptosis via various cell signaling pathways (Lu et al. 2000; Brash 1997). Inducers of apoptosis also include other types of intracellular and extracellular stimuli including DNA damage (such as thymine dimers and pyrimidine photodimers), cell cycle disruption, hypoxia, detachment from surrounding tissue, and loss of trophic signaling (Sun et al. 2004). Apoptosis is regulated through at least two well-recognized pathways, both involving intracellular caspase activation. The first of these is called the "intrinsic pathway" and is mediated through mitochondria, while the second is known as the "extrinsic pathway" and is mediated through cell surface death receptors (Lokshin and Zakeri 2004; Danial and Korsmeyer 2004). These "death receptors" belong primarily to a tumor necrosis factor receptor/nerve growth factor receptor [TNFR/NGFR] subfamily containing cysteine-rich domains in their extracytoplasmic region. Apoptotic cell death can be basically divided into four phases: *triggering* (by mitochondria or death receptors), *signaling* (by proteins or protein kinases [PK]), *execution* involving caspases and nucleases, and *burial* including phagocytosis (Eguchi 2001). More recently, a caspase-independent apoptotic pathway has also been gaining some prominence in apoptosis studies (Susin et al. 1999).

Evidence for the role of apoptotic deregulation in carcinogenesis comes from several different sources. Epidemiological studies with squamous cell carcinoma patients have revealed that homozygous genetic polymorphisms in DNA repair proteins, resulting in a greater susceptibility to apoptotic sunburn, are strongly correlated with skin cancer (Han et al. 2004; Nelson et al. 2002). Quantitative histological studies in the rat liver model have revealed that the rate of apoptosis tends to increase from normal to preneoplastic to malignant cells (Schulte-Hermann et al. 1995). Comparative studies with the rat bladder have also suggested that apoptosis is closely linked to chemically induced carcinogenesis (Zhang and Takenaka 2007). Scientific animal studies have shown that simply increasing the basal frequency of apoptosis in murine skin cells can be linked to the development of squamous cell carcinomas in transgenic mice (van Hogerlinden et al. 1999).

Furthermore, numerous mutations occur in tumor suppressor genes involved in the induction of apoptosis such as p53, which is the most frequently mutated in human cancers (Levine 1997). It transactivates genes that mediate apoptosis and functions in DNA repair, senescence, and cell cycle arrest. P53 deficiency leads to inappropriate survival of cells with DNA damage thereby predisposing them to neoplasia (Mihara et al. 2003). Numerous oncogenes may also be activated to inhibit the inherent controls of apoptosis such as Bcl-2, NF-κB, PI3-K, Ras, Myc, and Flip (Johnstone et al. 2002). The fact that mutations in genes that control apoptotic pathways are common in most cancers further emphasizes the importance of apoptosis in carcinogenesis (Sun et al. 2004).

Certain carcinogens like UV rays exert some of their carcinogenic effects via the generation of reactive oxygen species (ROS) in the cell (Lu et al. 1997). This also is true for X-rays, as well (Liao et al. 1999). Certain oncogenic proteins such as Ras also produce elevations in ROS upon stimulation. Many genes and proteins that respond to conditions of oxidative stress within the cell subsequently trigger apoptosis. Because mitochondria are important regulators of cellular redox status, the induction of oxidative stress exhibits its effects upon these organelles to trigger the intrinsic apoptotic pathway via cytochrome c release and caspase cascade activation (Martin 2006; Sun et al. 2004).

In addition, a number of studies have shown that targeting apoptosis with selective apoptotic agents (such as polyphenols and certain phytochemicals) at different stages of carcinogenesis can be successful, adding further support for this thesis (Surh 1999; Surh 2003; Kaur et al. 2006). Polyphenols have also been reported to suppress tumorigenesis, partly through the induction of apoptosis in transformed cells. For example, they protect against chemically induced hepatic tumors in mice (Klaunig and Kamendulis 1999). However, the truly remarkable feature of certain bioactive agents like resveratrol, a polyphenol in grapes, is that they seem to selectively target cancer cells while sparing normal cells, and this has been corroborated in animal studies (Borek 2004). Similarly, polyphenols from tea induce apoptosis in skin tumors formed by UV exposure of mice susceptible to sunburn (Lu et al. 1997).

There are many models of cancer in humans. One such model is provided by albinos, who have little or no melanin (the pigment that protects against UV radiation and apoptotic sunburn) in their skin and, as a result, display an increased skin cancer incidence (Kromberg et al. 1989). This is confirmed in the albino hairless mouse model, which displays a high incidence of UV-induced cutaneous cancers, as well (de Gruijl and Forbes 1995).

Patients with the inherited disorder ataxia-telangiectasia [AT] provide another human cancer model since they are highly susceptible to certain cancers, particularly lymphomas (Taylor et al. 1996). The genetic mutation in AT patients occurs in and affects a PI3-kinase associated with DNA damage and has been found to be involved in the apoptotic response to X-ray irradiation (Liao et al. 1999). In addition, a constitutive elevation of IFNβ production has been demonstrated in fibroblasts derived from AT individuals (Siddoo-Atwal et al. 1996). An elevation of spontaneous apoptosis in AT lymphocytes has also been observed (Duchaud et al. 1996), thereby establishing a link between apoptotic potential and an increase in cancer risk. Cytokines like IFNα/β are known to stimulate the transcription of many proapoptotic proteins (Gao 2005). In addition, the transcriptional activator NF-κB, which mediates interferon (IFN) production, is constitutively activated in AT cells (Jung et al. 1995).

Moreover, there appears to be a role for the transcription factor NF-κB in carcinogenesis, as indicated by various genetic studies (Chen et al. 1999). However, the relationship between NF-κB activation and apoptosis is not always straightforward, since it does not necessarily stimulate apoptosis in all cell types. This may be due to the possibility that different NF-κB members mediate different signals or, more likely, that the role of NF-κB in apoptosis depends on the target cell type. This is evidenced by the complex and distinctive lymphocyte cell death model (Rieux-Laucat et al. 2003) in which NF-κB is apoptotic, while it is anti-apoptotic in hepatocytes and synovial cells (Miagkov et al. 1998; Wang et al. 1998).

Apoptosis is involved in the cell homeostasis in tissues, and, although increased cell proliferation is necessary, it is certainly not sufficient for cell transformation to take place. Normally, in multicellular organisms, a dynamic equilibrium exists between cell birth and death, to maintain constant cell numbers throughout adult life. This homeostasis depends on an integrated balance between apoptosis and mitosis, such that these two activities are counterbalanced and equivalent. This homeostatic balance may contribute a critical cell defense mechanism toward various genotoxic agents such as carcinogens (James et al. 1998).

The increased proliferation in preneoplastic lesions is often accompanied by a parallel increase in apoptosis. A permanent loss in homeostatic equilibrium between cell proliferation and death may be a critical determinant in the transition to tumorigenesis. Support for this view comes from the islet B cells of a multistage mouse model of carcinogenesis, in which the incidence of apoptosis increased in parallel with increasing proliferation during tumor promotion. Malignancy was associated instead to a dramatic drop in apoptotic rate, without a corresponding decrease in proliferation rate (Naik et al. 1996). Also, in the natural human skin cancer model (*not* involving the application of inflammatory skin irritants or tumor promoters found in various suntan oils), there is always a fresh layer of new epidermis underlying the peeling or apoptosing cells, following sunburn. Tumor formation only seems to occur once the cancer cells have become resistant to apoptosis while continuing to proliferate. In fact, acquired resistance to apoptosis appears to be a pivotal event in immortalization and their transition to malignancy (James et al. 1998).

In summary, various laboratory studies on animals and certain human data (Siddoo-Atwal 2015) are suggestive that tumor formation requires at least two discrete events to take place in response to a carcinogen. The first involves an elevation of apoptosis in a particular tissue due to a genetic predisposition, stress, or mutation. The second confers resistance to apoptosis in that same tissue, resulting in the formation of an abnormal growth due to a dysregulation of cell number homeostasis. Moreover, there is some evidence to suggest that both these events can be reversible when treated with a selective apoptotic agent, and, hence, they may be either genetic or epigenetic in nature (see Figs 7.1 and 7.2).

Thus, according to this new model, apoptosis becomes an important focus of study and key determinant of carcinogenic potential for any particular chemical being studied, especially in normal, non-transformed cells, derived from the target tissue.

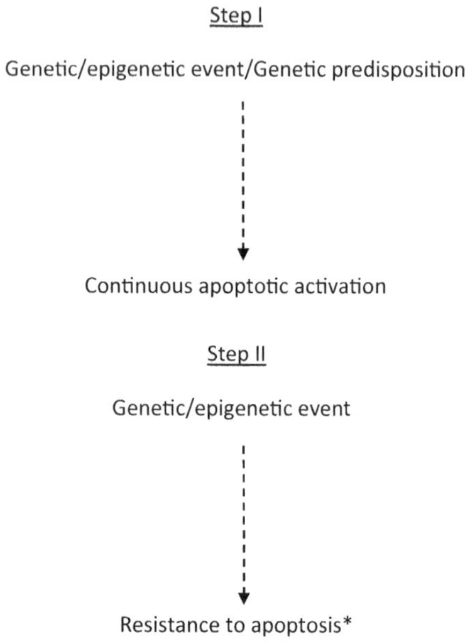

Fig. 7.1 Two-stage model of tumor formation (*apoptosis = programmed cell death)

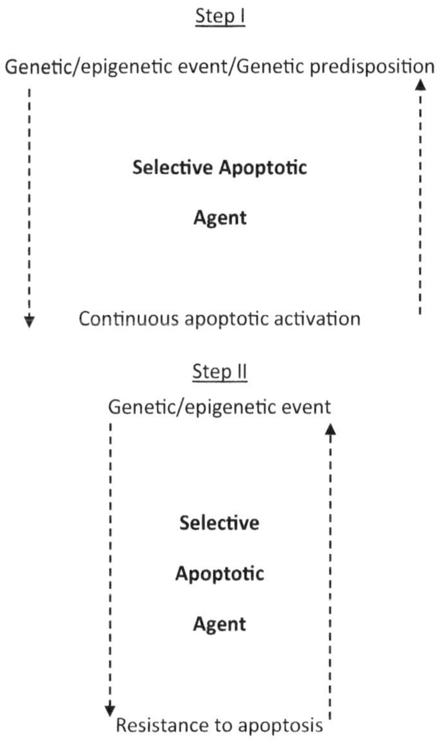

Fig. 7.2 Two-stage model of tumor formation and action of selective apoptotic agent

7.3.1 DDT and Its Metabolites

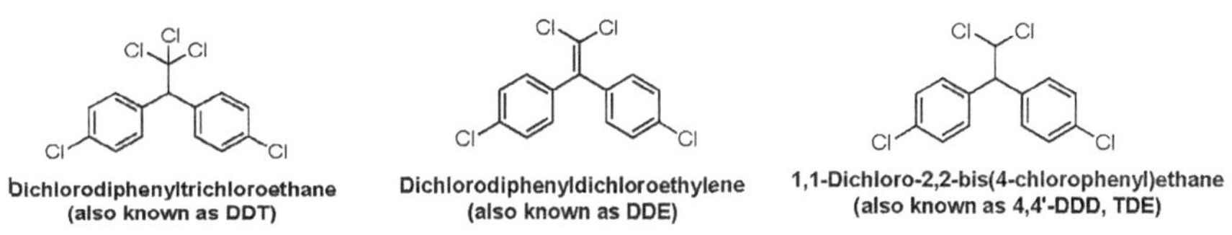

The insecticide DDT (dichlorodiphenyltrichloroethane) was first introduced in the United States in the 1950s for malarial vector control. Generally, DDT is not acutely toxic to humans but was mainly banned for ecological reasons, beginning from Sweden in 1970. It was banned in Russia and many other countries shortly after (Turusov et al. 2002). However, certain African, Latin American, and Asian countries (notably India) still employ it for vector control. Malaria is one of the leading causes of human deaths in tropical and subtropical regions, worldwide. Notably in South Africa, a malaria-endemic country, there were disastrous consequences when it was temporarily stopped (Jaga and Dharmani 2003).

The main concern is that DDT and its metabolites DDE (dichlorodiphenyldichloroethylene), and DDD (1,1-dichloro-2,2-bis(4-chlorophenyl)ethane), can persist for decades in humans, animals, plants, water, air, and soil. Specifically in humans, dietary and/or environmental exposure results in their accumulation in adipose tissue, blood serum, and breast milk. While DDT concentrations in breast milk have decreased over the last decades in most countries, DDE can still be found in breast milk in Ukraine (Gladen et al. 1999), due to local DDT chemical manufacture, and in parts of Punjab in India (personal communication from Balwinder Singh, Punjab Agricultural University), presumably due to ongoing vector control activities. Recent research suggests that DDT may cause preterm birth and early weaning of infants, which would offset some of its benefits for vector control (Rogan and Chen 2005).

DDT also accumulates in the tissues of other living organisms. It is toxic to freshwater and marine microorganisms, fish, amphibians, and birds. This pesticide has been associated with dramatic declines in many animal and bird populations (Beard 2006).

Biochemically, DDT and some of its metabolites are known to bind to estrogen receptors and to elicit estrogen-like effects (Kupfer and Bulger 1976). This is the mechanism via which they function as endocrine disruptors (Crews et al. 2000). However, breast cancer risk assessment is complicated by the ever-changing hormonal history in women due to menarche, pregnancy, lactation, and menopause. Thus, there is much conflicting data about the correlation between DDT blood levels and breast cancer (Cohn et al. 2007). Genetic variables may also be involved (Snedeker 2001; Schecter et al. 1997).

Epidemiologically, there is evidence for a role of DDT in pancreatic cancer under conditions of heavy and prolonged exposure. An association has been found in chemical manufacturing workers exposed to the pure product (Garabrant et al. 1992). However, no such clear trend has been observed in workers applying this pesticide in the field, where the exposure levels are likely to be lower (Andreotti et al. 2009). Genetic factors also appear to play a role. In a hospital study in Spain, for example, only those individuals with a K-Ras mutation were found to display an association between serum concentrations of DDT and pancreatic cancer (Porta et al. 1999). A weak association between serum DDE levels and pancreatic cancer has also been reported, due to environmental and occupational exposure to organochlorine chemicals (Hoppin et al. 2000).

In addition, there is some epidemiological evidence for an association between DDT and DDE and liver cancer in certain human populations. An elevation of DDE concentrations in the adipose tissue of Caucasian men has been correlated with an increased cancer risk (Cocco et al. 2000). The risk of liver cancer was also significantly elevated in Asian [Chinese] men with high blood DDT levels, while blood DDE levels were not associated with a higher risk (McGlynn et al. 2006). These types of studies highlight the possible role, in the etiology of cancer, of genetic polymorphisms and biochemical differences in various ethnic groups.

Hepatoblastomas have been induced in mice following both limited and lifetime treatment with DDT in a dose-dependent manner. DDE exposure also results in a high incidence of liver tumors in mice, while DDD moderately increases incidence in male mice only. The main significance of these animal studies is to show that DDT and its metabolites are cytochrome P450 inducers, which result in the generation of free radicals during their detoxification and thereby contribute to oxidative stress in the cell (Turusov et al. 2002). Their metabolism in the liver via this pathway may be a possible mechanism for triggering apoptosis and carcinogenesis in hepatocytes.

DDT and DDE are able to induce apoptosis in human mononuclear cell cultures. This particular study was performed in children exposed to the insecticide, who were found to have higher blood levels of DDT, DDE, and DDD, as compared to controls (Perez-Maldonado et al. 2004, 2006). DDT and DDE are also potent activators of AP-1 activity in human endometrial adenocarcinoma cell lines (Frigo et al. 2002). Since the transcription factors NF-κB and AP-1 function in the activation of the FasL-dependent apoptotic pathway in certain cell types, this may indicate a potential for apoptotic stimulation. This would suggest a possible mechanism of carcinogenesis for this endocrine disruptor. However, more evidence of the apoptotic potential of DDT and its metabolites is required in normal non-transformed human cell lines, specifically in hepatocytes and pancreatic cells.

7.3.2 *Chlorpyrifos*

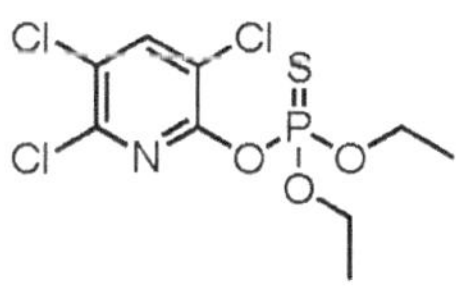

O,O-diethyl O-3,5,6-trichloropyridin-2-yl phosphorothioate (also known as Chlorpyrifos)

Chlorpyrifos (O,O-diethyl O-3,5,6-trichloropyridin-2-yl phosphorothioate) is a broad-spectrum chlorinated organophosphate insecticide. It has a large variety of crop and non-crop uses including termite control and is one of the most widely used insecticides in the world. The main degradation route appears to be via aerobic and anaerobic metabolism, while its persistence is significantly greater in the cold and dark conditions of the Arctic. Chlorpyrifos is less persistent under tropical conditions due to microbial degradation, photodegradation, and volatility at high temperatures.

This chemical has been associated with genotoxicity, mutagenicity, immunotoxicity, carcinogenicity in humans, reproductive toxicity, developmental neurotoxicity, and endocrine disruption. Ecotoxicity has been demonstrated in birds, mammals, terrestrial invertebrates, fish, and aquatic vertebrates. Its aquatic toxicity is most relevant in the context of the Arctic ecosystem (Watts 2012).

Epidemiological surveys, part of the US Agricultural Health Study, have shown statistically significant increases in rates of brain cancer (specifically glioma), lymphohematopoietic cancers, and leukemia in exposed male pesticide applicators (Lee et al. 2004). Smaller studies have also found a significant association, in chlorpyrifos-exposed individuals, with Hodgkin's (Karunanayake et al. 2012) and non-Hodgkin's lymphomas (Waddell et al. 2001). In addition, there is epidemiological evidence indicating an association between exposure and lung and rectal cancers (Lee et al. 2007).

Studies from the Kashmir apple-growing region of India implicated chlorpyrifos and mancozeb, as two of a range of chemicals employed in the area, in the increased incidence of primary malignant brain tumors and familial gliomas (Bhat et al. 2010; Baba et al. 2012). Chlorpyrifos is the more likely candidate for causative factor, in this particular instance, based on the other available data. Nevertheless, the neurodegenerative and neurotoxic effects of mancozeb may also contribute to the disease state.

The carcinogenicity of chlorpyrifos has not been demonstrated in rat models. This may be due to major metabolic differences in biochemical detoxification pathways. For example, major biochemical differences in PCB metabolism exist between humans and rats. However, developmental neurotoxicity has been observed in neonatal rats (Slotkin et al. 2001), and chlorpyrifos is known to interfere with the development of the mammalian brain and nervous system (Qiao et al. 2002). It is toxic to immature neurons and glial cells (Monnet-Tschudi et al. 2000). Genotoxicity has also been demonstrated, particularly in male rats, as evidenced by micronucleus formation and other DNA damage (Sandhu et al. 2013).

Chlorpyrifos induces cell apoptosis and nuclear DNA fragmentation associated with apoptosis in cultured HeLa cells (Diqiu Li et al. 2015). It has been shown to stimulate apoptosis in primary human hepatocytes, in combination with two other chemicals (Nawaz et al. 2013). Chemical modifications affecting the nucleus and cell membranes have been observed via Raman microspectroscopy in single human keratinocytes exposed to low doses of chlorpyrifos (Pema et al. 2011). Increased chromosomal aberrations have been reported in exposed human lymphocytes (Abbassy et al. 2006). It has also induced apoptosis and necrosis in cultured human peripheral blood lymphocytes, in vitro (Prabhavathy Das et al. 2006).

7.3.3 *Mancozeb*

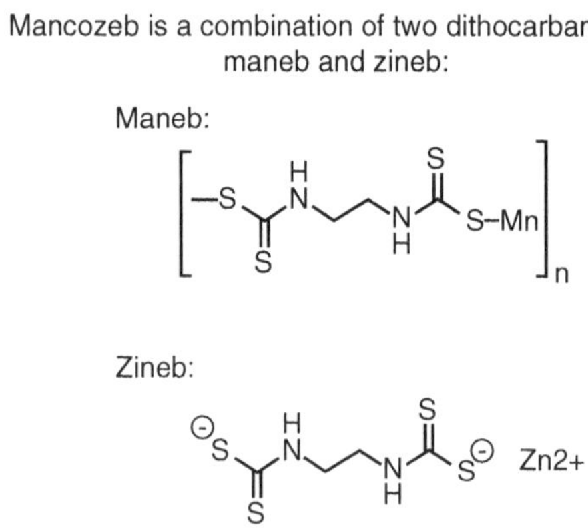

Manganese/zinc ethylene-bis-dithiocarbamate is a dithiocarbamate fungicide. It is widely used on a variety of horticultural crops and wheat (Belpoggi et al. 2002). It contains zinc and manganese, which are both essential elements for humans, but the latter is neurotoxic when excess amounts accumulate in the body (Watts 2014). The main metabolite is ethylenethiourea (ETU) which is known to cause thyroid cancer in rats. Biochemically, ETU acts by decreasing thyroxine and increasing thyroid-stimulating hormones, as well as being a thyroid peroxidase inhibitor (Houeto et al. 1995).

Chemical residues are found in food and drinking water. Manganese can be even detected in the hair of exposed women (Mora et al. 2014). Mancozeb has been associated with acute toxicity, chronic toxicity, neurological damage, cancer, genotoxicity, endocrine disruption, reproductive anomalies, birth defects, and modulation of the immune system. Environmental effects include toxicity to fish and aquatic vertebrates, reproductive disruption in birds and mammals, and toxicity in certain insects (US EPA 2005). Ecological effects include accumulation of manganese in the soil and accumulation of ETU in surface water (EURO Com 2009).

Epidemiologically, mancozeb has been correlated with an increased thyroid cancer risk in men in the agricultural regions of Minnesota, where its use is prevalent according to local crop diversification patterns. This American state is uniquely situated for conducting such a study since it is divided into four agricultural regions, due to its unusual geology. Each of these four regions is limited to growing specific crops, and, therefore, specific pesticides are in use in each area. Generally, thyroid cancer occurs more frequently in women. Thus, the increased thyroid cancer mortality observed in these specific areas of Minnesota is likely to be related to environmental or occupational exposure (Schreinemachers et al. 1999). In another epidemiological survey, an increased incidence of birth defects in farming families and the general public was also found in this same region (Garry et al. 1996). In addition, mancozeb has been associated with leukemia and other cancers (Mills et al. 2005; Watts 2014).

Experimental studies in rats on the carcinogenicity of mancozeb have resulted in multiple carcinomas including malignant tumors of the thyroid gland (Belpoggi et al. 2002). An overall increase in the incidence of head and neck carcinomas, in particular, suggest a susceptibility of the head region. This propensity may also be correlated with some of the neurotoxic effects of this chemical in rats (Domico et al. 2006). These include peripheral nerve damage resulting in abnormal gait and loss of muscle mass (US EPA 2005). Prenatal exposure also alters the developing brain in mice (Miranda-Contreras et al. 2005). Neurodegeneration and behavioral changes were observed in other species, as well (Negga et al. 2011).

Cell culture studies have revealed that mancozeb induces apoptosis in cultured human lymphocytes, as measured by flow cytometric methods. The release of ROS plays a critical role in the initiation of mancozeb-induced apoptosis in two ways. Firstly, it stimulates the mitochondria-mediated pathway involving cytochrome c release and activation of the caspase cascade. Secondly, it acts via NF-κB activation, expression of Fas-L, and triggering of the Fas-L-dependent apoptotic pathway (Srivastva et al. 2012). Fas is a member of the TNFR/NGFR apoptotic death receptor family, and Fas-L (Fas ligand) can be co-stimulated and activated by NF-κB and AP-1 in T lymphocytes (Kasibhatla et al. 1998). Mancozeb also induces a proapoptotic effect in rat fibroblasts and leukocytes, suggesting a mechanism for carcinogenesis and the pathogenesis of neurodegenerative disease (Calviello et al. 2006).

7.4 Conclusion

DDT is likely to be a risk factor for pancreatic cancer in humans under conditions of heavy exposure, such as chemical manufacture and certain genetic predispositions, such as the K-Ras mutation. DDT and its metabolites may also pose a risk for liver cancer. Animal studies support a role for DDT, DDE, and DDD in the induction of liver cancer in mice, which detoxify organochlorine chemicals via the cytochrome P450 biochemical pathway, like in humans.

Although there is some evidence for the apoptotic potential of DDT and its metabolites, more studies should be conducted on specific, normal human cell types. Further investigations should also be carried out on the possible effects of DDT and DDE found in blood samples of children and in breast milk in India and other places. Studies showing preterm birth and early weaning may be an indicator that children are more susceptible to the effects of these chemicals than adults. The possible correlation with leukemia and other childhood cancers should be investigated since apoptotic potential has been demonstrated in human mononuclear cells in vitro. At the same time, efficient pesticides without the negative ecological effects of DDT should be sought.

Chlorpyrifos is highly toxic to aquatic life, and there have been ecological incidents demonstrating its toxicity to terrestrial species. It has also been detected in human breast milk in India. Epidemiological evidence suggests an association with brain cancer (gliomas), lung cancer, lymphohematopoietic cancers, and leukemia. Although animal studies have not resulted in carcinogenesis, this may be due to biochemical differences between humans and laboratory animals. However, developmental neurotoxicity has been demonstrated in neonatal rats. Moreover, genotoxicity and mutagenicity have also been reported in rats and other species. Cell culture studies have revealed that chlorpyrifos has antiandrogenic (Viswanath et al. 2010) and estrogenic effects (Andersen et al. 2002; Kojima et al. 2004). Cell apoptosis, nuclear DNA fragmentation, chromosomal aberrations, and chemical modifications have also been observed in culture.

Thus, there is a strong possibility that chlorpyrifos is a carcinogen and is implicated in the formation of brain tumors in agricultural workers employing this pesticide. A suitable alternative should be found. A study of childhood cancer incidence is also strongly suggested in the areas of Punjab where chlorpyrifos has been found in breast milk (although, the latest results show a decline) and in the regions of Kashmir where chlorpyrifos use is prevalent among orchard farmers.

Mancozeb has been associated with various kinds of toxicity in humans. It has been found to have detrimental environmental and ecological effects that are far-reaching. Epidemiologically, it has been correlated with increased thyroid cancer risk in men and other cancers. Experiments in animal models, such as rats, have also resulted in malignant tumors of the thyroid gland. Neuronal damage has been observed in rats, mice, and other species. Apoptosis has been demonstrated in cultured human lymphocytes. Thus, the conclusion strongly suggested in this case that mancozeb is a dangerous chemical with carcinogenic and neurotoxic potential in humans. It also has long-term adverse effects upon the environment and other species. A substitute fungicide (preferably a biopesticide), possibly derived from the antifungal components of eucalyptus or manuka oils, should be developed for commercial purposes (Siddoo-Atwal and Atwal 2012).

In conclusion, the greatest irony of ecology lies in the basic principle by which it binds humans inextricably to the lowest forms of life on the planet and the realization that, by attempting to exterminate them, we may actually be exterminating ourselves.

Furthermore, the consequences are far too devastating when children become the unwitting victims of the world's relentless need to increase food production. Plausible alternatives like powerful natural pesticides must be found. Population control is also important and can play a significant role in limiting and managing the agricultural and farming needs of a country.

Concomitant with the development of new, less harmful biopesticides, there should be implementation of educational programs for farmers about proper pesticide use and application practices. A reduction in the excessive use of certain suspect pesticides, at plant growth stages and on fruits that they are not recommended for, may also result in a decrease in chemical-related cancer incidence.

In this century, chemical companies are increasingly likely to bear the responsibility for rigorous testing of their products *prior* to marketing. A culture of self-censorship and good citizenship should be demanded by the general public, with the aid of media globalization. Moreover, a statistically significant DNA damage and cell death (apoptosis) assay in a panel of cell cultures derived from different human tissue types may become the standard test for carcinogenicity. Due to the increasing acknowledgement of the importance of this parameter in carcinogenesis by the scientific community, the hallmark of skin carcinogenesis is finding ever wider application in the field of cancer research, as a whole.

7.5 Addendum: Glyphosate (Organophosphate/Organophosphorus)

Glyphosate is a broad-spectrum, nonselective systemic herbicide that targets broadleaf plants and grasses. It is one of the most heavily used herbicides worldwide. Chemically, it is an organophosphate or organophosphorus compound, but it does not affect the nervous system in the same way as organophosphate pesticides. Glyphosate acts by blocking a specific enzymatic pathway, the shikimic acid pathway, that is required for making certain proteins necessary for plant growth.

This herbicide binds tightly to soil where it can persist for up to 6 months depending upon various factors like climate and soil type. Then, it is broken down by bacteria in the soil. Due to its affinity for soil, glyphosate is not so likely to get into the groundwater. However, glyphosate and its degradation product, aminomethylphosphonic acid (AMPA), can be mobile, and both of these chemicals are usually detected together in the environment. They occur widely in soils and sediment, rivers, and streams and less frequently in lakes, ponds, wetlands, soil water, and groundwater (Battaglin et al. 2014). On its own, glyphosate has low toxicity toward fish and wildlife, but in combination with other chemicals, it may be toxic. It has been linked to reproductive and developmental effects in mammals.

Glyphosate can be absorbed through the skin, eyes, and nose during application. It may also be ingested, following application in the absence of good hygiene. It passes through the body relatively quickly, and the majority of it exits through the urine and feces (National Pesticide Information Centre – General Fact Sheet, Glyphosate).

Epidemiologically, glyphosate has been associated with a very specific type of cancer [non-Hodgkin's lymphoma, NHL] in at least three independent case-control studies in humans (De Roos et al. 2003; Hardell and Eriksson 1999, Hardell et al. 2002; McDuffie et al. 2001). However, in another large prospective cohort study, no association was found between glyphosate and any solid tumors or lymphoid malignancies overall, including NHL and its subtypes. Nevertheless, there was some evidence of an increased risk of acute myeloid leukemia (AML) among the highest exposed group of individuals (Andreotti et al. 2017). Certain epidemiological studies have also suggested an association with multiple myeloma incidence (De Roos et al. 2005). Thus, taken together, this data appears to suggest an association between glyphosate and hematological malignancies, namely, lymphoma, leukemia, and myeloma.

Glyphosate has been found to disrupt cell cycle regulation in marine invertebrate models such as sea urchin embryos. Dysregulation of the cell cycle is often observed in tumor cells and can result in genomic instability (Marc et al. 2004). One of the effects of a failure of cell cycle checkpoints is to trigger cell death or apoptosis.

Glyphosate-based herbicides have also been shown to produce teratogenic effects on vertebrates by impairing retinoic acid signaling. These defects include impaired neural crest development and craniofacial malformations (Paganelli et al. 2010).

Glyphosate increased the frequency of micronucleus formation in mice in vivo (Manas et al. 2009). It also increases the frequency of chromosomal aberrations in the bone marrow cells of Swiss albino mice (Prasad et al. 2009). Another interesting study with mice indicated that glyphosate has tumor-promoting potential similar to 12-O-tetradecanoylphorbol-13-acetate (TPA) in a skin carcinogenesis model (George et al. 2010). Moreover, a study with glyphosate demonstrated that maternal exposure induced a variety of functional abnormalities in enzyme activities in pregnant rats and their fetuses (Daruich et al. 2001).

One long-term study using environmentally relevant concentrations of a glyphosate-based herbicide reported a significant increase in the incidence of mammary tumors in rats (Seralini et al. 2014). In this regard, it is interesting to note that glyphosate is estrogenic at relatively high concentrations and can mimic estrogen. However, the typical exposure levels of humans are considered to be well below these concentrations and, therefore, may not constitute a breast cancer risk factor (Mesnage et al. 2017). In another study, glyphosate was reported to exert proliferative effects in human hormone-dependent breast cancer cells, even at environmentally relevant concentrations, thereby confirming its potential estrogenic activity (Thonprakaisang et al. 2013).

In addition, a glyphosate-based herbicide (Roundup™) induced necrosis and apoptosis in mature rat testicular cells, at high doses in vitro. At lower environmental doses, it displayed an endocrine impact by decreasing testosterone levels by 35% (Clair et al. 2012). In a similar study, Roundup disrupted male reproductive functions by triggering calcium-mediated cell death in rat testis and Sertoli cells (De Liz Oliveira Cavalli et al. 2013). Glyphosate also induced cell death through apoptotic and autophagic mechanisms in differentiated rat PC12 cells, in a dose- and time-dependent manner (Gui et al. 2012). A commercial formulation of glyphosate increased caspase-3-like activity and annexin V-positive cells, which are both indicative of apoptosis, in mouse 3 T3-L1 fibroblasts (Martini et al. 2012).

Glyphosate-based herbicides are endocrine disruptors and cytotoxic in human liver carcinoma cells (HepG2) (Gasnier et al. 2009). Glyphosate formulations have also been observed to trigger apoptosis via caspase 3 and caspase 7 activation, in these same cells (Chaufan et al. 2014). Another interesting finding is that human-derived epithelial cells are more susceptible than cells from internal organs to the cytotoxic and DNA-damaging properties of the herbicide, as determined by an elevated frequency of micronuclei and nuclear budding (Koller et al. 2012).

Low dilutions of glyphosate formulations have been found to induce apoptosis in three different human cell types, including umbilical, embryonic, and placental cells, via activation of enzymatic caspases 3 and 7. These results have been confirmed via DNA fragmentation, nuclear shrinkage, and nuclear fragmentation assays (Benachour and Seralini 2009). In human erythrocytes, hemolysis was only observed at high (potentially toxic) concentrations in vitro (Kwiatkowska et al. 2014).

Although epidemiology indicates a correlation between glyphosate and lympho-hematopoietic malignancies, animal studies provide little supporting evidence for a corresponding association with cancer. However, various animal models demonstrate that glyphosate can cause disruption of cell cycle regulation and DNA damage such as chromosomal aberrations. Both these types of biological effects can trigger cellular apoptosis. Apoptosis has also been observed in response to glyphosate, and its commercial chemical preparations, in a variety of animal models. Moreover, a definite sensitivity to this chemical has been observed in pregnant rats and their fetuses. Therefore, the stimulation of apoptosis in three different human cell types, namely, umbilical, embryonic, and placental, may point to a specific susceptibility of pregnant women and their fetuses to the detrimental effects of glyphosate (especially when coupled with its estrogenic activity). Furthermore, apoptosis assays should be conducted in various types of normal, non-transformed human white blood cells corresponding to the predominant cancers associated with this chemical at environmental doses, in order to test if there is any basis for an apoptotic model of carcinogenesis. At the very least, glyphosate appears to display endocrine effects at lower concentrations which may impact mammalian hormone metabolism, reproductive health, and fertility. It may also act as a tumor promoter in the presence of other chemicals that constitute carcinogenic stimuli.

Finally, in one intriguing study, honey bee (*Apis mellifera*) larvae were exposed to various pesticides. All the applied pesticides triggered an increase in apoptosis in treated, compared to untreated, larvae. Glyphosate was found to induce apoptosis in all the larval tissues tested including midgut, salivary glands, and ovaries via DNA fragmentation labelling and phosphatidylserine (PS) localization (Gregorc and Ellis 2011). The potential implications of glyphosate and other pesticide use on commercial apiculture demonstrated by this insect study seem to be far-reaching.

In fact, one of my father's favorite quotes was "If the bee disappeared off the face of the Earth, man would have only four years left to live" (*Albert Einstein*).

Acknowledgments The author would gratefully like to acknowledge the help of Professor Rajinder Peshin of Sher-e-Kashmir University of Agricultural Sciences and Technology of Jammu, India, and Drs. A. K. Dhawan and Balwinder Singh of Punjab Agricultural University Ludhiana, India, for sharing their data. Professor Paul J. Hergenrother, University of Illinois, is specially acknowledged for preparing the chemical structures and correct chemical names.

[This paper is dedicated to the memory of Dr. A. S. Atwal, the founding father of Apiculture and Ecology in modern India, and also to that of his student-successor, Dr. G. S. Dhaliwal.]

References

Abbassy, M.A., Belal, A.H., El-Seehy, M.A., Mossa, A.H. (2006). In vitro genotoxic effect of two organophosphorus and two pyrethroid insecticides on human lymphocytes. *Journal of Agricultural & Environmental Sciences, 5*(1) (Alexandria University, Egypt):45–56.

Andersen, H. R., Vinggaard, A. M., Rasmussen, T. H., Gjermandsen, I. M., & Bonefeld-Jorgensen, E. C. (2002). Effects of currently used pesticides in assays for estrogenicity, androgenicity, and aromatase activity *in vitro*. *Toxicology & Applied Pharmacology, 179*, 1–12.

Andreotti, G., Freeman, L. E., Hou, L., Coble, J., Rusiecki, J., Hoppin, J. A., Silverman, D. T., & Alavanja, M. C. (2009). Agricultural pesticide use and pancreatic cancer risk in the agricultural health study cohort. *International Journal of Cancer, 124*, 2495–2500.

Andreotti, G., Koutros, S., Hofmann, J.N., Sandler, D.P., Lubin, J.H., Lynch, C.F., Lerro, C.C., De Roos, A.J., Parks, C.G., Alavanja, M.C., Silverman, D.T., Freeman, L.E.B.(2017). Glyphosate use and cancer incidence in the agricultural health study. *JNCI: Journal of the National Cancer Institute*, djx233, https://doi.org/10.1093/jnci/djx233. Published: 09 November 2017.

Atwal, A. S., & Siddoo-Atwal, C. (2012). A possible role for honey bee products in the detoxification of mycotoxins. *Acta Horticulturae, 963*, 237–245 I International symposium on mycotoxins in nuts and dried fruits.

Baba, S. H., Wani, M. H., Zargar, B. A., Wani, S. A., & Kubrevi, S. S. (2012). Pesticide delivery system in apple growing belt of Kashmir valley. *Agricultural Economics Research Review, 25*, 435–444.

Battaglin, W. A., Meyer, M. T., Kuivila, K. M., & Dietze, J. E. (2014). Glyphosate and its degradation product ampa occur frequently and widely in U.S. soils, surface water, groundwater, and precipitation. *Journal of the American Water Resources Association, 50*(2), 275–290.

Beard, J. (2006). DDT and human health. *Science of the Total Environment, 355*(1–3), 78–89.

Belpoggi, F., Soffritti, M., Guarino, M., Lambertini, L., Cevolani, D., & Maltoni, C. (2002). Results of long-term experimental studies on the carcinogenicity of ethylene-bis-dithiocarbamate (mancozeb) in rats. *Annals of the New York Academy of Sciences, 982*, 123–136.

Benachour, N., & Seralini, G. E. (2009). Glyphosate formulations induce apoptosis and necrosis in human umbilical, embryonic, and placental cells. *Chemical Research in Toxicology, 22*(1), 97–105.

Bhat, A. R., Wani, M. A., Kimani, A. R., & Raina, T. H. (2010). Pesticides and brain cancer linked in orchard farmers of Kashmir. *Indian Journal of Medical & Paediatric Oncology, 31*(4), 110–120.

Borek, C. (2004). Dietary antioxidants and human cancer. *Integrative Cancer Therapy, 3*, 333–341.

Brash, D. E. (1997). Sunlight and the onset of skin cancer. *Trends in Genetics, 13*, 410–414.

Calviello, G., Piccioni, E., Boninsegna, A., Tedesco, B., Maggiano, N., Serini, S., Wolf, F. I., & Palozza, P. (2006). DNA damage and apoptosis induction by the pesticide mancozeb in rat cells: Involvement of the oxidative mechanism. *Toxicology and Applied Pharmacology, 211*(2), 87–96.

Chambers, H. W. (1992). Organophosphorus compounds: An overview. In J. E. Chambers & P. E. Levi (Eds.), *Organophosphates chemistry, fate, and effects*. San Diego: Academic Press.

Chaufan, G., Coalova, I., & Ríos de Molina Mdel, C. (2014). Glyphosate commercial formulation causes cytotoxicity, oxidative effects, and apoptosis on human cells: Differences with its active ingredient. *International Journal of Toxicology, 33*(1), 29–38.

Chen, F., Castranova, V., Shi, X., & Demers, L. M. (1999). New insights into the role of NF-κB, a ubiquitous transcription factor in the initiation of diseases. *Clinical Chemistry, 45*(1), 7–17.

Clair, E., Mesnage, R., Travert, C., & Séralini, G. É. (2012). A glyphosate-based herbicide induces necrosis and apoptosis in mature rat testicular cells *in vitro*, and testosterone decrease at lower levels. *Toxicology in Vitro, 26*(2), 269–279.

Cocco, P., Kazerouni, N., & Zahm, S. H. (2000). Cancer mortality and environmental exposure to DDE in the United States. *Environmental Health Perspectives, 108*, 1–4.

Cohn, B. A., Wolff, M. S., Cirillo, P. M., & Sholtz, R. I. (2007). DDT and breast cancer in young women: New data on the significance of age at exposure. *Environmental Health Perspectives, 115*(10), 1406–1414.

Crews, D., Willingham, E., & Skipper, J. K. (2000). Endocrine disruptors: present issues, future directions. *The Quarterly Review of Biology, 75*, 243–260.

Danial, N., & Korsmeyer, S. (2004). Cell death: Critical control points. *Cell, 116*, 205–219.

Daruich, J., Zirulnik, F., & Gimenez, M. S. (2001). Effect of the herbicide glyphosate on enzymatic activity in pregnant rats and their fetuses. *Environmental Research, 85*(3), 226–231.

De Gruijl, F. R., & Forbes, P. D. (1995). UV-induced skin cancer in a hairless mouse model. *BioEssays, 17*(7), 651–661.

De Liz Oliveira Cavalli, V. L., Cattani, D., Heinz Rieg, C. E., Pierozan, P., Zanatta, L., Benedetti Parisotto, E., Wilhelm Filho, D., Mena Barreto Silva, F. R., Pessoa-Pureur, R., & Zamoner, A. (2013). Roundup disrupts male reproductive functions by triggering calcium-mediated cell death in rat testis and sertoli cells. *Free Radical Biology and Medicine, 65*, 335–346.

Denmeade, S. R., & Isaacs, J. T. (1996). Programmed cell death (apoptosis) and cancer chemotherapy. *Cancer Control, 3*(4), 303–309.

De Roos, A. J., Zahm, S. H., Cantor, K. P., Weisenburger, D. D., Holmes, F. F., Burmeister, L. F., & Blair, A. (2003). Integrative assessment of multiple pesticides as risk factors for non-Hodgkin's lymphoma among men. *Occupational and Environmental Medicine, 60*, E11 Available: http://oem.bmjjournals.com/cgi/content/full/60/9/e11, 11e.

De Roos, A. J., Blair, A., Rusiecki, J. A., Hoppin, J. A., Svec, M., Dosemeci, M., Sandler, D. P., & Alavanja, M. C. (2005). Cancer incidence among glyphosate-exposed pesticide applicators in the agricultural health study. *Environmental Health Perspectives, 113*(1), 49–54.

Domico, L. M., Zeevalk, G. D., Bernard, L. P., & Cooper, K. R. (2006). Acute neurotoxic effects of mancozeb and maneb in mesencephalic neuronal cultures are associated with mitochondrial dysfunction. *Neurotoxicology, 27*(5), 816–825.

Duchaud, E., Ridet, A., Stoppa-Lyonnet, D., Janin, N., Moustacchi, E., & Rosselli, F. (1996). Deregulated apoptosis in ataxia-telangiectasia: Association with clinical stigmata and radio sensitivity. *Cancer Research, 56*, 1400–1404.

Eguchi, K. (2001). Apoptosis in autoimmune diseases. *Internal Medicine, 40*(4), 275–284.

El-Abaseri, T. B., Putta, S., & Hansen, L. A. (2006). Ultraviolet irradiation induces keratinocyte proliferation and epidermal hyperplasia through the activation of the epidermal growth factor receptor. *Carcinogenesis, 27*(2), 225–231.

European Commission. (2009). Review report of the active substance mancozeb SANCO/4058/2001 – rev 4.4.

Frigo, D. E., Burrow, M. E., Mitchell, K. A., Chiang, T.-C., & McLachlan, J. A. (2002). DDT and its metabolites alter gene expression in human uterine cell lines through estrogen receptor-independent mechanisms. *Environmental Health Perspectives, 110*(12), 1239–1245.

Gao, B. (2005). Cytokines, STATS and liver disease. *Cellular & Molecular Immunology, 2*(2), 92–100.

Garabrant, D. H., Held, J., Bryan, L., Peters, J. M., & Mack, T. M. (1992). DDT and related compounds and risk of pancreatic cancer. *Journal of the National Cancer Institute, 84*, 764–771.

Garry, V. F., Schreinemachers, D., Harkins, M. E., & Griffith, J. (1996). Pesticide appliers, biocides, and birth defects in rural Minnesota. *Environmental Health Perspectives, 104*, 394–399.

Gasnier, C., Dumont, C., Benachour, N., Clair, E., Chagnon, M. C., & Séralini, G. E. (2009). Glyphosate-based herbicides are toxic and endocrine disruptors in human cell lines. *Toxicology, 262*(3), 184–191.

George, J., Prasad, S., Mahmood, Z., & Shukla, Y. (2010). Studies on glyphosate-induced carcinogenicity in mouse skin: A proteomic approach. *Journal of Proteomics, 73*(5), 951–964.

Gladen, B. C., Monaghan, S. C., Lukyanova, E. M., Hulchiy, O. P., Shkyryak-Nyzhnyk, Z. A., Sericano, J. L., & Little, R. E. (1999). Organochlorines in breast milk from two cities in Ukraine. *Environmental Health Perspectives, 107*, 459–462.

Gregorc, A., & Ellis, J. D. (2011). Cell death localization *in situ* in laboratory reared honey bee (*Apis mellifera* L.) larvae treated with pesticides. *Pesticide Biochemistry and Physiology, 99*(2), 200–207.

Gui, Y. X., Fan, X. N., Wang, H. M., Wang, G., & Chen, S. D. (2012). Glyphosate induced cell death through apoptotic and autophagic mechanisms. *Neurotoxicology and Teratology, 34*(3), 344–349.

Hall, R. H. (1992). A new threat to public health: Organochlorines and food. *Nutrition and Health, 8*, 33–43.

Han, J., Colditz, G. A., Samson, L. D., & Hunter, D. J. (2004). Polymorphisms in DNA double-strand break repair genes and skin cancer risk. *Cancer Research, 64*, 3009–3013.

Hardell, L., & Eriksson, M. (1999). A case-control study of non-Hodgkin lymphoma and exposure to pesticides. *Cancer, 85*, 1353–1360.

Hardell, L., Eriksson, M., & Nordstrom, M. (2002). Exposure to pesticides as risk factor for non-Hodgkin's lymphoma and hairy cell leukemia: Pooled analysis of two Swedish case-control studies. *Leukemia and Lymphoma Journal, 43*, 1043–1049.

Hoppin, J. A., Tolbert, P. E., Holly, E. A., Brock, J. W., Korrick, S. A., Altshul, L. M., Zhang, R. H., Bracci, P. M., Burse, V. W., & Needham, L. L. (2000). Pancreatic cancer and serum organochlorine levels. *Cancer Epidemiology, Biomarkers & Prevention, 9*, 199–205.

Houeto, P., Bindoula, G., & Hoffman, J. R. (1995). Ethylenebisdithiocarbamates and ethylenethiourea: Possible human health hazards. *Environmental Health Perspectives, 103*, 568–573.

Jaga, K., & Dharmani, C. (2003). Global surveillance of DDT and DDE levels in human tissues. *International Journal of Occupational Medicine and Environmental Health, 16*(1), 7–20.

James, S. J., Muskhelishvili, L., Gaylor, D. W., Turturro, A., & Hart, R. (1998). Upregulation of apoptosis with dietary restriction: Implications for carcinogenesis. *Environmental Health Perspectives, 106*(1), 307–312.

Johnstone, R., Rueflir, A., & Lowe, S. (2002). Apoptosis: A link between cancer genetics and chemotherapy. *Cell, 108*, 153–164.

Jung, M., Zhang, Y., Lee, S., & Dritschilo, A. (1995). Correction of radiation sensitivity in ataxia telangiectasia cells by a truncated I kappa B-alpha. *Science, 268*, 1619–1621.

Karunanayake, C. P., Spinelli, J. J., & McLaughlin, J. R. (2012). Hodgkin lymphoma and pesticides exposure in men: A Canadian case-control study. *Journal of Agromedicine, 17*(1), 30–39.

Kasibhatla, S., Brunner, T., Genestier, L., Echeverri, F., Mahboubi, A., & Green, D. R. (1998). DNA damaging agents induce expression of fas ligand and subsequent apoptosis in t lymphocytes via the activation of of NF-κB and AP-1. *Molecular Cell, 1*, 543–551.

Kaur, M., Agarwal, R., & Agarwal, C. (2006). Grape seed extract induces anoikis and caspase-mediated apoptosis in human prostate carcinoma LNCaP cells: Possible role of ataxia-telangiectasia mutated -p53 activation. *Molecular Cancer Therapeutics, 5*, 1265–1274.

Klaunig, J., & Kamendulis, L. (1999). Mechanisms of cancer chemoprevention in hepatic carcinogenesis: Modulation of focal lesion growth in mice. *Toxicological Sciences, 52*, 101–106.

Kojima, H., Katsura, E., Takeuchi, S., Niyama, K., & Kobayashi, K. (2004). Screening for estrogen and androgen receptor activities in 200 pesticides by *in vitro* reporter gene assays using Chinese hamster ovary cells. *Environmental Health Perspectives, 112*, 524–531.

Koller, V. J., Fürhacker, M., Nersesyan, A., Mišík, M., Eisenbauer, M., & Knasmueller, S. (2012). Cytotoxic and DNA-damaging properties of glyphosate and roundup in human-derived buccal epithelial cells. *Archives of Toxicology, 86*(5), 805–813.

Kromberg, J. G. R., Castle, D., Zwane, E. M., & Jenkins, T. (1989). Albinism and skin cancer in Southern Africa. *Clinical Genetics, 36*, 43–52.

Kupfer, D., & Bulger, W. H. (1976). Interactions of chlorinated hydrocarbons with steroid hormones. *Federation Proceedings, 35*, 2603–2608.

Kwiatkowska, M., Huras, B., & Bukowska, B. (2014). The effect of metabolites and impurities of glyphosate on human erythrocytes (*in vitro*). *Pesticide Biochemistry and Physiology, 109*, 34–43.

Lee, W. J., Blair, A., Hoppin, J. A., Lubin, J. H., Rusiecki, J. A., Sandler, D. P., Dosemeci, M., & Alavanja, M. C. (2004). Cancer incidence among pesticide applicators exposed to chlorpyrifos in the agricultural health study. *Journal of the National Cancer Institute, 96*(23), 178–179.

Lee, W. J., Sandler, D. P., Blair, A., Samanic, C., Cross, A. J., & Alavanja, M. C. R. (2007). Pesticide use and colorectal cancer risk in the agricultural health study. *International Journal of Cancer, 121*, 339–346.

Levine, A. (1997). P53, the cellular gatekeeper for growth and division. *Cell, 88*, 323–331.

Li, D., Huang, Q., Lu, M., Zhang, L., Yang, Z., Zong, M., & Tao, L. (2015). The organophosphate insecticide chlorpyrifos confers its genotoxic effects by inducing DNA damage and cell apoptosis. *Chem, 135*, 387–393.

Liao, W.-C., Haimovitz-Friedman, A., Persaud, R. S., McLoughlin, M., Ehleiter, D., Zhang, N., Gatei, M., Lavin, M., Kolesnick, R., & Fuks, Z. (1999). Ataxia telangiectasia-mutated gene product inhibits DNA damage-induced apoptosis via ceramide synthase. *Journal of Biological Chemistry, 274*(25), 17908–17917.

Lokshin, R., & Zakeri, Z. (2004). Apoptosis, autophagy, and more. *International Journal of Biochemistry and Cell Biology, 36*, 2405–2419.

Lu, Y. P., Lou, Y. R., Xie, J. G., Yen, P., Huang, M. T., & Conney, A. H. (1997). Inhibitory effect of black tea on the growth of established skin tumours in mice: Effects on tumour size, apoptosis, necrosis, and bromodeoxyuridine incorporation into DNA. *Carcinogenesis, 18*, 2163–2169.

Lu, Y.-P., Lou, Y.-R., Li, X. H., Xie, J. G., Brash, D., Huang, M.-T., & Conney, A. H. (2000). Stimulatory effect of oral administration of green tea or caffeine on ultraviolet light-induced increases in epidermal wild-type p53, p21(WAF1/CIP1), and apoptotic sunburn cells in SKH-1 mice. *Cancer Research, 60*, 4785–4791.

Mañas, F., Peralta, L., Raviolo, J., Ovando, H. G., Weyers, A., Ugnia, L., Cid, M. G., Larripa, I., & Gorla, N. (2009). Genotoxicity of glyphosate assessed by the comet assay and cytogenetic tests. *Environmental Toxicology and Pharmacology, 28*(1), 37–41.

Marc, J., Mulner-Lorillon, O., & Bellé, R. (2004). Glyphosate-based pesticides affect cell cycle regulation. *Biology of the Cell, 96*(3), 245–249.

Martin, K. R. (2006). Targeting apoptosis with dietary bioactive agents. *Experimental Biology and Medicine, 231*, 117–129.

Martini, C. N., Gabrielli, M., & Vila Mdel, C. (2012). A commercial formulation of glyphosate inhibits proliferation and differentiation to adipocytes and induces apoptosis in 3T3-L1 fibroblasts. *Toxicology in Vitro, 26*(6), 1007–1013.

McDuffie, H. H., Pahwa, P., McLaughlin, J. R., Spinelli, J. J., Fincham, S., Dosman, J. A., Robson, D., Skinnider, L. F., & Choi, N. W. (2001). Non-Hodgkin's lymphoma and specific pesticide exposures in men: Cross-Canada study of pesticides and health. *Cancer Epidemiology Biomarkers and Prevention, 10*, 1155–1163.

McGlynn, K. A., Abnet, C. C., Zhang, M., Sun, X. D., Fan, J. H., O'Brien, T. R., Wei, W. Q., Ortiz-Conde, B. A., Dawsey, S. M., Weber, J. P., Taylor, P. R., Katki, H., Mark, S. D., & Qiao, Y. L. (2006). Serum concentrations of 1,1,1-trichloro-2,2-bis(p-chlorphenyl)ethane (DDT) and 1,1-dichloro-2,2-bis(p-chlorophenyl) ethylene (DDE) and risk of primary liver cancer. *Journal of the National Cancer Institute, 98*(14), 1005–1010.

Mesnage, R., Phedonos, A., Biserni, M., Arno, M., Balu, S., Corton, C. J., Ugarte, R., & Antoniou, M. N. (2017). Evaluation of estrogen receptor alpha activation by glyphosate-based herbicide constituents. *Food and Chemical Toxicology, 108A*, 30–42.

Miagkov, A. V., Kovalenko, D. V., Brown, C. E., Didsbury, J. R., Cogswell, J. P., Stimpson, S. A., Baldwin, A. S., & Makarov, S. S. (1998). NF-κB activation provides the potential link between inflammation and hyperplasia in the arthritic joint. *Proceedings of the National Academy of Sciences of the United States of America, 95*, 13859–13864.

Mihara, M., Erster, S., Zaika, A., Petrenko, O., Chittenden, T., Pancoska, P., & Moll, U. (2003). P53 has a direct apoptogenic role in the mitochondria. *Molecular Cell, 11*, 577–590.

Mills, P. K., Yang, R., & Riordan, D. (2005). Lymphohematopoietic cancers in the United Farm Workers of America (UFW), 1988–2001. *Cancer Causes & Control, 16*(7), 823–830.

Miranda-Contreras, L., Dávila-Ovalles, R., Benítez-Díaz, P., Peña-Contreras, Z., & Palacios-Prü, E. (2005). Effects of prenatal paraquat and mancozeb exposure on amino acid synaptic transmission in developing mouse cerebellar cortex. *Brain Research. Developmental Brain Research, 160*(1), 19–27.

Monnet-Tschudi, F., Zurich, M. G., Schilter, B., Costa, L. G., & Honegger, P. (2000). Maturation-dependent effects of chlorpyrifos and parathion and their oxygen analogs on acetylcholinesterase and neuronal and glial markers in aggregating brain cell culture. *Toxicology & Applied Pharmacology, 165*, 175–183.

Mora, A. M., Van Wendel de Joode, B., Mergler, D., Córdoba, L., Cano, C., Quesada, R., Smith, D. R., Menezes-Filho, J. A., Lundh, T., Lindh, C. H., Bradman, A., & Eskenazi, B. (2014). Blood and hair manganese concentrations in pregnant women from the infants' environmental health study (ISA) in Costa Rica. *Environmental Sciences & Technology, 48*(6), 3467–3476. https://doi.org/10.1021/es404279r?src=recsys-cor1.

Naik, P., Karrim, J., & Hanahan, D. (1996). The rise and fall of apoptosis during multistage tumorigenesis: Down-modulation contributes to tumour progression from angiogenic progenitors. *Genes and Development, 10*, 2105–2116.

National Pesticide Information Centre – General fact sheet, glyphosate. http://npic.orst.edu/factsheets/archive/glyphotech.html

Nawaz, A., Razpotnik, A., Rouimi, P., De Souza, G., & Cravedi, J.-P. (2013). Cellular impact of combinations of endosulfan, atrazine, and chlorpyrifos on human primary hepatocytes and HepaRG cells after short and chronic exposures. *Cell Biology and Toxicology, 30*(1), 17–29.

Negga, R., Rudd, D. A., Davis, N. S., Justice, A. N., Hatfield, H. E., Valente, A. L., Fields, A. S., & Fitsanakis, V. A. (2011). Exposure to Mn/Zn ethylene-bis-dithiocarbamate and glyphosate pesticides leads to neurodegeneration in *Caenorhabditis elegans*. *Neurotoxicology, 32*(3), 331–341.

Nelson, H. H., Kelsey, K. T., Mott, L. A., & Karagas, M. R. (2002). The XRCC1 Arg399Gln polymorphism, sunburn, and non-melanoma skin cancer: Evidence of gene-environment interaction. *Cancer Research, 62*, 152–155.

Paganelli, A., Victoria Gnazzo, V., Helena Acosta, V., Silvia, L., López, S. L., & Carrasco, A. E. (2010). Glyphosate-based herbicides produce teratogenic effects on vertebrates by impairing retinoic acid signaling. *Chemical Research in Toxicology, 23*(10), 1586–1595. https://doi.org/10.1021/tx1001749-cor1.

Pema, G., Lasalvia, M., D'Antonio, D., L'Abbate, N., & Capozzi, V. (2011). Identification of chemical modification in single human keratinocyte cells exposed to low doses of chlorpyriphos by Raman micro-spectroscopy. *Journal of Raman Spectroscopy, 42*(4), 603–611.

Perez-Maldonado, I. N., Díaz-Barriga, F., de la Fuente, H., González-Amaro, R., Calderón, J., & Yáñez, L. (2004). DDT induces apoptosis in human mononuclear cells *in vitro* and is associated with increased apoptosis in exposed children. *Environmental Research, 94*(1), 38–46.

Perez-Maldonado, I. N., Athanasiadou, M., Yáñez, L., González-Amaro, R., Bergman, A., & Díaz-Barriga, F. (2006). DDE-induced apoptosis in children exposed to the DDT metabolite. *Science of the Total Environment, 370*(2–3), 343–351.

Porta, M., Malats, N., Jariod, M., Grimalt, J. O., Rifà, J., Carrato, A., Guarner, L., Salas, A., Santiago-Silva, M., MCorominas, J., Andreu, M., & X-Real, F. (1999). Serum concentrations of organochlorine compounds and K-Ras mutation in exocrine pancreatic cancer. *Lancet, 354*, 2125–2129.

Prabhavathy Das, G., Shak, A. P., & Jamil, K. (2006). Estimation of apoptosis and necrosis caused by pesticides *in vitro* on human lymphocytes using DNA diffusion assay. *Drug Chemical Toxicology, 29*(2), 147–156.

Prasad, S., Srivastava, S., Singh, M., Shukla, Y. (2009). Clastogenic effects of glyphosate in bone marrow cells of Swiss albino mice. *Journal of Toxicology*, 2009 , Article ID 308985, 6 pages. https://doi.org/10.1155/2009/308985.

Qiao, D., Seidler, F., Padilla, S., & Slotkin, T. A. (2002). Developmental neurotoxicity of chlorpyrifos: What is the vulnerable period? *Environmental Health Perspectives, 11*, 1097–1103.

Rieux-Laucat, F., Le Deist, F., & Fischer, A. (2003). Autoimmune lymphoproliferative syndromes: Genetic defects of apoptosis pathways. *Cell Death and Differentiation, 10*, 124–133.

Rogan, W. J., & Chen, A. (2005). Health risks and benefits of DDT. *The Lancet, 366*(9487), 763–773.

Sandhu, M. A., Saeed, A. A., Khilji, M. S., Ahmed, A., Latif, M. S., & Khalid, N. (2013). Genotoxicity evaluation of chlorpyrifos: A gender related approach in regular toxicity testing. *Journal of Toxicological Sciences, 38*(2), 237–244.

Schecter, A., Toniolo, P., Dai, L. C., Thuy, L. T. B., & Wolff, M. S. (1997). Blood levels of DDT and breast cancer risk among women living in the north of Vietnam. *Archives of Environmental Contamination and Toxicology, 33*(4), 453–456.

Schreinemachers, D. M., Creason, J. P., & Garry, V. F. (1999). Cancer mortality in agricultural regions of Minnesota. *Environmental Health Perspectives, 107*(3), 205–211.

Schulte-Hermann, R., Bursch, W., Grasl-Kraupp, B., Mullauer, L., & Ruttkay-Nedecky, B. (1995). Apoptosis and multistage carcinogenesis in rat liver. *Mutation Research, 333*(1–2), 81–87.

Seralini, G. E., Clair, E., Mesnage, R., Gress, S., DeFarge, N., Malatesta, M., Hennequin, D., & De Vendomois, J. S. (2014). Republished study: Long-term toxicity of a roundup herbicide and a roundup-tolerant genetically modified maize. *Environmental Sciences Europe, 26*, 14.

Siddoo-Atwal, C. (2009). AT, apoptosis, and cancer: A viewpoint. *Indian Journal of Ecology, 36*(2), 103–110.

Siddoo-Atwal, C. (2011). A case-study of apoptotic sunburn with sunscreen. *Indian Journal of Ecology, 38*(2), 300–301.

Siddoo-Atwal, C. (2015). A case-study of abnormal growth formation and regression. Published online. *Science 2.0*, December 22 (2015); http://www.science20.com/cancer_science/chanda siddoo-atwal/ a_casestudy_of_abnormal_growth_formation_and_regression-162516

Siddoo-Atwal, C., Rosin, M. P., & Haas, A. L. (1996). Elevation of interferon β- inducible proteins in ataxia telangiectasia cells. *CancerResearch, 56*, 443–447.

Slotkin, T. A., Cousins, M. M., Tate, C. A., & Seidler, F. J. (2001). Persistent cholinergic presynaptic deficits after neonatal chlorpyrifos exposure. *Brain Research, 902*(2), 229–243.

Snedeker, S. M. (2001). Pesticides and breast cancer risk: A review of DDT, DDE, and dieldrin. *Environmental Health Perspectives, 109*(Suppl 1), 35–47.

Srivastva, A. K., Ali, W., Singh, R., Bhui, K., Tyagi, S., Al-Khedhairy, A. A., Srivastava, P. K., Musarrat, J., & Shukla, Y. (2012). Mancozeb-induced genotoxicity and apoptosis in cultured human lymphocytes. *Life Sciences, 90*(21–22), 815–824 https://www.sciencedirect.com/science/article/pii/S0024320512000021.

Sun, S., Hail, N., & Lotan, R. (2004). Apoptosis as a novel target for cancer chemoprevention. *Journal of the National Cancer Institute, 96*, 662–678.

Surh, Y. (1999). Molecular mechanisms of chemopreventive effects of selected dietary and medicinal phenolic substances. *MutationResearch, 428*, 305–327.

Surh, Y. J. (2003). Cancer chemoprevention with dietary phytochemicals. *NationalReviews Cancer, 3*, 768–780.

Susin, S. S., Lorenzo, H. K., Zamzami, N., Marzo, I., Snow, B. E., Brothers, G. M., Mangion, J., Jacotot, E., Constantini, P., Loeffler, M., Larochette, N., Goodlett, D. R., Aebersold, R., Siderowski, D. P., Penninger, J. M., & Kroemer, G. (1999). Molecular characterization of mitochondrial apoptosis-inducing factor. *Nature, 397*, 441–446.

Taylor, A. M., Metcalfe, J. A., Thick, J., & Mak, Y. F. (1996). Leukemia and lymphoma in ataxia telangiectasia. *Blood, 87*(2), 423–438.

Thongprakaisang, S., Thiantanawat, A., Rangkadilok, N., Suriyo, T., & Satayavivad, J. (2013). Glyphosate induces human breast cancer cells growth via estrogen receptors. *Food and Chemical Toxicology, 59*, 129–136.

Thornton, J. (reviewed by Stirling D - 2000). Pandora's poison: Chlorine, health, and a new environmental strategy. *Electronic Green Journal, 1*(16): 1–4.

Turusov, V., Rakitsky, V., & Tomatis, L. (2002). Dichlorodiphenyltrichloroethane (DDT): Ubiquity, persistence, and risks. *Environmental Health Perpectives, 110*(2), 125–128.

US EPA. (2005). Reregistration Eligibility Decision (RED) Mancozeb . https://www3.epa.gov/pesticides/chem_search/reg_actions/reregistration/red_PC-014504_20-Sep-05.pdf

Van Hogerlinden, M., Rozell, B. L., Ahrlund-Richter, L., & ToftgArd, R. (1999). Squamous cell carcinomas and increased apoptosis in skin with inhibited rel/nuclear factor-κB signaling. *Cancer Research, 59*, 3299–3303.

Viswanath, G., Chatterjee, S., Dabral, S., Nanguneri, S. R., Divya, G., & Roy, P. (2010). Anti-androgenic endocrine disrupting activities of chlorpyrifos and piperophos. *Journal of Steroid Biochemistry & Molecular Biology, 120*, 22–29.

Waddell, B. L., Zahm, S. H., Baris, D., Weisenburger, D. D., Holmes, F., Burmeister, L. F., Cantor, K. P., & Blair, A. (2001). Agricultural use of organophosphate pesticides and the risk of non-Hodgkin's lymphoma among male farmers (United States). *Cancer Causes & Control, 12*, 509–517.

Wang, C. Y., Mayo, M. W., Korneluk, R. G., Goeddel, D. V., & Baldwin, A. S. (1998). NF-kappaB antiapoptosis: Induction of TRAF1 and TRAF2 and c-IAP1 and c-IAP2 to suppress caspase-8 activation. *Science, 281*, 1680–1683.

Watts, M. (2012, August). Chlorpyrifos as a possible global persistent organic pollutant. *Pesticide Action Network North America.*

Watts, M. (2014). Mancozeb. *A Pesticide Action Network Asia and the Pacific Factsheet Series.*

WHO. (1988). *Dithiocarbamate pesticides, ethylenethiourea and propylenethiourea: A general introduction.* (Environmental health criteria, Vol. 78). Geneva: World Health Organization.

Zhang, X., & Takenaka, I. (2007). Apoptosis and carcinogenesis: Morphologic observations in the rat bladder treated N-Butyl-N-(4-hydroxybutyl)nitrosamine. *International Journal of Urology, 5*(3), 262–267.

Chapter 8
Climate Change: Impact on Biotic Stresses Afflicting Crop Plants

Chirantan Chattopadhyay, Ajanta Birah, and Bushan L. Jalali

Abstract Climate change is of immense concern in view of dependence on agriculture for food and nutritional security of Indian growing population. Effect of climate change on insect pests and diseases of agricultural crops is multidimensional. Magnitude of this impact could vary with the species and their growth patterns. Changing pattern of disease and insect pest scenario due to climate change has warranted the need for improved novel agricultural practices and use of eco-friendly approaches for sustainable crop production. We need to critically revisit the efficacy of current chemical, physical and biological control tactics, including pest-resistant cultivars under climate change, and to include future climate scenarios in all research aimed at developing new tools and strategies. Research on host response and adaptation should be launched to understand better as to how forthcoming change could influence crop pests, coupled with pest risk analyses done regularly, based on host–pathogen/pests interactions.

Keywords Climate change · Diseases · Insect pests · Crop production

Abbreviations

CIAI	Central Island Agricultural Research Institute
CROPSAP	Crop Pest Surveillance and Advisory Project
DRMR	Directorate of Rapeseed-Mustard Research

C. Chattopadhyay (✉)
Uttar Banga Krishi Viswavidyalaya, Pundibari, West Bengal, India
e-mail: chirantan_cha@hotmail.com

A. Birah
ICAR: National Research Centre for Integrated Pest Management, New Delhi, India
e-mail: ajantabirah69@gmail.com

B. L. Jalali
Haryana Agricultural University, Hisar, Haryana, India
e-mail: bljalali@gmail.com

R. Peshin, A. K. Dhawan (eds.), *Natural Resource Management: Ecological Perspectives*, Sustainability in Plant and Crop Protection,
https://doi.org/10.1007/978-3-319-99768-1_8

DST	Department of Science and Technology
GBNV	Groundnut Bud Necrosis Virus
HAU	Haryana Agricultural University
ICAR	Indian Council of Agricultural Research
IIOR	Indian Institute of Oilseeds Research
IIPR	Indian Institute of Pulses Research
NCIPM	National Research Centre for Integrated Pest Management
NICRA	National Innovations in Climate Resilient Agriculture
NRDC	National Research Development Corporation
UBKV	Uttar Banga Krishi Viswavidyalaya
UGC	University Grants Commission

8.1 Introduction

Climate change has direct impact on crop production and productivity. It is of immense concern to India in view of the country dependence on agriculture for food and nutritional security of its ever-growing population. High coefficients of variation in yields of maize (>60%), rice (25–38%), wheat (32%) and soybean (26–34%) are explained due to climate (temperature and precipitation) variations, with special reference to central India as compared to safer regions, namely, West Bengal (Ray et al. 2015). Changing pest scenario due to climate change has warranted the need for future studies on such models which can predict the severity of important pathogens of major crops, in real-field conditions. These changes have direct effect on growth and multiplication, spread and severity/infestation of many plant pathogens/insect pests, which in turn are affecting the pattern of incidence of pests and diseases (West et al. 2015; Lamichhane et al. 2015). Population dynamics of pests is dependent upon temperature and humidity, among several weather factors. Small changes in temperature can result in changed virulence as well as appearance of new pests in a region. Areas presently uncongenial for pests may become suitable, due to rise in temperatures. However, it needs to be borne in mind that insects (including insect pests) and microorganisms (including crop pathogens) could adapt to slow changes in the environment (viz. increase in temperature), and thus their favourable thermal range could also shift. Any change in them, depending upon their base value, can significantly alter the scenario, which ultimately may result in yield losses.

The effect of climate change on insect pests and diseases of agricultural crops is multidimensional. Magnitude of this impact could vary with the type of species and their growth patterns. It may be assumed that the vegetation tolerating high temperature, salinity and having high CO_2-use efficiency could perform better than other species. Intergovernmental Panel on Climate Change predicted, in its 1995 report, that doubling the level of CO_2 could possibly increase yields in some crops by 30%. However, observations on changing insect pest and disease incidence over the twentieth century suggest that growing agricultural production and trade has been affected by their wide dissemination. There is some evidence for a latitudinal bias in range shifts that indicates a global warming signal (Bebber 2015).

The increased production could be offset partly or entirely by an emerging spectrum of insect pests, pathogens or weeds. It is, therefore, important to consider all the biotic components under the changing pattern of climate.

8.2 Focus on Climate Change and Disease Scenario

In India, limited efforts have been made in this area for any disease and crop (Subba Rao et al. 2007; Chattopadhyay and Huda 2009). Presently, most of the work related to climate change vis-à-vis plant diseases and insect pests is going on in rice (*Oryza sativa*, for rice blast, bacterial leaf blight), wheat (*Triticum aestivum*, for *Puccinia*, *Septoria*) and horticultural (*Meloidogyne* spp.) crops. In 2011 there was an upsurge in false smut incidence on rice in West Bengal (Ghosh et al. 2012). The disease reached a very high severity in 2013, with additional areas of Jharkhand, Orissa and Bihar. Along with the established sheath blight (Biswas 2011), it is expected to cause serious problems in rice production in the coming years. Shorter winters may affect the oil yields of the rapeseed-mustard crops and grain filling of other crops grown during a similar period. Root rot is an emerging threat for rapeseed-mustard production system, recently reported from the fields in some pockets of the country (Meena et al. 2010). Apart from the already established *Sclerotinia* rot, it is likely to increase with severe winters (Kumar et al. 2013; West et al. 2015). In India, severity of yellow wheat rust reached up to 100% with prevalence being 30–40% in the Terai and northern hills in 2010–2013. Along with powdery mildews, the diseases are predicted to rise further with shortening, milder winters (Kumar et al. 2013; West et al. 2015).

Spot blotch (*Septoria* sp.) has also become very important on wheat in recent times, which has been attributed to climate changes. It seems that there is severe occurrence of Indian cassava mosaic virus in Kerala, due to a shift in climatic conditions. A new report of African cassava mosaic virus and Sri Lankan cassava mosaic virus is attributed to increase in temperature and CO_2 levels. The groundnut bud necrosis virus (GBNV) has been on the rise on several field and horticultural crops across the country.

At Kanpur the rise in temperature passed the tolerance limit of the vectoring mite of pigeon pea sterility mosaic virus. Sucking insect pests and mites of crop plants are predicted to rise due to increase in temperature and precipitation (Chattopadhyay et al. 2012; West et al. 2015). The climate variability may have also influenced *Phytophthora* blight incidence at Kanpur and Pantnagar, in mutually, opposite directions (Kumar et al. 2012).

Since the first report of *Stemphylium* blight on chickpea from Bangladesh in 1987, the disease has gradually moved to Nepal. In India it has been found infesting both lentil and chickpea in the North-Eastern Plain Zone, Terai region, namely, Pantnagar and Pusa (Bihar) (Ghosh et al. 2012). Contrastingly, *Ascochyta blight*, which used to be a major problem in chickpea in the North-Western Plains of India, has almost vanished from the area, which could also be due to substantial reduction in the area under the crop in the region and shift in the same towards southern India.

Phyllody is being noticed on mung bean with up to 8% incidence (Mohapatra and Chattopadhyay 2012). There are preliminary indications for increase in sheath blight, sheath rot, false smut diseases of rice, fruit rot and die back of chilli, bitter gourd and potato, including postharvest losses due to rise in temperature (Biswas 2011). Further examples are described in Table 8.1.

Table 8.1 Impact of climate changes on plant diseases

Component of climate change/pathogen/disease	Impact
Elevated CO_2 level (550 ppm CO_2)	
Biotrophic fungi such as rust	Promotes growth and development, due to enhanced carbohydrate contents
Downy mildew (*Peronospora manshurica*), brown spots (*Septoria glycines*) and sudden death syndrome (*Fusarium virguliforme*) of soybean	Altered expression of diseases
Powdery mildew in barley (*Blumeria graminis*)	Increased resistance to powdery mildew
Blast (*Pyricularia oryzae*) and sheath blight (*Rhizoctonia solani*)	More susceptible to injury
Powdery mildew and anthracnose (*Colletotrichum gloeosporioides*)	Increased pathogen reproduction
Rice blast (*Magnaporthe grisea*), wheat scab (*Fusarium* spp.), stripe rust (*Puccinia striiformis*) and powdery mildew (*Blumeria graminis*)	Increased infestation
Rhizoctonia solani, *Sclerotium rolfsii*, *Fusarium oxysporum* f. sp. *ricini*, *Macrophomina phaseolina*, *Alternaria tenuissima*	Increased fungal biomass with increased sclerotial bodies (*R. solani*). Increased cellulase activity (*Trichoderma viride*, *T. harzianum*, *F. o.* f. sp. *ricini* and *M. phaseolina*)
Effect of increased temperatures	
Stripe rust isolates (*Puccinia striiformis*)	Increased aggressiveness
Common bunt (*Tilletia caries*) and Karnal bunt (*Tilletia indica*) in wheat	Increased infection
Dry root rot (*Rhizoctonia bataticola*) of chickpea	Increased infestation
Phytophthora blight (*Phytophthora drechsleri* f. sp. *cajani*) of pigeon pea	Increased infestation
Needle blight (*Dothistroma septosporum*)	Increasing northward
Leaf rust (*Puccinia recondita*) in wheat, broomrape (*Orobanche cumana*) in sunflower, black shank (*Phytophthora nicotianae*) in tobacco and bacterial blight (*Xanthomonas oryzae* pv. *oryzae*) in rice	Temperature sensitivity
More cyst formation by *Trichoderma* in higher temperature (48 °C) for better survival (Rao et al. 2015)	

Source: Gautam et al. (2013) and CRIDA Annual Report (2014)

8.3 Focus on Climate Change and Insect Pest Scenario

Climate affects the population dynamics and distribution systems of invertebrate pests like insects and mites. Studies under National Innovations in Climate Resilient Agriculture (NICRA) have shown a faster development of *Tetranychus urticae* under elevated (550 ppm) CO_2 (Rao et al. 2015). Climate change may affect the distribution of the insect pests, physiology, abundance and phenology, and the major factors comprise CO_2 concentration, temperature, natural enemies, precipitation and their host plant (Lastuvka 2009). Severe weather events such as strong rainstorms, elevated temperatures or high wind also influence their continued existence. Climate changes may include shifts in species distributions, changes in life cycles with phenology, rise in number of generations and population growth rates, alterations in crop-pest synchrony, change in migratory behaviour, natural enemy-pest interaction and changes in interspecific interactions (Root and Hughes 2005). Extinction of some species and changes in community structure are also normal (Thomas et al. 2004). In India, Bihar hairy caterpillar (*Spilarctia obliqua*) was surprisingly observed in increasing trends on mustard crop, in recent times. Oilseeds *Brassicas* have been affected by the painted bug (*Bagrada cruciferarum*) in the western part of India, whereas in eastern India, it has been affected by saw fly (*Athalia proxima*) (Chattopadhyay et al. 2012). In rice, gall midge (*Orseolia oryzae*) was a major problem earlier in the Chhattisgarh state, which gradually slowed down due to several factors including midge-tolerant varieties/cultivars and climate changes. Finally, a resurgence of gall midge was noted at Kanker, Chhattisgarh, ascribed to a climate change and to the appearance of new biotype of the midge. This pest is being seen as a major future threat on rice crop.

Gao et al. (2009) examined interactions across three trophic levels, cotton (*Gossypium hirsutum*), aphid (*Aphis gossypii*) and its coccinellid predator (*Propylaea japonica*), as influenced by crop cultivars and CO_2 concentrations. Plant carbon/nitrogen ratios, condensed tannin and gossypol content were significantly higher while nitrogen content was lower in plants exposed to elevated CO_2 levels, as compared to ambient CO_2 concentrations. No significant difference in the survival and lifetime fecundity of *P. japonica* was observed between CO_2 concentration treatments and cultivars. However, larval durations of the predatory beetle were significantly longer when fed on aphids from the elevated CO_2 concentrations. It was speculated that *A. gossypii* may become a more serious pest under an environment with elevated CO_2 concentrations, due to increased survival of aphid and longer time for development of its coccinellid predator. Mango midges infesting particularly the inflorescence are becoming a serious problem in Maharashtra and Uttar Pradesh (CROPSAP 2013), which is in line with future projections about dipteran pests.

Studies under National Initiative for Climate Resilient Agriculture NICRA (ICAR) since 2011 have indicated that rise in ambient temperature due to climate changes could result in 2–4 additional generations of lepidopteran pests (viz. *Helicoverpa armigera*) with shortened life cycle by 5–6 days across locations, which could possibly lead to increased crop losses (Sharma et al. 2010; Rao et al. 2014). On the contrary, it has also been observed in the recent times that *Helicoverpa* infestations have been on a lower ebb than before, which possibly could be explained based on fitness levels of the pest, vis-á-vis a changed climate. This could also be true for the lepidopteran yellow stem borer (*Scirpophaga incertulas*) that damages rice. South American tomato leaf miner, *Tuta absoluta*, was documented as a new invasive pest into India during Nov–Dec 2014 at Bengaluru and in March 2015 at Hyderabad (NCIPM 2015). Outbreak of rice black bug (*Scotinophara lurida*) has been noticed since 2006 in Aduthurai (NCIPM 2015).

Overall, there is a change in crop-wise pest scenario with several fresh introductions, which may be due to increased human activities that are triggering major pest outbreaks, that may be inviting more sprays of chemicals coupled with higher risks of chemical pesticide residues as well (Bebber 2015). Elevated CO_2 (550 ppm) have been favourable to faster development of *Bactrocera dorsalis* (Sridhar et al. 2014; Rao et al. 2015). There are indications of increased predatory activity on pests having influenced high infestation of *Nilaparvata lugens* on rice crops in Asia (Karuppaiah and Sujayanand 2012; Ali et al. 2014). Extreme events like Hudhud cyclone have been ascribed as reason for the rise in populations of both *N. lugens* and *Sogatella furcifera* in rice fields in Chhattisgarh, coastal regions of Andhra Pradesh (Rao et al. 2015). Apprehensions have been expressed towards adverse impact of climate change on pollinators and lac productivity also (Srivastava et al. 2014; Mohanasundaram et al. 2014). Many other relevant examples are described in Table 8.2.

We have seen measurable changes in insect pest and pathogen scenarios attributed to changes in climate patterns. However, establishing such correlations through research endeavours remains to be done in India. Due to variation in climatic situations and fitness of insect pests and pathogens to such changes, there is the possibility of their migration to their comfort zones, which could thereby trigger invasions to newer regions/countries and establishment therein. There is no clear indication on injury profile on attainable yields and changes thereof in relation to any change in pest scenario. Therefore, we need to be sure about the fitness status of any pest vis-á-vis the changed climatic situation, apart from our readiness to detect the issue. Hence, there is a need for monitoring pre-border pest risk, vis-á-vis of the possibilities of incursions, for carrying out the necessary policy and preparedness thereof. This is apart from monitoring changing national emerging/invasive pest problems and levels of their severity on regular basis, for appropriate prioritisation. The pest diagnostic systems has to be simple, autonomous, rapid and enrichment-free, field operable, inexpensive, real-time, sensitive and specific with abilities for early detection. Thus, quarantine as part of pest risk analysis for all pests existing globally and not existing in India should be the first order of the Integrated Pest Management (IPM), so that management of pests starts even before their entry to the country. Policy-makers and authorities need to realise the importance of any invasion, the cost involved to manage and eradicate them. Thus, in globalised economy, monitoring of human and cargo-linked movements of insect pests, pathogens and nematodes through containers (dusts, dirt therein) should be enabled. This is a necessary

Table 8.2 Impact of climate change on insect pests

Components of climate change	Name of the insect pests/ pathogens	Impact	References
Elevated temperature	*Spodoptera litura*	Developmental duration decreased with increase in temperature for immature stages of *S. litura*, till 30 °C, but increased at 35 °C in egg, 3rd and 4th larval instars and pupal stage, indicating a nonlinear response at extreme temperatures	Rao et al. (2015)
Temperature dependent	Maize aphid (*Rhopalosiphum maidis*)	Nymphal development and adult longevity showed a linear decreasing trend till 25 °C and a nonlinear response above this temperature. Fecundity highest at 20 °C and reduced by >50% at 30 °C	
Elevated CO_2	Peanut aphids (*Aphis craccivora*)	The number of nymphs laid per female was significantly higher at CO_2 (550 ppm) > than ambient. Increased mean fecundity and significant reduction in longevity and development time	
	Many foliage feeding Lepidopterans in agricultural and forest species	Enhanced feeding by insects in order to obtain sufficient nitrogen for their metabolism; slower development; increased length of life stages; more foliage feeding than the normal	Fand et al. (2012)
	Gypsy moth *Lymantria dispar* in apple	Reduced larval weight gain; increased larval feeding; prolonged development	
High temperature and water stress	Midge *Stenodiplosis sorghicola* and spotted stem borer *Chilo partellus* in sorghum	Breakdown of resistance against target insect pests. Heavy loss in yields, due to increased pest damage	
	Bollworms *Heliothis virescens*, *Helicoverpa armigera* and *Helicoverpa punctigera* on cotton	Negative impacts on transgene expression in *Bt* cotton. Reduced production of *Bt* toxins. Enhanced susceptibility of crops to insect pests	
Recent abnormal weather patterns	Sugarcane woolly aphid *Ceratovacuna lanigera*	30% yield losses	Joshi and Viraktamath (2004) and Srikanth (2007)
		Reduced cane recovery	

(continued)

Table 8.2 (continued)

Components of climate change	Name of the insect pests/ pathogens	Impact	References
	Rice plant hoppers	Crop failure over more than 33,000 ha paddy area	IARI News (2008, 2009)
	Nilaparvata lugens (Stal) and *Sogatella furcifera*		
Recent abnormal weather patterns changed cropping environment (introduction of *Bt* cotton)	Mealybug, *Phenacoccus solenopsis* in cotton, vegetables and ornamentals	Heavy yield (30–40%) loss to the cotton	Dhawan et al. (2007)
		Increased cost of crop protection due to overuse of chemical pesticides	
	Papaya mealy bug *Paracoccus marginatus*	Significant yield loss to the papaya growers	Tanwar et al. (2010)
Efficacy of insect biocontrol by fungi	Leaf webber of mango (*Orthaga exvinacea*)	Increased infestation	Rao et al. (2014)
Reliability of economic threshold levels	Leaf miner of mango (*Acrocercops syngramma*)	Increased infestation	
Insect diversity in ecosystems. Parasitism	Stem borer (*Batocera rufomaculata*)	Increased infestation	
	Fruit borer (*Deanolis albizonalis*/*Citripestis eutraphera*)	Increased infestation	
	Early shoot borer (*Chlumetia transversa*)	Increased infestation	
	Scale insects (*Aspidiotus destructor*)	Increased infestation	
	Mealy scale (*Chloropulvinaria polygonata*)	Increased infestation	
	Mango scale (*Aulacaspis tubercularis*)	Increased infestation	
	Mealy bug (*Rastrococcus iceriyodes*, *Ferrisia virgata*)	Increased infestation	
	Thrips (*Thrips tabaci*, *Scirtothrips dorsalis*)	Increased infestation	Bergant et al. (2005) and Rao et al. (2015)

(continued)

Table 8.2 (continued)

Components of climate change	Name of the insect pests/ pathogens	Impact	References
	Corn earworms	Altitudes wise range expansion and increased overwintering survival in the United States	Fand et al. (2012)
	Heliothis zea (Boddie) and *Helicoverpa armigera* in maize		
	European corn borer	Northward shifts in the potential distribution up to 1220 km are estimated to occur An additional generation per season	
	Ostrinia nubilalis		
	Old world bollworm	Phenomenal increase in the United Kingdom from 1969 to 2004 and outbreaks at the northern edge of its range in Europe	
	Helicoverpa armigera		
	Cottony cushion scale	Cottony cushion scale *Icerya purchasi*	
	Icerya purchasi		
	Oak processionary moth *Thaumetopoea processionea*	Northward range extension from central and southern Europe into Belgium, Netherlands and Denmark	
	Cottony camellia scale, *Chloropulvinaria floccifera*	More abundant in the United Kingdom, extending its range northwards in England and increasing its host range in the last decade	
	Cotton bollworm/pulse pod borer *Helicoverpa armigera* on cotton, pulses, vegetables	Expansion of geographic range in Northern India	
		Adult flights/migratory behaviour	

policy and requires preparedness to manage such invasions at border (through regulations, inspections) and also post-border (movement within country, pest-surveillance, incidence management) conditions. A system should be set in place for emergency situations for pest management, with actions, roles and responsibilities identified in advance and duly linked with an information flow, in a decision framework from pre-border checks and pest risk analysis in relation to present and future climates. This should enable biosecurity (safeguarding of biological resources from external threats) to be perceived for better prevention and preparedness to ensure a biosecure country.

8.4 Novel Plant Protection Approaches

Changing pattern of disease and insect pest scenario due to climate change has warranted the need for improved novel agricultural practices and use of eco-friendly approaches for sustainable crop production. A knowledge gap exists in understanding key parameters such as pathogen survival and the timing of spore release to infect subsequent crops, as different responses to climate by the pathogen and crop could lead to variations for the incidence of infections. Concerted research efforts must be intensified to understand impacts of climate changes on soil microbes, particularly those that mitigate root diseases. Generally, warmer conditions will increase severity of root and stem rots, and they will advance with earlier crop growth in view of higher temperatures. Yield losses could increase due to both biotic (diseases/insect-pests) and abiotic (higher, earlier transpiration stress caused by heat or drought) stresses. Effects of increased carbon dioxide concentrations on dynamics of plant pathogens also require further research. Increased carbon dioxide may trigger denser crop canopies, which will encourage a range of foliar diseases (West et al. 2015). Due to changes in climate, there would be a shift in seasons, cropping patterns, insect pest and disease scenarios. Therefore, the choice of crop management practices based on the real-time situation is very crucial. In such scenarios, weather-based insect pests and disease monitoring, coupled with rapid diagnostics, would play a significant role for reliable information on the level of pest condition and for their coordinated management at regional, national and global levels. There is a need to adopt novel approaches to respond to the resurgence of diseases/insect pests under changed climates. Thus, there is an urgent need to devise better ways for data compilation (viz. e-pest surveillance, use of satellite-based information for tracking pest mobility) with knowledge about the migratory behaviour of crop pests, duly aided by international cooperation at regional and global scale. This will facilitate improved understanding of criticalities of pests' impact through agroecological analyses (including alternate hosts of pests), yield losses and economics. Ultimately, improved forewarning capabilities will be achieved for pest risk under enhanced level of unpredictability in temporal and spatial interactions between weather, cropping systems and pests, thus enabling a closer to reality forecast of doom and gloom scenarios. Educating farmers to keep records of farming practices undertaken date-wise may enable experts to guide on appropriate and timely decision-makings, apart from equipping policy-makers, crop health institutions and field functionaries with updated information.

IPM strategies would undoubtedly be the pragmatic solution to cope with any complex insect pest/disease scenario. It includes multiple approaches such as the use of healthy disease-free seeds, with durable disease resistance, and various types of cropping systems that promote the conservation of natural/native biocontrol agents. Early warning systems and monitoring for forecasting disease/insect pest epidemics/epizootics should also be developed for important pathogens/insect pests, which have a direct link on crop production. Plant-origin biopesticides such as neem (*Azadirachta indica*) oil, karanja (*Pongamia pinnata*) seed extract and neem cake may also facilitate in mitigation of climate changes, because they may help in the reduction of nitrous oxide emission by nitrification inhibitors. Biological control studies in several field crops are lacking and need to be addressed to reduce the reliance on chemical pesticides and consequent development of pest resistance and/or resurgence thereof. Holistically, there is a need for improving resilience of production systems to withstand expected shocks from increased pest prevalence, for which research and developmental efforts are required, including pre-emptive breeding for shock-tolerant varieties (Dinesh et al. 2015). Unfortunately, appropriate expertise has been declining over the years in several sectors, more so in the area of diagnostics and taxonomy of fungi, bacteria, nematodes, mites, insects and weeds. Overall, under extreme uncertainties, extension workers should be prepared for worst-case scenarios that uncover crops resilience to pests, apart from capability to face disasters through all possible means of cooperation.

8.5 Conclusion

Climatic changes provide challenges and opportunities for Indian agriculture. Though their effect is vast, only limited research on their impact on plant diseases/insect pests has been done in real field conditions. However, some assessments have been carried out in few countries on fewer crops and particular pathogens/insect pests, under field condition to counter current as well as upcoming problems. Emphasis must shift from impact assessment to developing adaptation and mitigation strategies, with special attention towards outbreaks. We need to critically revisit the efficacy of current chemical, physical and biological control tactics, including pest-resistant cultivars under climate changes, and to include future climate scenarios in all research aimed at developing new tools and strategies. Research on host response and adaptation should be launched to understand better how forthcoming changes could influence crop pests, coupled with regular pest risk analyses, based on host–pathogen/pests' interactions. India is fortunate enough to have such a diverse climate suitable to grow various types of crop plants, with varied pest population, which can help counter the problems in a changed climate scenario, enabling farmers to cope with such uncertainties with confidence.

References

Ali, M. P., Huang, D., Nachman, G., Ahmed, N., Begum, M. A., & Rabbi, M. F. (2014). Will climate change affect outbreak patterns of plant hoppers in Bangladesh? *PLoS One, 9*(3), e91678. https://doi.org/10.1371/journal.pone.0091678.

Bebber, D. P. (2015). Range-expanding pests and pathogens in a warming world. *Annual Review of Phytopathology, 53*. https://doi.org/10.1146/annurev-phyto-080614-120207.

Bergant, K., Trdan, S., Dragan, Z., Nidarc, I. C., Zalika, C., Repins, E. K., Luc, K. A., & Bogataj, K. (2005). Impact of climate change on developmental dynamics of *Thrips tabaci* (Thysanoptera: Thripidae): Can it be quantified? *Environmental Entomology, 34*, 755–766.

Biswas, A. (2011). Plant diseases in the perspective of climate change. *SATSA Mukhapatra-Annual Technical Issue, 15*, 84–86.

Central Research Institute for Dryland Agriculture (CRIDA). (2014). *Annual report* (Report 2014). Central Research Institute for Dryland Agriculture (CRIDA), Santoshnagar, Hyderabad, India.

Chattopadhyay, C., & Huda, A. K. S. (2009, December 17–19). Changing climate forcing alteration in cropping pattern to trigger new disease scenario in oilseeds and pulses in Indian subcontinent. *Abstracts, national symposium on climate change, plant protection and food security interface*, Kalyani, p. 70.

Chattopadhyay, C., Bhattacharya, B. K., Kumar, V., Kumar, A., & Meena, P. D. (2012). Impact of climate change on pests and diseases of oilseeds *Brassica* – The scenario unfolding in India. *The Journal of Oilseed Brassica, 2*, 48–55.

CROPSAP (Crop Pest Surveillance and Advisory Project) team of 'E' pest surveillance. (2013). *Pests of fruits (banana, mango and pomegranate) 'E' Pest Surveillance and Pest Management Advisory* (Ahuja, D.B. Ed.). New Delhi/Pune: National Centre for Integrated Pest Management/ State Department of Horticulture, Commissionerate of Agriculture. 67 p.

Dhawan, A. K., Singh, K., Saini, S., Mohindru, B., Kaur, A., Singh, G., & Singh, S. (2007). Incidence and damage potential of mealybug, *Phenacoccus solenopsis* Tinsley, on cotton in Punjab. *Indian Journal of Ecology, 34*, 110–116.

Dinesh, D., Bett, B., Boone, R., Grace, D., Kinyangi, J., Lindahl, J., Mohan, C. V., Ramirez, V. J., Robinson, R., Rosenstock, T., Smith, J., & Thornton, P. (2015). *Impact of climate change on African agriculture: Focus on pests and diseases.* CCAFS Info Note. Copenhagen, Denmark: CGIAR Research Program on Climate Change, Agriculture and Food Security (CCAFS) downloaded on 26 July 2015 from https://ccafs.cgiar.org/publications/impact-climate-change-african-agriculture-focus-pests-and-diseases#.VbUObfmqqko at 2217 IST.

Fand, B. B., Kamble, A. L., & Kumar, M. (2012). Will climate change pose serious threat to crop pest management-A critical review. *International Journal of Scientific and Research Publications, 2*, 1–14.

Gao, F., Zhu, S., Sun, Y., Du, L., Parajulee, M., Kang, L., & Ge, F. (2009). Interactive effects of elevated CO_2 and cotton cultivar on tri-trophic interaction of *Gossypium hirsutum*, *Aphis gossypii* and *Propylaea japonica*. *Environmental Entomology, 37*, 29–37.

Gautam, H. R., Bhardwaj, M. L., & Kumar, R. (2013). Climate change and its impact on plant diseases. *Current Science, 105*, 1685–1691.

Ghosh, A., Das, A., Bhattacharya, B. K., Kumar, V., Kumar, A., Meena, P. D., & Chattopadhyay, C. (2012). Impact of climate change on pests and diseases – the scenario unfolding in Indian agriculture. *SATSA Mukhopatra Annual Technical Issue, 16*, 15–29.

IARI. (2008). Brown plant hopper outbreak in rice. *News 24*. New Delhi: Indian Agricultural Research Institute.

IARI. (2009). Brown plant hopper outbreak in rice. *News 25*. New Delhi: Indian Agricultural Research Institute.

Joshi, S., & Viraktamath, C. A. (2004). The sugarcane woolly aphid, *Ceratovacuna lanigera* Zehntner (Hemiptera: Aphididae): Its biology, pest status and control. *Current Science, 87*, 307–316.

Karuppaiah, V., & Sujayanand, G. K. (2012). Impact of climate change on population dynamics of insect pests. *World Journal of Agricultural Sciences, 8*, 240–246.

Kumar, A., Chattopadhyay, C., Singh, K. N., Vennila, S., & Rao, V. U. M. (2012). Trend analysis of climate variables for pigeonpea growing area. In *Abstract, international conference on statistics and computation,* pp. 145–148.

Kumar, A., Kumar, V., Bhattacharya, B. K., Singh, N., & Chattopadhyay, C. (2013). Integrated disease management: Need for climate-resilient technologies. *Journal of Mycology and Plant Pathology, 43*, 28–36.

Lamichhane, J. R., Barzman, M., Booij, K., Boonekamp, P., Desneux, N., Huber, L., Kudsk, P., Langrell, S. R. H., Ratnadass, A., Ricci, P., Sarah, J. L., & Messéan, A. (2015). Robust cropping systems to tackle pests under climate change: A review. *Agronomy for Sustainable Development, 35*, 443–459.

Lastuvka, Z. (2009). Climate change and its possible influence on the occurrence and importance of insect pests. *Plant Protection Science, 45*, S53–S62.

Meena, P. D., Awasthi, R. P., Chattopadhyay, C., Kolte, S. J., & Kumar, A. (2010). *Alternaria blight*: A chronic disease in rapeseed-mustard. *Journal of Oilseed Brassica, 1*, 1–11.

Mohanasundaram, A., Monobrullah, M., Sharma, K. K., Singh, R. K., Meena, S. C., & Verma, S. (2014, October 16–18). Effect of weather parameters on production of summer rangeeni lac crop at Ranchi, Jharkhand. In *Compendium-cum-abstract, international symposium on new-dimensions in agrometeorology for sustainable agriculture*, GBPUAT, Pantnagar (UA), Association Agrometeorologists, p. 329.

Mohapatra, S. D., & Chattopadhyay, C. (2012, January 10–13). Occurrence of mungbean phyllody and its management. *Abstracts. 3rd global conference on plant pathology for food security*, Org. by ISMPP and MPUAT, Udaipur, p. 259.

NCIPM (National Research Centre for Integrated Pest Management). (2015). *Annual report* (2014–2015). New Delhi: ICAR- National Research Centre for Integrated Pest Management.

Rao, T. M., Jayanthi, P. D. K., Verghese, A., Arthikirubha, A., Sowmya, B. R., & Bhatt, R. M. (2014). Emerging pests of mango under climate change scenario. *Technical bulletin No.18*, IIHR, Bengaluru.

Rao, M. S., Swathi, P., Rama Rao, C. A., Rao, K. V., Raju, B. M., Srinivas, K., Manimanjari, D., & Maheshwari, M. (2015). Model and scenario variations in predicted number of generations of *Spodoptera litura* Fab. on peanut during future climate change scenario. *PLoS ONE, 10*(2), e0116762. https://doi.org/10.1371/journal.pone.0116762.

Ray, D. K., Gerber, J. S., MacDonald, G. K., & West, P. C. (2015). Climate variation explains a third of global crop yield variability. *Nature Communications, 6*, 5989. https://doi.org/10.1038/ncomms6989.

Root, T. L., & Hughes, L. (2005). Present and future phenological changes in wild plants and animals. In T. Lovejoy & L. Hannah (Eds.), *Climate change and biodiversity* (pp. 61–69). New Haven: Yale University Press.

Sharma, H. C., Srivastava, C. P., Durairaj, C., & Gowda, C. L. L. (2010). Pest management in grain legumes and climate change. In S. S. Yadav, D. L. McNeil, R. Redden, & S. A. Patil (Eds.), *Climate change and management of cool season grain legume crops* (pp. 115–140). Dordrecht: Springer Science & Business Media.

Sridhar, V., Verghese, A., Vinesh, L. S., Jayashankar, M., & Kamala Jayanthi, P. D. (2014). CLIMEX simulated predictions of oriental fruit fly, *Bactrocera dorsalis* (Hendel) (Diptera: Tephritidae) geographical distribution under climate change situations in India. *Current Science, 106*, 1702–1710.

Srikanth, J. (2007). World and Indian scenario of sugarcane woolly aphid. In N. Mukunthan, J. Srikanth, B. Singaravelu, T. Rajula Shanthy, R. Thiagarajan, & D. Puthira Prathap (Eds.), *Woolly aphid management in sugarcane*. Extension Publication, Sugarcane Breeding Institute, Coimbatore, 154: 1–12.

Srivastava, P., Chauhan, D., & Khan, M. S. (2014, October 16–18). Climate changes: A threat to insect pollinators in litchi crop at Pantnagar. In *Compendium-cum-abstract, international symposium on new-dimensions in agrometeorology for sustainable agriculture*, GBPUAT, Pantnagar (UA), Association Agrometeorologists, pp. 323–324.

Subba Rao, A. V. M., Agarwal, P. K., Huda, A. K. S., & Chattopadhyay, C. (2007). Using InfoCrop – A user friendly crop simulation model for mustard. In *Climate and crop disease risk management – An international initiative in Asia Pacific region* (p. 16). Hyderabad: CRIDA.

Tanwar, R. K., Jeyakumar, P., & Vennila, S. (2010). Papaya mealybug and its management strategies. *Technical bulletin* 22. New Delhi: National Centre for Integrated Pest Management.

Thomas, C. D., Cameron, A., Green, R. E., Bakkenes, M., Beaumont, L. J., Collingham, Y. C., Erasmus, B. F. N., Ferreira de Siqueira, M., Grainger, A., Hannah, L., Hughes, L., Huntley, B., Jaarsveld, A. S., Midgley, G. F., Miles, L., Ortega-Huerta, M. A., Peterson, A. T., Phillips, O. L., & Williams, S. E. (2004). Extinction risk from climate change. *Nature, 427*, 145–148.

West, J. S., Fitt, B. D. L., Townsend, J. A., Stevens, M., Edwards, S. G., Turner, J. A., Ellerton, D., Flind, A., King, J., Hasler, J., Werner, C. P., Tapsell, C., Holdgate, S., Summers, R., Angus, B., & Edmonds, J. (2015). Impact of climate change on diseases in sustainable arable crop systems: CLIMDIS. *HGCA Project report No. 539*, Agriculture and Horticulture Development Board, UK, 67 p.

Chapter 9
Role of Microbes in Sustainable Agriculture

Ambalal N. Sabalpara and Lalit Mahatma

Abstract Agriculturists have multifaceted challenges as natural resources are constantly shrinking, available resources are becoming worse everyday, pathogens are evolving fast, and expectations of consumers are progressively increasing. Recent technical developments in agriculture saw an increased use of chemicals affecting the microbial ecosystem, whereby soil rapidly loses vitality. Existing technologies have reached a plateau, and hitherto it is extremely difficult to further increase food production. Under these circumstances, sustainability in crop productions cannot be attained without the sustaining role of the microbial populations in soil. Microbes perform multiple functions of supplying nutrients; controlling diseases, insects, nematodes, and weeds; and recycling by waste degradation. The role of microbes in sustainable agriculture is discussed in this chapter.

Keywords Microbes · Sustainable agriculture · Plant nutrient supply · Plant disease management · Insects · Nematodes · Weeds · Waste

Abbreviations

AEFB	Aerobic endospore-forming bacteria
BNF	Biological nitrogen fixation
Bt	*Bacillus thuringiensis*
C:N ratio	Carbon/nitrogen ratio
FYM	Farmyard manure

A. N. Sabalpara
Navsari Agricultural University, Navsari, Gujarat, India
e-mail: sabalpara_2008@yahoo.com

L. Mahatma (✉)
Department of Plant Pathology, N.M. College of Agriculture, Navsari Agricultural University, Navsari, Gujarat, India
e-mail: mahatmalalit@yahoo.co.in

R. Peshin, A. K. Dhawan (eds.), *Natural Resource Management: Ecological Perspectives*, Sustainability in Plant and Crop Protection,
https://doi.org/10.1007/978-3-319-99768-1_9

GV	Granulosis virus
NPV	Nuclear polyhedrosis virus
PGPR	Plant growth-promoting rhizobacteria
PSB	Phosphate-solubilizing bacteria
PSM	Phosphate-solubilizing microorganisms

9.1 Introduction

Sustainability in crop production is of utmost importance to feed the ever-increasing population of the world. The world's population is projected to reach 8.5 billion by 2030 (Anonymous 2017a). Under this changing scenario, agriculturists have multifaceted challenges as the natural resources are constantly shrinking, available resources are progressively deteriorating, pathogens are evolving fast, and expectations of consumers are progressively increasing. Recent developments achieved by agricultural sciences may have interfering effects on the microbial ecosystem. The use of chemical fertilizers alone has increased by 233% after the Green Revolution (Khunt et al. 2014). Excessive use of chemicals increased the yield of different agricultural commodities; however, they seriously affected the environment (Mahmood et al. 2016). Production and use of chemical fertilizers, especially nitrogenous fertilizers, has increased the level of nitrous oxide in the atmosphere. Improvements in nitrogen use efficiency and nitrogen inhibitors can substantially reduce emissions of N_2O and help in protecting the environment.

Biofertilizers are the safest and cheapest source of different nutrients used in agriculture. The World Health Organization estimated that there are three million cases of pesticide poisoning each year and up to 220,000 deaths, primarily in developing countries (Anonymous 2017b). Pesticide exposure can cause a range of neurological health effects such as memory loss, loss of coordination (Lah 2011), reduced speed of response to stimuli, reduced visual ability, altered or uncontrollable mood and general behavior, and reduced motor skills. Other possible health effects include asthma, allergies, and hypersensitivity. Pesticide exposure is also linked to cancer, hormone disruption, and problems with reproduction and fetal development (Mahmood et al. 2016; Anonymous 2017b). Apart from these effects, the existing technologies appear to have reached a plateau, and hitherto it is extremely difficult to further increase food production. Overall, the main side effect of these chemicals is represented by the disturbance of the ecosystem. Therefore, there is an urgent need for exploring nonconventional resources, not only to increase the demand of an ever-increasing population but also to sustain our ecosystem and protect it from further degradation. Microorganisms have tremendous potential to support life without causing any deleterious effect. Sustainability in agriculture cannot be attained without sustaining the microbial population in the soil under the present circumstances. Microbes can supply the essential elements required by plants, have the potential to control pathogens causing diseases, and may ameliorate abiotic stress conditions under the pressure of changing climatic effects.

9.2 Microorganisms as Biofertilizers

A number of microorganisms, acting symbiotically, associative symbiosis, or without association with plants, fix atmospheric nitrogen, with high utilization efficiency. These microorganisms, if utilized properly, have potential to reduce chemical nitrogen application by 50%, thereby helping in the reduction of toxic N_2O deposition in the atmosphere. Similarly, the process of manufacturing phosphate and potash fertilizers also adversely affects the environment and soil microclimate. Growth of different flora and fauna is also affected adversely. Formulations made up with bacteria have potential to replace 50% of phosphorous and potash fertilizers and may help in the reduction of atmosphere's deleterious effects induced by chemical fertilizers (Mahatma et al. 2016a, b). Not only the major but also the minor nutrients are made available to the plants by some bacteria, thereby maintaining overall nutrient balance in crops. Application of biofertilizers and subsequent amendment of organic matter over an extended period of time help in establishing these microorganisms in the soil ecosystem. These species may be endophytic or rhizosphere colonizers. They also act as plant growth-promoting rhizobacteria (PGPR) to help plants withstand various biotic and abiotic stresses.

9.2.1 *Nitrogen-Fixing Microorganisms*

The atmosphere contains about 78% of nitrogen, but plants or animals cannot utilize it directly. Nitrogen fixing microorganisms can fix atmospheric nitrogen in soil by the process known as biological nitrogen fixation (BNF). Different nitrogen-fixing species may be grouped into symbiotic nitrogen fixers (belonging to genera *Rhizobium*, *Azorhizobium*, *Frankia*, etc.), nonsymbiotic nitrogen fixers (*Azotobacter*, *Klebsiella*, etc.), and associative symbiotic nitrogen fixers (*Azospirillum* and *Gluconacetobacter*).

9.2.1.1 Symbiotic Nitrogen Fixers

Bacteria collectively known as rhizobia are found associated with leguminous root nodules or occasionally on stem nodules. Inside nodules, rhizobia live as bacteroides and fix atmospheric nitrogen to ammonia that is provided to the plant. In return, the hosting plant provides shelter and nutrients to the bacteria. *Rhizobium* spp. inoculation is a popular practice to replace nitrogenous fertilizers in legumes (Gupta 2004). Oxygen level inside nodules is usually regulated by a specific protein called leghaemoglobin, produced only after plant infection with *Rhizobium* spp. The roots supply organic material for the growth of the bacteroides in a symbiotic relationship with the bacterial nitrogen supply.

It has been reported that reduction of nitrogenous synthetic fertilizers by using nitrogen-fixing bacterial inoculation is one of the most efficient steps in sustainable agriculture (Zahran 1999). Symbiotic association between *Rhizobium* and leguminous plants can fix up to 100–200 kg N/ha/year in one crop season. This can leave substantial nitrogen behind for the subsequent crop (Tilak et al. 2005). This association is also important in nature to reduce soil erosion and weeds, to enrich soil organic nitrogen (Carsky et al. 2001). Application of *Pseudomonas* spp. and *Bacillus* spp. as co-inoculants with *Rhizobium* spp. in chickpea accelerates plant growth, nodulation, and nitrogen fixation (Mohammadi et al. 2010). Rhizobium–legume symbiosis is influenced by the rhizosphere application of naringenin, which increases nodule numbers. There was significant improvement in the number of nodules in mung bean (*Vigna radiata*) with the treatment of 15 μM concentration of naringenin over control (Patel and Mahatma 2017).

Another excellent case of symbiosis comprises several floating water ferns (*Azolla* spp.) found in tropical and temperate ecosystems. There are seven important species reported, namely, *Azolla caroliniana*, *A. mexicana*, *A. microphylla*, *A. filiculoides*, *A. rubra*, *A. nilotica*, and *A. pinnata*. These pteridophytes have the ability to fix atmospheric nitrogen through symbiosis with blue-green algae (*Nostoc* and *Anabaena* spp.). Therefore, they are considered as an important potential source of nitrogen for wetland rice (*Oryza sativa*). The contribution of nitrogen from *Azolla* spp. to wetland rice plants has been found to be maximum when incorporated into the soil as green manure. Rice yields could increase by 0.5–2 t/ha by applying *Azolla* (Gupta 2004).

Another distinct and noteworthy association between microbes and plants occurs in the actinomycetes *Frankia* spp., involving a broad spectrum of plants belonging to eight families, collectively called actinorhizal plants. In addition, nitrogen-fixing cyanobacteria (mainly *Nostoc*) have also been found to colonize different plant organs, either intracellularly in the family Gunneraceae or, extracellularly, in liverworts, hornworts, *Azolla*, and *Cycadaceae*.

9.2.1.2 Associative Symbiotic Nitrogen Fixers

Azospirillum is a heterotrophic, associative symbiotic nitrogen fixer belonging to the Rhodospirillaceae family. *Azospirillum lipoferum*, *A. brasilense*, and *A. amazonese* are important species in soil. *Azospirillum* spp. directly benefits plants, improving shoot and root development and increasing the rate of water and mineral uptake by roots. It is recommended for maize, sugarcane, sorghum, wheat, pearl millet, etc. It can fix 20–40 kg/ha/yr of nitrogen in soil. Apart from this, *Azospirillum* produces plant growth-promoting substances. Maize (*Zea mays*) inoculated with *Azospirillum* may get 10–20% of nitrogen that is used for plant. Considering only the cultivation of this crop, annual savings may reach millions of nitrogen tons (Vose and Ruschel 1981).

Other associative, endophytic nitrogen-fixing bacteria belong to the genus *Gluconacetobacter* (*Acetobacter*), initially described in Brazil, associated with sugarcane plant, and named *Acetobacter diazotrophicus*. Then the species was renamed *Gluconacetobacter diazotrophicus*. This bacterium may reduce by 50% the plant's nitrogen requirement in sugarcane crop (Mahatma et al. 2016a). *Gluconacetobacter* has also been found to be associated with nonsugar-rich plants (Sevilla and Kennedy 2000).

9.2.1.3 Nonsymbiotic Nitrogen Fixers

The genus *Azotobacter* includes aerobic, heterotrophic, cyst forming, capsulated, free-living nitrogen bacteria. It belongs to the Azotobacteriaceae family. *Azotobacter chroococum*, *A. vinelandii*, *A. insignis*, and *A. beijerinckii* are the most common species found in soil. However, a number of *Azotobacter* rarely exceed 10^5 cfu/g of soil, due to the presence of antagonistic organisms and limited nutrients. Apart from nitrogen fixation, *Azotobacter* produces antifungal antibiotics and plant growth-promoting substances. It can promote seed germination and plant vigor. According to field trial, *Azotobacter*-treated (10^7cfu/ml) melon and papaya (*Carica papaya*) seedlings were 68% and 103% heavier than seedling without inoculation. Mahatma et al. (2016b) observed that different isolates of *Azotobacter* have different growth and yield effects on finger millet (*Eleusine coracana*). It may reduce by 50% nitrogen requirement in this crop. *Azotobacter* spp. have been recommended as biofertilizers in a number of crops including rice, maize, wheat, sugarcane, pearl millet, vegetable crops, horticultural crops, forest trees, etc.

9.2.2 Phosphate-Solubilizing Microorganisms

Phosphorous is the second-most important nutrient of plants, absorbed as $H_2PO_4^-$ in acidic pH or as HPO_4^{2-} in alkaline pH. After the application of phosphorous fertilizers in soil, chemical reactions convert available form of phosphorous in insoluble and unavailable forms.

Highly alkaline and acidic soils reduce phosphate availability. In alkaline soils, calcium is the dominant cation that will react with phosphate. A general sequence of reactions in alkaline soils is the formation of dibasic calcium phosphate dihydrate, octacalcium phosphate, and hydroxyapatite. The formation of each product results in a decrease in phosphate solubility and availability. When pH is acidic, plant-available phosphorus becomes increasingly tied up in aluminum phosphates. As soils become more acidic ($pH < 5$), phosphorus is fixed in iron phosphates. The amorphous Al and Fe phosphates gradually change into compounds that resemble crystalline variscite (an Al phosphate) and strengite (a Fe phosphate).

Fixed insoluble phosphate in soil can be solubilized by a group of bacteria known as phosphate-solubilizing microorganisms (PSM), which have biochemical mechanisms converting insoluble unavailable phosphate to soluble available phosphate through organic acid production. Bacteria of genera *Bacillus*, *Pseudomonas*,

Rhizobium, *Enterobacter*, and fungi belonging to genera *Aspergillus* or *Penicillium* are considered as potent PSM (Whitelaw 2001). However, bacteria are more efficient in phosphate solubilization than fungi (Alam et al. 2002). Therefore, application of PSB inoculants in agriculture has been preferred to save expensive chemical phosphatic fertilizers. PSB populations in rhizosphere soil will always be higher as compared to non-rhizospheric soil, suggesting synergistic effect of PSB and roots. Application of PSB along with single super-phosphate and rock phosphate may reduce phosphatic fertilizer inputs by 25% and 50%, respectively (Sundara et al. 2002; Mahatma et al. 2016a). PSB are recommended for all crops.

9.2.3 Phosphate Absorbers

Mycorrhizae are symbiotic associations between fungi and plants root. Fungi absorb nutrients such as phosphate, zinc, calcium, etc. from nearby soil that are supplied to roots, while in turn the plant provides shelter and nutrients to fungi. *Acaulospora mellea*, *Funneliformis mosseae*, *Gigaspora margarita*, *Glomus etunicatum*, and *Scutellospora calospora* are predominant species, especially in South Gujarat (Tandel and Mahatma 2016). Mycorrhizae form symbiotic associations in the majority of crops, except plants belonging to families of Chenopodiaceae, Amaranthaceae, Caryophyllaceae, Polygonaceae, Brassicaceae, Commelinaceae, Juncaceae, and Cyperaceae (Mishra et al. 2013).

9.2.4 Potash Solubilizers/Mobilizers

Potassium is the third most important nutrient for plants and the fourth most abundant nutrient, constituting about 2.5% of the lithosphere. Three forms of potassium are found in soil, namely, soil minerals, non-exchangeable, and available forms. Soil contains only 1–2% of available potassium, the rest being soil minerals or non-exchangeable forms that are unavailable for plants growth. The total pool of soil K is extremely complex and can be solubilized by bacteria through the production of acids that make K available to the plant. *Frateuria aurantia*, a species of Proteobacteria, is one of the potential microorganisms that can mobilize K in the rhizosphere of different crops (Chandra and Greep 2006).

9.2.5 Zinc Solubilizers

Zinc is a micronutrient for plant growth that may be present in soil in an available or unavailable form. Several bacterial species like *B. subtilis*, *Thiobacillus thiooxidans*, and yeasts (*Saccharomyces* sp.) can be used as zinc solubilizers. *Bacillus sp.*

can be applied as zinc solubilizer biofertilizers along with cheaper insoluble zinc oxide, zinc carbonate, and zinc sulfide, to replace costly zinc sulfate (Mishra et al. 2013).

9.2.6 Plant Growth-Promoting Rhizobacteria (PGPR)

PGPR are a group of bacteria that stimulate plant growth and increase crop yields. Due to increasing awareness of PGPR in agricultural production, scientists are trying to explore these microorganisms to increase plant growth through field applications (Burr et al. 1984). PGPR include several bacterial genera, e.g., *Bacillus*, *Pseudomonas*, *Rhizobium*, *Bradyrhizobium*, *Enterobacter*, *Alcaligenes*, *Arthrobacter*, *Azotobacter*, *Cellulomonas*, *Flavobacterium*, *Streptomycetes*, and *Xanthomonas* (Weller 1988). They affect plant growth by nutrient uptake through direct or indirect mechanisms. Some of these bacteria have the capacity to secrete plant growth-promoting chemicals. However, efficiency depends upon soil physical and chemical properties. Surprisingly, bacteria showed better activity in nutrient-deficient soils as compared to nutrient-rich soils (Egamberdiyeva 2007). The bacterial inoculation on cotton plants significantly increased seed yield, plant height, and microbial populations in soil (Anjum et al. 2007). Use of PGPR with phosphate-enriched compost in an integrated approach improved growth, yield, and nodulation in chickpea (Shahzad et al. 2008). Several PGPR isolated from farm waste compost showed synergistic effects on pearl millet growth (Hameeda et al. 2006).

9.3 Microorganisms as Biopesticides

9.3.1 Management of Soilborne Diseases

Biopesticide is a broad term covering all microorganisms and/or products used for management of diseases and insect pests in agriculture. Presence of microorganisms that suppress growth of a pathogenic microorganism was first discovered by Sanford in 1926. He observed that the incidence of potato scab caused by *Streptomyces scabies* was drastically reduced in soil when a population of a nonpathogenic strain of *S. scabies* increased, changing the soil from "pathogen conducive" to "pathogen suppressive." Soon after, Millard and Taylor (1927) reported the control of scab grown in sterilized soil and inoculated with *S. scabies*, through simultaneous inoculation of soil with a vigorous saprophytic species, *S. praecox*. This observation led to focusing on different soil-inhabiting saprophytic microorganisms for management of devastating diseases. In 1932, Weindling showed that *Trichoderma viride*, a common saprophytic fungus, was able to parasitize the mycelia of other fungi. Later, he showed that the lethal action of *T. viride* was due to the secretion of an antibiotic substance which he called "gliotoxin." Still later, in 1945 Brain and McGowan isolated another antibiotic substance from *T. viride*, which they named "viridin" (see Mehrotra 1993). These discoveries raised hopes of controlling plant diseases by biological means. In 1951, a new approach was initiated by Bliss in the USA, who was working on root disease of *Citrus* spp. caused by *Armillaria mellea*. He suggested that if the soil was fumigated with carbon disulfide, the disease-inciting fungus was killed not directly by the fumigant but indirectly by the dominance of *T. viride*, population of which increased after fumigation. Garrett (1958) tested the effect of fumigation and the antibiotic effect of *T. viride* and accessed the relative parts played by these two components. He found that about 30% of *Armillaria mellea* was killed due to the dominance of *T. viride*, while the rest was directly killed by the fumigant. He also pointed out that, in the thicker plant roots, *T. viride* was more instrumental in the process of killing than the fumigant, which might not be able to reach easily. Since then, many groups from all over the world showed efficacy of different saprophytic microorganisms for management of different plant pathogens, at different levels (Mehrotra 1993). Detailed accounts of different microorganisms extensively used for management of various pathogens are given in Table 9.1. The most fascinating part of the biopesticide disease management is that they work on wide ranges of fungi and bacteria irrespective of crop.

9.3.2 Management of Seed-Borne Diseases

Seed is a primary source of inoculum and vehicle for the long-distance spread of many diseases. The pathogen that survives on the seeds can be controlled effectively by the seed treatment. Pathogens responsible for purple blotch of onion (*Alternaria porri*), *Phomopsis* of brinjal/eggplant (*Phomopsis vexans*), and anthracnose of beans and many other crops (*Colletotrichum lindemuthianum*) can be effectively controlled by seed treatments with *Trichoderma harzianum*. Sigatoka of banana (*Mycosphaerella musicola*) and bacterial leaf spot of tomato (*Xanthomonas* sp.) can be effectively controlled by seed treatments with *Bacillus subtilis*.

9.3.3 Management of Airborne Diseases

There are numerous constraints in using biopesticides for the management of airborne diseases. However, *Chaetomium* sp. and *Athelia bombacina* are an excellent example of biopesticides that suppress *Venturia inaequalis* production of ascospores and conidia in fallen and growing leaves, respectively (Carisse et al. 2000). Antibiotics from *Chaetomium* sp. can diffuse passively on the leaf surface and inhibit the infection by *V. inaequalis* (Bouderau and Andrews 1987). Biocontrol efficacy of *Chaetomium* sp. increases in combination with biological fertilizers, effective strains of cellulose-degrading fungi, and specific fungi for plant growth stimulants (Soytong

Table 9.1 Diseases, pathogens, and nematodes managed by different organisms

Disease/nematode	Pathogen/scientific name
Diseases/nematode managed by *Bacillus subtilis*	
Sigatoka of banana	*Mycosphaerella musicola*
Lesion nematode of banana	*Pratylenchus coffeae*
Burrowing nematode of banana	*Radopholus similis*
Anthracnose of mango	*Colletotrichum gloeosporioides*
Purple blotch of onion	*Alternaria porri*
Root-knot of vegetables	*Meloidogyne incognita*
Diseases managed by *Pseudomonas fluorescens*	
Cucumber wilt	*Fusarium oxysporum*
Carnation wilt	*F. oxysporum*
Cucumber damping off	*Pythium aphanidermatum*
Radish wilt	*F. oxysporum* f. sp. *raphani*
Tomato wilt	*F. oxysporum* f. sp. *lycopersici*
Root rot	*Aphanomyces euteiches*
Sugar beet root rot	*Aphanomyces cochliodes*
Cucumber wilt	*F. oxysporum*
Cucumber damping off	*P. aphanidermatum*
Nematodes managed by *Paecilomyces lilacinus*	
Root-knot nematode	*Meloidogyne* spp.
Potato golden nematode	*Globodera pallida*
Cyst nematode	*Heterodera* spp.
Citrus nematode	*Tylenchulus semipenetrans*
Burrowing nematode	*Radopholus similis*
Reniform nematode	*Rotylenchulus reniformis*
Diseases managed by *Trichoderma* spp.	
Sigatoka of banana	*Mycosphaerella musicola*
Purple blotch of onion	*Alternaria porri*
Cucumber wilt	*F. oxysporum*
Carnation wilt	*F. oxysporum*
Cucumber damping off	*P. aphanidermatum*
Tomato wilt	*F. oxysporum* f. sp. *lycopersici*
Cucumber wilt	*F. oxysporum*
Cucumber damping off	*P. aphanidermatum*

and Ratanacherdchai 2005). *Tuberculina maxima*, commonly known as purple mold, parasitizes the white pine blister rust fungus *Cronartium ribicola*. *Darluca filum* and *Verticillium lecanii* parasitize carnation rust *Uromyces caryophyllinus* and brown rust of wheat *Puccinia recondita* (Spencer and Parasitic 1981). *Ampelomyces quisqualis* parasitizes several genera of powdery mildew pathogen *Oidium*, *Erysiphe*, *Sphaerotheca*, *Podosphaera*, *Uncinula*, and *Leveillula* (Sztejnberg et al. 1989; Falk et al. 1995). *Tilletiopsis* sp. parasitizes the cucumber powdery mildew fungus *Sphaerotheca fuliginea*. *Nectria inventa* and *Gonatobotrys simplex* parasitize two pathogenic species of *Alternaria*. Spraying of *Cladosporium herbarum* or *Penicillium* sp. almost completely suppressed the subsequent infection of developing tomato fruits by *Botrytis cinerea*. Sprays with *Trichoderma* in the field also reduced Botrytis rot of strawberries (*Fragaria × ananassa*) and of grapes (*Vitis vinifera*) at the time of harvest and in storage. Sclerotinia head rot of sunflower was reduced significantly by releasing into the field honeybees that had been previously contaminated heavily with spores of the biocontrol fungi *Trichoderma* spp., which the honeybees delivered promptly to the flowers (Agrios 2005).

9.3.4 Postharvest Diseases

The time elapse between harvesting and consumption may be, in many cases, significant. Therefore, the harvest should remain disease-free until consumption. Many postharvest diseases originate in the field and are expressed either in transit or during storage. Application of chemicals for management of postharvest diseases is considered the most disagreeable, as the chances of product directly entering the food chain are higher. Postharvest rots of several fruits can be reduced considerably by spraying them with spores of antagonistic fungi and saprophytic yeasts, at different stages of fruit development, or by dipping the fruit in the inoculum. Yeast such as *Cryptococcus* and *Candida* spp. can effectively reduce the infection by *Penicillium expansum*, preventing postharvest decay of apples (He et al. 2003). Also, significant reduction of citrus green mold caused by *Penicillium digitatum* can be obtained by treating the fruit with antagonistic yeasts or the fungal antagonist *Trichoderma viride*. Preharvest and postharvest Botrytis rot of strawberries can be reduced by several sprays of *Trichoderma* spores on strawberry blossoms and young fruit. Penicillium rot of pineapple (*Ananas comosus*) was reduced considerably by spraying the fruit with nonpathogenic strains of the pathogen. Similarly, several antagonistic yeasts protected grapes and tomatoes from Botrytis, Penicillium, and Rhizoctonia rots. *Aureobasidium pullulans* controls decay of apple fruit caused by *Botrytis cinerea* and *Penicillium expansum*. In addition to controlling the decay, *A. pullulans* caused a transient increase in beta 1,3-glucanase, chitinase, and peroxidase activities, starting 24 h after the treatment and reaching maximum levels in the following 48 and 96 h. An increase in beta 1,3-glucanase, chitinase, and peroxidase activity was also triggered by wounding, although the level of increase was markedly lower than in treated fruit. The ability of *A. pullulans* to increase activities of beta 1,3-glucanase, chitinase, and peroxidase in addition to its known capacity to outcompete the pathogen for nutrients and space may be the basis of its biocontrol activity (Ippolito et al. 2000).

The yeast *Candida saitoana* controlled postharvest decay of apple fruit by inducing systemic resistance in the fruit while at the same time increasing chitinase and beta 1,3-glucanase activities. In addition, *Candida oleophila* was approved for postharvest decay control in citrus and apples, under the trade name Aspire™ (Agrios 2005).

9.4 Mechanisms of Action of Biopesticides

The mechanisms by which antagonistic microorganisms affect pathogen populations are not always clear, but they are generally attributed to one of these four effects:

1. Direct parasitism or lysis and death of the pathogen
2. Competition with the pathogen for nutrients
3. Direct toxic effects on the pathogen by antibiotic substances released by the antagonist
4. Indirect toxic effects on the pathogen by volatile substances, such as ethylene, released by the metabolic activities of the antagonist

Many antagonistic microorganisms are naturally present in crop soils and exert a certain degree of biological control over one or many pathogens, regardless of human activities. We, however, have been attempting to increase the effectiveness of antagonists either by introducing new and larger populations of, e.g., *Trichoderma harzianum* and *Pasteuria penetrans*, in fields where they are lacking and/or by adding soil amendments. The latter serve as nutrients for, or otherwise stimulate growth of, the antagonistic microorganisms, increasing their pathogen inhibitory activity.

9.5 Management of Insects

Insects can be controlled by using fungi, bacteria, viruses, and protozoans. The most successful insect pathogen used for control is the bacterium *Bacillus thuringiensis* (Bt). There are different species of bacteria with potential to kill insect from different taxa. In a commercial product, different strains are either mixed together to allow a broader spectrum of activities against the pests the product is prepared for. Bt subsp. *kurstaki* and *aizawai* are used for lepidopteran larvae, whereas Bt subsp. *tenebrionis* is used for coleopteran larvae. Other strains are specific to mosquitoes (Bt subsp. *israelensis*).

Other than Bt, insect viruses, namely, baculoviruses (nuclear polyhedrosis virus, NPV) and granulosis virus (GV), are the most promising for insect control, particularly of lepidoptera and diptera, because of their specificity. NPVs have been successfully used for management of devastating pests like *Heliothis* and *Spodoptera* spp. on cotton, fruit, and vegetable crops in several countries (Ramanujam et al. 2014). Entomopathogenic fungi, namely, *Beauveria bassiana*, *Metarhizium anisopliae*, *Lecanicillium lecanii*, and *Isaria fumosorosea*, are the most common mycoinsecticides used for the various groups of pests in different crop ecosystems. Fungal pathogens have certain advantages in pest control programs over other insect pathogens such as bacteria and viruses. Mass production techniques of fungi are much simpler, easier, and cheaper than those used for Bt and NPVs. Fungi, unlike bacteria or viruses, directly infect their hosts through the cuticle and do not require ingestion for infection. By this way also the sucking insects may be infected by these entomopathogenic fungi. In some insect orders, nymphal or larval stages are more often infected than the adult stages; in others the reverse may be the case. Entomopathogenic fungi of the genus *Aschersonia* are specific for whitefly and scale insects. They can be used as biological control agents against silverleaf whitefly, *Bemisia argentifolii*, and greenhouse whitefly, *Trialeurodes vaporariorum* (Meekes 2001). *Nomuraea rileyi* is a specific pest antagonist, infecting only lepidopteran larvae (Ignoffo 1981; Carruthers and Soper 1987).

9.6 Management of Nematodes

Plant-parasitic nematodes cause serious crop losses worldwide and are among the most important agricultural pests. Their management is more difficult than that of other pests because nematodes mostly inhabit soil and usually attack the underground parts of the plants. Nematodes in soil are subject to infections by bacteria and fungi. This creates the possibility of using soil microorganisms to control plant-parasitic nematodes. Bacteria are numerically the most abundant organisms in soil, and some of them, for example, fungi such as *Purpureocillium* (former *Paecilomyces*) *lilacinum*, *Metarhizium anisopliae*, and *Beauveria bassiana* or bacteria such as *Pasteuria penetrans* and other *Pasteuria* sp., *Pseudomonas*, and *Bacillus*, have shown great potential for biological control. Members of the genus *Pasteuria* are obligate, mycelial, endospore-forming bacterial parasites of plant-parasitic nematodes and water fleas (Tian et al. 2007). Aerobic endospore-forming bacteria (AEFB) (mainly *Bacillus* spp.) and *Pseudomonas* spp. are among the dominant populations in the rhizosphere that are able to antagonize nematodes. Organic amendments in the soil increase efficacy of different microbes. In general, efficacy against nematodes increases as per cent N in amendments increases and as C:N ratio decreases (Rodriguez-Kabana et al. 1987).

9.7 Management of Weeds

Microbial weed control represents an innovative means to manage troublesome weeds and utilize the naturally occurring biological herbicides produced by soil microorganisms. These compounds kill or hinder the growth of weeds so that beneficial plant species can gain a competitive advantage. The vast diversity of microorganisms in our environment is largely untapped, and the potential discovery and characterization of these microbial compounds represents an opportunity to complement chemical herbicides or to reduce soil erosion or degradation due to tillage for weed control (Stubbs and Kennedy 2012). *Puccinia chondrillina* has been used to control *Chondrilla juncea* (Barton 2004), and *Colletotrichum gloeosporioides* has been used for control of *Senna obtusifolia* (Boyette 2006). Use of biological herbicides requires a shift in thinking from the use of chemical herbicides, as biological control will most likely not eliminate a weed problem as quickly or as thoroughly as some herbicides on an annual basis. The goal is to inhibit the weed pest below an economically damaging threshold over a long time period, in order for beneficial species to gain a competitive advantage (Ghosheh 2005).

Table 9.2 Different types of agricultural substrates utilized for mass multiplication of biopesticides

Type of substrate	Substrates to be utilized
Agricultural wastes	Sugarcane bagasse and trash, groundnut trash, rice straw and husk, wheat bran and husk, bajra husk, dry weed, banana pseudostem, dried banana leaves, dried mango leaves, chickpea husk, kitchen waste, etc.
Organic matters	Well-decomposed farm yard manure, seasoned press mud, poultry manure, neem, mustard, linseed, groundnut and castor cakes, fresh cow dung, vermicompost, etc.
Green manuring crops	*Gliricidia sepium*, *Pongamia glabra*, *Azadirachta indica*, *Peltophorum ferrugineum*, *Leucaena leucocephala*, *Calotropis gigantea*, *Cassia fistula*, *Sesbania grandiflora*, etc.

9.8 Waste Management

Waste is any substance, solution, mixture, or article of which no direct use is envisaged but which is transported for reprocessing, dumping, elimination by incineration, or other methods of disposal. Substantial amounts of wastes generated from various sources, including domestic establishments, accumulated in natural environment are an excellent site for colonization and multiplication of microorganisms including fungi. Preparation of value-added compost having rich inoculums of *Trichoderma* spp. offers triple advantages of waste management, recycling of organic carbon, and preparation of cost-effective value-added product (Sabalpara 2014). Many different types of agricultural substrates are utilized for the mass multiplication of biopesticides and preparation of their value-added product (Table 9.2).

9.9 Model Methods for the Mass Multiplication of Bioagents

9.9.1 Multiplication of T. viride in Seasoned Press Mud

A model protocol for the mass multiplication of *T. viride* in seasoned press mud at the farm level was developed and popularized in south Gujarat. The method involved uniformly mixing 9-day-old culture of T. *viride* prepared in potato dextrose broth into 120 kg press mud. Water was sprinkled intermittently to keep it moistened. This was covered by gunny bags to permit air movement and trap moisture under shade. Within 25 days, nucleus culture for further multiplication becomes ready. The same was added to 8 tons of press mud, mixed thoroughly, and incubated for 8 days under shade condition before applied in the field. In this way, 8000 times more inoculum was added in the soil than the recommended doses of biopesticides, which rapidly established showing quick and visible effect. Similarly, other substances could also be effectively used for the multiplication of different bioagents at the mass level.

9.9.2 Multiplication of *Pseudomonas* sp. in Coconut Water

Coconut water is most commonly available in the tropics and is a very good source of nutrients for most bioagents. For fast multiplication of microorganisms, particularly *Pseudomonas* spp., a raw coconut fruit is selected, and the stem-end portion is cut. The cut surface is sterilized with 0.1% sodium hypochlorite solution. In the same volume a 0.1 ml *Pseudomonas* sp. culture (2×10^9 cfu/ml) inoculum is injected and incubated for 24 h, after which the preparation becomes ready and can be applied in the field by diluting appropriately. Final quantity depends on the amount of water present in the coconut. Generally, one fruit is mixed in 200 kg organic matter and incubated to be applied in one acre area.

9.9.3 Multiplication of *T. viride* in Sugarcane Trash

To prepare value-added product from the sugarcane trash, a pit or trench of convenient width and length near the boundary of sugarcane field is prepared. One kg culture (*T. viride*), 7.5 kg urea, and 50–75 kg of fresh cow dung per ton of trash are taken. Frequent watering must be done for maintaining the moisture content. The compost becomes ready within 10–12 weeks. This trash compost can be prepared by both pit and heap methods. Trash compost has the nutrient content of 0.8% N, 0.25% P, and 0.7% K with C:N ratio of 22:1. Trash can be also composted along with press mud.

9.9.4 Multiplication of Biopesticides in Paddy Straw

Goyal and Sindhu (2011) suggested a methodology for multiplication of biopesticides in paddy straw. For the same, a pit of 1 m × 1 m × 2 m is prepared in the shade. The pit is filled with 1 ton of unchopped paddy straw soaked in 0.1% urea solution. Cultures of *Aspergillus awamori*, *Paecilomyces fusisporous*, and *Trichoderma viride* (100 g each) are mixed in 1 L water containing 0.1% urea, and the same is mixed in paddy straw. Optimum watering is done to maintain 60% moisture and turning at 15, 30, and 60 days. Enriched compost becomes ready after 90 days for field application.

9.9.5 Mass Multiplication of *Trichoderma* sp. on Banana Waste

Balasubramanian et al. (2008) proposed the mass multiplication protocol of *Trichoderma* sp. on banana waste. For the same banana waste, urea, rock phosphate, culture of *Bacillus polymyxa*, *P. sajor caju*, and *T. viride* are used. A pit of different banana waste, namely, sheath, pseudostem, and core, is chopped in the length of 5–8 cm. A pit is prepared, and different ingredients are placed in five different layers. Each layer contains 1 ton banana waste, 5 kg urea, 125 kg rock phosphate, and 1 L broth culture of *Bacillus polymyxa*, *P. sajor caju*, and *T. viride*. Five different layers are prepared similarly and mixed thoroughly. Banana waste is decomposed within 45 days, and enriched culture is made available for field use.

9.9.6 Multiplication of *Pseudomonas* spp. in Sugarcane Juice and Banana Pseudostem Sap

Mass multiplication of *Pseudomonas* spp. may be done on sugarcane juice and banana pseudostem. For this, 10 L sugarcane/banana pseudostem sap is inoculated with 10 ml culture of *Pseudomonas* sp. and incubated for 3 days for maturation. Then the same is mixed in another 990 ml matured juice and incubated for 2 days before it is applied in the field. This can be stored at room temperature (up to 6 months).

References

Agrios, G. N. (2005). *Plant pathology* (5th ed.). Burlington: Elsevier Academic.

Alam, S., Khalil, S., Ayub, N., & Rashid, M. (2002). *In vitro* solubilization of inorganic phosphate by Phosphate Solubilizing Microorganism (PSM) from maize rhizosphere. *International Journal of Agriculture and Biology, 4*, 454–458.

Anjum, M. A., Sajjad, M. R., Akhtar, N., Qureshi, M. A., Iqbal, A., Jami, A. R., & Hasan, M. (2007). Response of cotton to Plant Growth Promoting Rhizobacteria (PGPR) inoculation under different levels of nitrogen. *Journal of Agricultural Research, 45*, 135–143.

Anonymous. (2017a). http://www.un.org/sustainabledevelopment/blog/2015/07/un-projects-world-population-to-reach-8-5-billion-by-2030-driven-by-growth-in-developing-countries/ website. Accessed 30 Oct 2017.

Anonymous. (2017b). http://www.toxipedia.org/display/toxipedia/Effects+of+Pesticides+on+Human+Healthwebsite. Accessed 30 Oct 2017.

Balasubramanian, C., Udayasoorian, P., Prabhu, C., & Kumar, G. S. (2008). Enriched compost for yield and quality enhancement in sugarcane. *Journal of Ecobiology, 22*, 173–176.

Barton, J. (2004). How good are we at predicting the field host-range of fungal pathogens used for classical biological control of weeds? *Biological control: Theory and Applications in Pest Management, 31*(1), 99–122.

Bouderau, M. A., & Andrews, J. H. (1987). Factors influencing antagonism of *Chaetomium globosum* to *Venturia inaequalis:* A case study in failed biocontrol. *Phytopathohology, 77*, 1470–1475.

Boyette, C. (2006). Adjuvants enhance the biological control potential of an isolate of *Colletotrichum gloeosporioides* for biological control of sicklepod (*Senna obtusifolia*). *Biocontrol Science and Technology, 16*(9–10), 1057–1066.

Burr, T. J., Caesar, A. M., & Schrolh, N. (1984). Beneficial plant bacteria: Critical review. *Plant Science, 2*, 1–20.

Carisse, O., Philion, V., Rolland, D., & Bernier, J. (2000). Effect of fall application of fungal antagonists on spring ascospore production of the apple scab pathogen, *Venturia inaequalis*. *Phytopathology, 90*, 31–37.

Carruthers, R. I., & Soper, R. S. (1987). Fungal diseases. In J. R. Fuxa & Y. Tanada (Eds.), *Epizootiology of insect diseases* (pp. 357–416). New York: Wiley.
Carsky, R. J., Becker, M., & Hauser, S. (2001). Mucuna cover crop fallow systems: Potential and limitations, In G. Tian, F. Ishida, & Keatinge (Eds.), *Sustaining soil fertility in West Africa*, (Special Publication No 58). Madison: Soil Science Society of America.
Chandra, K., & Greep, S. (2006). *Potash mobilizing bacteria Frateuria aurantia, Regional Centre of Organic Farming*. Bangalore: Government of India.
Egamberdiyeva, D. (2007). The effect of plant growth promoting bacteria on growth and nutrient uptake of maize in two different soils. *Applied Soil Ecology, 36*, 184–189.
Falk, S. P., Gadoury, D. M., Pearson, R. C., & Seem, R. C. (1995). Partial control of grape powdery mildew by the mycoparasite *Ampelomyces quisqualis*. *Plant Disease, 79*, 483–490.
Garrett, S. D. (1958). Inoculums potential as a factor limiting lethal action by *Trichoderma viride* Fr. On *Armillaria mellea* (Fr.) Quelet. *Transactions of the British Mycological Society, 41*, 157–164.
Ghosheh, H. (2005). Constraints in implementing biological weed control: A review. *Weed biology and management, 5*(3), 83–92.
Goyal, S., & Sindhu, S. S. (2011). Composting of rice straw using different inocula and analysis of compost quality. *Microbiology Journal, 1*, 126–138.
Gupta, A. K. (2004). *The complete technology book on biofertilizers and organic farming*. New Delhi: National Institute of Industrial Research.
Hameeda, B., Rupela, O., Reddy, G., & Satyavani, K. (2006). Application of plant growth-promoting bacteria associated with composts and macrofauna for growth promotion of Pearl millet (*Pennisetum glaucum* L). *Biology and Fertility of Soils, 43*, 221–227.
He, D., Zheng, X. D., Yin, Y. M., Sun, P., & Zhang, H. Y. (2003). Yeast application for controlling apple postharvest diseases associated with *Penicillium expansum*. *Botanical bulletin of Academia Sinica, 44*, 211–216.
Ignoffo, C. M. (1981). The fungus *Nomuraea rileyi* as a microbial insecticide. In H. D. Burges (Ed.), *Microbial control of pests and plant diseases* (pp. 513–538). New York: Academic.
Ippolito, A., Ghaouth, A. E., Wilson, C. L., & Wisniewski, M. (2000). Control of postharvest decay of apple fruit by *Aureobasidium pullulans* and induction of defense responses. *Postharvest Biology and Technology, 19*, 265–272.
Khunt, M. D., Solanki, V. A., Sabalpara, A. N., & Mahatma, L. (2014). Role of biofertilizers in plant nutrient management under present scenario. In U. C. Bhale (Ed.), *Major constraints and verdict of crop productivity*. New Delhi: Astral International Private Limited.
Lah, K. (2011) Effects of pesticides on human health. In Toxipedia. Available from http://www.toxipedia.org/display/toxipedia/Effects+of+Pesticides+on+Human+Health.
Mahatma, L., Makwana, K. V., & Sabalpara, A. N. (2016a). Enhancement of sugarcane production and productivity by the biofertilizers with graded chemical fertilizers. *Indian Journal of Sugarcane Technology, 31*(01), 6–9.
Mahatma, L., Naik, B. M., Mehta, B. P., Solanky, K. U., Chaudhary, P. P., & Sabalpara, A. N. (2016b). Field efficacy of different isolates of *Azotobacter croococum* for improving the yield of finger millet (*Eleusine coracana* (L.) Gaertn). *World Journal of Pharmaceutical and Life Sciences, 2*(5), 285–291.
Mahmood, I., Imadi, S. R., Shazadi, K., Gul, A., & Hakeem, K. R. (2016). Effects of pesticides on environment. In K. R. Hakeem et al. (Eds.), *Plant, soil and microbes*. Cham: Springer International Publishing. https://doi.org/10.1007/978-3-319-27455-3_13.
Meekes, E. T. M. (2001). Entomopathogenic fungi against whitefly: Tritrophic interactions between *Aschersonia* species, *Trialeurodes vaporariorum* and *Bemisia argentifolii*, and glasshouse crops. ISBN: 90-5808-443-4.
Mehrotra, R. S. (1993). *Plant pathology*. New Delhi: Tata McGraw Hill Publishing Company.
Millard, W. A., & Taylor, C. B. (1927). Antagonism of microorganisms as the controlling factor in the inhibition of scab by green manuring. *Annals of Applied Biology, 14*, 202–215.

Mishra, D. J., Singh, R., Mishra, U. K., & Shahi, S. K. (2013). Role of bio-fertilizer in organic agriculture: A review. *Research Journal of Recent Sciences, 2*, 39–41.

Mohammadi, K., Ghalavand, A., Aghaalikhani, M., Heidari, G. R., & Sohrabi, Y. (2010). Introducing the sustainable soil fertility system for chickpea (*Cicer arietinum* L.). *African Journal of Biotechnology, 10*(32), 6011–6020.

Patel, A. H., & Mahatma, L. (2017). Enhancement of nodulation efficiency of Mungbean rhizobia. *International Journal of Current Microbiology and Applied Sciences, 6*(10), 2581–2585. https://doi.org/10.20546/ijcmas.2017.610.303.

Ramanujam, B., Rangeshwaran, R., Sivakmar, G., Mohan, M., & Yandigeri, M. S. (2014). Management of insect pests by microorganisms. *Proceedings of the Indian National Science Academy, 80*(2), 455–471.

Rodriguez-Kabana, R., Jones G. M., & Chet, I. (1987). Biological control of nematodes: Soil amendments and microbial antagonists *plant and soil* Vol. 100, No. 1/3. In *Proceedings of international symposium: Plant and soil: Interfaces and interactions Wageningen*, The Netherlands 6–8 August 1986, (pp. 237–247).

Sabalpara A. N. (2014). Mass multiplication of biopesticides at farm level. In Presidential address *delivered at the 35th ISMPP annual conference, Dr PDKV Akola, MS, 8 Jan 2014.*

Sevilla, M., & Kennedy, C. (2000). Colonization of rice and other cereals by *Acetobacter diazotrophicus*, an endophyte of sugarcane. In J. K. Ladha & P. M. Reddy (Eds.), *The quest for nitrogen fixation in rice. Proceedings of the third working group meeting on assessing opportunities for nitrogen fixation in rice*. Makati City: International Rice Research Institute.

Shahzad, S. M., Khalid, A., Arshad, M., Khalid, M., & Mehboob, I. (2008). Integrated use of plant growth promoting bacteria and P-enriched compost for improving growth, yield and nodulating of Chickpea. *Pakistan Journal of Botany, 40*, 1735–1441.

Soytong, K., & Ratanacherdchai, K. (2005). Application of mycofungicide to control late blight of potato. *Journal of Agricultural Technology, 1*(1), 19–32.

Spencer, D. M., & Parasitic, P. T. (1981). Effects of *Verticillium lecanii* on two rust fungi. *Transactions of the British Mycological Society, 77*, 535–542.

Stubbs, T. L., & Kennedy, A. C. (2012). *Microbial weed control and microbial herbicides*. In R. Alvarez-Fernandez (Ed.), *Herbicides environmental impact studies and management approaches* (pp. 135–166). Croatia: InTech, Rijeka.

Sundara, B., Natarajan, V., & Hari, K. (2002). Influence of phosphorus solubilizing bacteria on the changes in soil available phosphorus and sugarcane yields. *Field Crops Research, 77*, 43–49.

Sztejnberg, A., Galper, S., Mazar, S., & Lisker, N. (1989). *Ampelomyces quisqualis* for biological and integrated control of powdery mildews in Israel. *Journal of Phytopathology, 124*, 285–295.

Tandel, M. H., & Mahatma, L. (2016). Standardization of mass multiplication protocol for arbuscular mycorrhizal fungi isolated from the South Gujarat. *Advances in Life Sciences, 5*(21), 9681–9685.

Tian, B., Yang, J., & Zhang, K. Q. (2007). Bacteria used in the biological control of plant-parasitic nematodes: Populations, mechanisms of action, and future prospects. *FEMS Microbiology Ecology, 61*, 197–213.

Tilak, K. V. B. R., Ranganayaki, N., Pal, K. K., De, R., Saxena, A. K., Nautiyal, C. S., Mittal, S., Tripathi, A. K., & Johri, B. N. (2005). Diversity of plant growth and soil health supporting bacteria. *Current Science, 89*, 136–150. toxipedia.org/display/toxipedia/Effects+of+Pesticides +on+Human+Health. Accessed 16 May 2016.

Vose, P. B., & Ruschel, A. P. (1981). *Associative N2 –Fixation*. Boca Raton: CRC press.

Weller, D. M. (1988). Biological control of soil borne plant pathogens in the rhizosphere with bacteria. *Annual Review of Phytopathology, 26*, 379–407.

Whitelaw, M. A. (2001). Growth promotion of plants inoculated with phosphate solubilizing fungi. *Advances in Agronomy, 69*, 99–151.

Zahran, H. H. (1999). *Rhizobium*-legume Symbiosis and nitrogen fixation under severe conditions and in an arid climate. *Microbiology and Molecular Biology Reviews, 63*, 968–989.

Chapter 10
Exploiting Chemical Ecology for Developing Novel Integrated Pest Management Strategies for Africa

Zeyaur R. Khan, Charles A. O. Midega, Jimmy Pittchar, and John A. Pickett

Abstract Push-pull, a novel approach for integrated management of insect pests, weed and soil fertility, was developed through the exploitation of chemical ecology and agro-biodiversity to address agricultural constraints facing millions of resource-poor African farmers. The technology was developed by selecting appropriate plants that naturally emit signalling chemicals (semiochemicals) and influence plant-plant and insect-plant interactions. Plants highly attractive for egg laying by lepidopteran cereal stemborer pests were selected and employed as trap crops, to draw pests away from the main cereal crops. Among these, *Pennisetum purpureum* produced significantly higher levels of volatile cues (stimuli), used by gravid stem borer females to locate host plants, than maize (*Zea mays*) or sorghum (*Sorghum bicolor*). Despite its attractiveness to stemborer moths, *P. purpureum* supported minimal survival of the pests' immature stages. Plants that repelled stem borer moths, notably *Melinis minutiflora* and forage legumes in the genus *Desmodium*, were selected as intercrops, which also attracted natural enemies of the pests through emission of (E)-β-ocimene and (E)-4,8-dimethyl-1,3,7-nonatriene. *Desmodium* intercrop suppressed parasitic weed, *Striga hermonthica*, through an allelopathic mechanism. Their root exudates contain novel flavonoid compounds which stimulate suicidal germination of *S. hermonthica* seeds and dramatically inhibit its attachment to the host roots. We identified and selected new drought- and temperature-tolerant trap [*Brachiaria* (*B. brizantha* × *B. ruziziensis*) cv. *mulato*] and intercrop plants (*Desmodium*, e.g. *D. intortum*) suitable for drier agroecologies. The new trap and intercrop plants also have appropriate chemistry in controlling stemborers, a new invasive pest, fall armyworms and parasitic striga weeds. Opportunities for semiochemical delivery by companion plants, including plant-plant signalling and early herbivory alert, are explored for developing future smart integrated pest management (IPM) strategies.

Z. R. Khan (✉) · C. A. O. Midega · J. Pittchar
International Centre of Insect Physiology and Ecology (icipe), Nairobi, Kenya
e-mail: zkhan@icipe.org; cmidega@icipe.org; jpittchar@icipe.org

J. A. Pickett
Cardiff University, School of Chemistry, Cardiff, UK
e-mail: PickettJ4@cardiff.ac.uk

R. Peshin, A. K. Dhawan (eds.), *Natural Resource Management: Ecological Perspectives*, Sustainability in Plant and Crop Protection,
https://doi.org/10.1007/978-3-319-99768-1_10

Keywords Biodiversity and basic science · Chemical ecology · Push-pull technology · Novel pest management · Fall armyworm · Climate change

Abbreviations

BBSRC	Biotechnology and Biological Sciences Research Council
BIRE	Biological Interactions in The Root Environment
DFID	Department for International Development
FAW	Fall armyworm
GC-EAG	Coupled gas chromatography–electroantennography
GDP	Gross domestic product
HIPVs	Herbivore-induced plant volatiles
IPM	Integrated pest management
OPV	Open pollinated varieties
SDC	Swiss Agency for Development and Cooperation
SDG	Sustainable Development Goals
SSA	Sub-Saharan Africa

10.1 Introduction

Agricultural development in sub-Saharan Africa (SSA) is constrained by biophysical factors, low capacities and institutional and policy bottlenecks, while development assistance to agriculture has declined to only 4% of public expenditure (World Bank 2008; IFAD 2011). Although agriculture contributes to over 25% of the gross domestic product (GDP) and more than half of export earnings, the per capita food production in Africa has declined over the past two decades. Rural Africa is characterized by continuing stagnation and often deterioration, low crop and livestock productivity, low farm incomes and the rising vulnerability of resource-poor smallholder farmers, who constitute the majority and whose basic source of livelihood is agriculture (World Bank 2008). More than 265 million rural poors face higher levels of hunger and poverty. Longer-term development challenges include dependence on a few primary commodities, poor human capacity, increasing migration to urban areas, low employment, especially of the youth and women, and climate changes (World Bank 2007).

10.2 Low Agricultural Productivity and Other Compounding Effects in SSA

Several factors interrelated with the low productivity remain a concern, not least the environmental degradation of the natural resource bases, the human population pressure on productive resources, leading to food insecurity, undernutrition, poverty, high morbidity and human migrations. Intensive land use, without sustainable investments in soil fertility enhancement, has transformed most of the natural landscape characterized by a decline in the overall quality of soil and vegetation, reducing agricultural yields. Soil erosion further leads to land degradation. Land clearance for agriculture is causing serious deforestation (World Resources Institute 2007), and there is progressive depletion of soil nutrients, particularly of nitrogen and soil organic carbon (Solomon et al. 2010). African soils are poor in organic matter due to continuous cropping and poor farming practices and in need of an agronomic innovation that continuously improves soil health (Sanchez 2002; Oswald 2005; Rodenburg et al. 2005).

This situation, driven further by a significantly increased demand for food because of population growth, urbanization and changing food consumption patterns, raises the need for sustainable intensification of production systems (Pretty et al. 2011), which includes both the management of pests and soil health and the reduction of land degradation. Effects of climate changes are expected to have greater impacts on sustainable agricultural development in SSA with production constraints expected to increase during the next few decades, as agriculture intensifies to meet the extra food demand by the growing population. The resource-constrained smallholder farmers living in arid and semiarid regions, who practise mixed crop–livestock systems, are particularly badly affected. SSA is projected to have more than 500 million food insecure people by 2020.

Cereals remain the major food and cash crops for the majority of the resource-poor smallholders in SSA, grown alongside livestock in mixed farming systems. Production of main cereals, maize, *Zea mays*, and sorghum, *Sorghum bicolor*, is severely reduced by a complex of biotic constraints, namely, stemborer pest complexes, parasitic striga weeds and more recently fall armyworm invasion, as well as abiotic factors, such as water stress and degraded soils. The most economically significant insect pests are Lepidopteran stemborers of the families Noctuidae and Crambidae, e.g. the indigenous *Busseola fusca* (Noctuidae) and the invasive *Chilo partellus* (Crambidae), as well as the fall armyworm (FAW) *Spodoptera frugiperda* (Lepidoptera: Noctuidae) (Midega et al. 2018). Over 20 stemborer species attack cultivated gramineae in SSA (Kfir et al. 2002), causing between 30% and 80% yield losses of cereal crops (Kfir et al. 2002).

10.2.1 A New Challenge: The Fall Armyworm Invasion in Africa

In addition, losses of up to 100% to maize and other crops have been reported by smallholder farmers (<2 ha), due to the recent FAW invasion of Africa (Midega et al. 2018). FAW is a native pest in tropical and sub-tropical Americas (Todd and Poole 1980). More than 28 countries in Africa are already affected, and the pest is extending rapidly to other countries (Goergen et al. 2016; Cock et al. 2017). The pest adds to the diversity of Lepidopterans of cereal crops and signals increased about its negative impacts on agricultural production and food security in Africa. FAW invasion has adverse economic impacts on smallholder farmers as it directly increases their capital costs through increased labour and inputs needed, specialised knowledge required to deal with the pest, the inability of agricultural systems to respond to sudden invasions and the overall effect of higher production costs and yield losses on household incomes.

FAW is a polyphagous pest with a wide host range of economically important grasses such as maize, rice, sorghum and sugarcane, as well as other crops, including cabbage, beet, peanut, soybean, alfalfa, onion, cotton, pasture grasses, millet, tomato, potato and cotton. It is prolific, ovipositing egg masses in batches of 100–200 eggs (Sparks 1979; Johnson 1987; CABI 2017). Eggs hatch in 2 to 4 days under optimum temperatures. Adult moths can survive 2 to 3 weeks during which females mate multiple times, producing up to 1000 eggs each. Their larval stages have six instars, the first of which is most voracious, consuming the most plant material. The larvae eat different parts of the plant, mainly young whorls, ears and tassels, depending on the larval age, stage of development and the host plant type. On maize, young larvae eat leaves at night, leaving a 'window pane' effect, and hide in the plant funnel during the day. Larval feeding often kills the growing point, causing 'dead heart'. At the reproductive stage of maize, the larvae also attack reproductive organs, feeding on tassels and/or boring into the ears (Midega et al. 2018). The larger caterpillars also act as cutworms by entirely sectioning the stem base of maize plantlets. As larvae grow older, they hide inside the funnel, limiting the effect of pesticide applications and natural enemies. FAW is a sporadic and long-distance migratory pest whose adult moths are able to fly over 100 km in a single night (Goergen et al. 2016). It has a significant economic impact, up to 100% crop loss (CABI 2017), finding most farming communities ill prepared to manage the invasive pest.

Single control methods are relatively costly, unsustainable or ineffective. A number of control methods have been tried, including:

i) Application of pesticides (chemical control)
ii) Use of microbial organisms that attack FAW in its native range, for example, *Beauveria bassiana* and *Spodoptera frugiperda* multiple *nucleopolyhedrovirus* (SfMNPV)
iii) Use of predatory insects and parasitic wasps (parasitoids)
iv) Use of genetically modified crops containing Bt genes that are resistant to FAW
v) Mass trapping of male moths using pheromones, preventing them from mating
vi) Integrated pest management (IPM) – a combination of methods minimizing pesticides use (CABI 2017)

Past management of FAW has mainly relied on application of synthetic chemical pesticides, with dismal results, depending on farmers' knowledge, consistency of use, purchasing power and choice of pesticide products (Midega et al. 2018). Although chemical insecticides have been shown to provide control of the pest (Young 1979), cases of resistance to some key a.i. have been reported (Yu 1992; Al-Sarar et al. 2006). Dispersion of FAW larvae lower into the maize plant canopy keeps them out of reach of topical insecticide applications (Cook et al. 2004, Midega et al. 2018). Moreover, pesticides are not affordable for the vast majority of smallholder famers in Africa, and their incorrect use has resulted in poisoning farmers and the environment, as well.

10.2.2 The Need for Integrated Pest Management (IPM) of Stemborers and Fall Armyworm

Suitable and cost-effective integrated pest management (IPM) strategies need therefore to be developed, specifically for smallholder farmers in Africa (Midega et al. 2018). However, most available literature on FAW control relates to agricultural systems in its native Americas, which differ from those in Africa. Therefore, control methods that have been effective in FAW native habitats may not be effective in Africa. However, evidence from Latin America indicates that an IPM approach may be necessary in which pesticides use is minimised and alternative approaches, such as exploiting the pest's natural enemies and crop monitoring, be consistently applied (CABI 2017). Climatic conditions in Africa support prolific reproduction of FAW, which is expected to result in increasingly severe damage to crops (Goergen et al. 2016), as the invasive pest is likely to have few natural enemies. Conventional control methods have limited effectiveness, as explained above, notably the difficultly in application of pesticides and development of resistance by the pest to some insecticides and transgenic technologies such as Bt-maize. Therefore, an integrated management approach for fall armyworm that fits within the mixed cropping nature of the African farming systems is necessary, for the resource constrained farmers. Thus, an IPM technology like the push-pull (www.push-pull.net) that exploits natural processes, including the use of natural enemies, appears as the most promising (see paragraph 3).

10.2.3 Parasitic Striga Weeds

Parasitic weeds in the genus *Striga* (Scrophulariaceae), commonly known as striga, further severely constrain cereal production in SSA (Oswald et al. 2001; Khan et al. 2014). There are at least 22 species of *Striga* of which *Striga hermonthica* and *Striga asiatica* have been identified as the most socio-economically important in cereal cultivation, in much of SSA (Gressel et al. 2004; Gethi et al. 2005). Striga infestation weakens host cereal plants by competing for and absorbing its supply of moisture, photosynthates and minerals (Tenebe and Kamara 2002). The weed quickly adapts to its environment (Bebawi and Metwali 1991) and germinates in response to specific chemical cues present in root exudates of its hosts or certain non-host plants (Yoder 1999, Parker and Riches 1993). Striga roots mesh with the host root system and also injects phytotoxins into roots (Frost et al. 1997, Gurney et al. 1999; Gurney et al. 2006). Striga infestations cause significant reductions in host plant height, biomass and grain yields (Gurney et al. 1999). Striga weed parasitization causes up to 100% cereal yield losses. Conditions such as degraded environments, low soil fertility, higher soil temperature and low rainfall (Gurney et al. 2006) exacerbate striga attacks in subsistence farming systems.

10.2.4 Unsustainable Pest and Weed Control Methods

Research and extension institutions in Africa often recommend the use of insecticides and herbicides in management of stemborers, FAW and striga, respectively, with dismal results. This is complicated by the resource-constrained nature of subsistence farmers in the region, impeding their ability to afford expensive chemicals. Insecticide use in Africa is limited, largely due to shortage of information, inaccessibility of appropriate and effective products and associated high costs (Midega et al. 2018). Transgenic plant technologies such as Bt maize have been tried (Frizzas et al. 2014). However, development of field resistance by stemborers and FAW to transgenic crops has been documented, including resistance to Cry1F maize in Puerto Rico (Storer et al. 2010). Majority of smallholders therefore do not attempt to manage stemborers, FAW or striga and consequently suffer high grain yield losses and food insecurity (Chitere and Omolo 1993; Oswald 2005; Midega et al. 2018).

Sustainable management of pests, weeds and resource degradation therefore needs sustainable intensification that maximizes soil quality and crop productivity, adopting a systems approach (social, economic and environmental) to agricultural development and developing solutions based on integrated analyses of specific agro-ecosystem conditions and farmer practices (IAASTD 2009; Pinstrup-Andersen 2010). Significant and sustainable increases of productivity require a more holistic approach, reflecting the multifunctionality of agriculture, which integrate a variety of resource-conserving technologies and practices – i.e. IPM, integrated soil fertility management (ISFM) and livestock integration (Pretty et al. 2006). In order to manage the production constraints, approaches that are compatible with the polycultural nature of low-input farming systems in Africa need to be developed by understanding the intricate biological interactions within ecosystems, farm landscapes and socio-economic conditions of smallholder farmers. One such approach is the 'push-pull' technology which is compatible with African socio-economic conditions as it does not rely on high external inputs, but biological management of local bio-resources.

10.3 Development of the 'Push-Pull' Technology

The push-pull technological innovation was developed by the International Centre of Insect Physiology and Ecology (ICIPE) to significantly increase cereal and livestock productivity, by addressing the interrelated problems caused by the above biotic constraints (notably insect pests and weeds), soil and environmental degradation, lack of livestock fodder, loss of biodiversity, increasing temperatures and water stress, through improved management strategies (Khan et al. 2014; Midega et al. 2015b). The technology is a polycropping innovation that holistically combines resource-conserving principles of IPM and ISFM, by using natural processes and locally available bio-resources (Cook et al. 2007; Hassanali et al. 2008). It was developed for smallholder farming systems based on the traditional African diversified cereal-legume-fodder intercropping practice. In this method, the perennial intercrop maintains continuous soil cover and provides live mulching, thus conserving soil moisture and improving arthropod abundance and biodiversity as well as the food web of natural enemies of stemborers (Khan et al. 2002, 2006a; Midega et al. 2015a).

The technology effectively controls the major insect pests of cereals in SSA, i.e. lepidopteran stemborers and FAW (Midega et al. 2018), and the devastating parasitic striga weeds, both of which can cause total yield loss to cereals (Khan et al. 2014). The technology relies on understanding natural biochemical processes and their underlying chemical ecology, agro-biodiversity and plant-plant and insect-plant interactions (Cook et al. 2007; Khan et al. 2014; Pickett and Khan 2016). It deploys inter- and trap crops in a mixed cropping system (Khan et al. 2006b) which release behaviour-modifying stimuli (plant chemicals) to manipulate the distribution and abundance of stemborers and beneficial insects.

ICIPE initially conducted a series of studies on the behaviour and chemical ecology of plants, insect pests and their natural enemies in order to identify companion plants that would naturally manipulate the behaviour of insects, while delivering additional benefits to the farmers (Khan et al. 2014). A stimulo-deterrent cropping strategy, known as 'push-pull', was thus developed by exploiting phytochemicals released by two sets of carefully chosen companion plants. These consist of repellent and trap plants, intercropped in between and planted around the main cereal crops (Khan et al. 2010). The intercrops, (e.g. *Desmodium* spp.) emit semiochemicals that repel ('push') insect pests from the main crop while attracting their natural enemies. The trap plant [e.g. Napier grass, *Pennisetum purpureum* or *Brachiaria* (*B. brizantha* × *B. ruziziensis*) cv. Mulato], grown as a border crop, attracts ('pull') the insect pests away from the main crop (Cook et al. 2007; Khan et al. 2010). The *Desmodium* intercrop was found to effectively suppress parasitic striga weeds (Khan et al. 2002). In a long-term research, ICIPE later established that the underpinning mechanism was an allelopathic effect of *Desmodium* root exudates which caused abortive germination of Striga seeds, thus preventing root-to-root attachment to the host plant and progressively depleting its seed bank. Furthermore, this strategy was found to improve soil health by fixing atmospheric nitrogen, improving carbon sequestration and conserving soil moisture (Midega 2015a; CABI 2017).

10.3.1 Exploiting Semiochemistry of Companion Plants for Stemborer Control

Cereal stemborers are polyphagous herbivores that feed on a range of host plants, including uncultivated grasses that serve as reservoirs (Khan et al. 1997a). Herbivorous arthropods are attracted to their host plants through olfactory detection of specific attractive semiochemicals which mediate changes in their behaviour or development (Nordlund and Lewis 1976; Dicke and Sabelis 1988) or to specific ratios of semiochemicals (Bruce et al. 2005) emitted by host plants or organisms, as well. Semiochemicals or their blends emitted by non-host plants conversely cause avoidance of those chemical compounds by the insects (Hardie et al. 1994).

Development of 'push-pull' was based on field surveys of more than 500 grass species (Poaceae, Cyperaceae and Typhaceae), as well as some leguminous crops in different agroecological zones in East Africa. Attractive and antagonistic plant species for use as trap and repellent intercrops, respectively, were thus identified. Potential trap crops were selected from plant species that were attractive but did not support development of the larval stages of the stemborer pest. Napier grass, *Pennisetum purpureum*, a forage crop, was found to produce significantly higher amounts of volatile organic compounds (VOCs) that attracted gravid stemborer females more than maize or sorghum (Birkett et al. 2006). The production of VOCs by Napier grass was found to be 100-fold in the first hour of nightfall (scotophase) (Chamberlain et al. 2006), the period during which stemborer moths seek host plants for oviposition (Päts 1991). This strategy influences the oviposition preference by gravid stemborer females. However, about 80% of the stemborer larvae did not survive (Khan et al. 2006a; 2007), as Napier grass produced a sticky sap which mechanically trapped them, exposing them to natural elements and enemies and causing high mortality. This resulted in significant reductions in stemborer infestation in maize, increasing yields by up to 1.5 t ha^{-1} (Khan et al. 2000).

Conversely, the fodder intercrops, *Melinis minutiflora* and legumes in the genus *Desmodium* (commonly known as desmodium) similarly repelled stemborer moths and provided effective control of these pests in intercrops with maize (Khan et al. 1997b, 2000). Both the *Desmodium* genus (silverleaf, *D. uncinatum*, and greenleaf, *D. intortum*) were found to produce repellent volatile semiochemicals during the damaging stage to plants of herbivorous insects, which repelled the stemborer moths (Khan et al. 2001). Also the intercropping of maize with silverleaf desmodium, *D. uncinatum*, was serendipitously found to suppress the emergence of the parasitic striga weed *S. hermonthica*, through allelopathic mechanism of the *Desmodium* root exudates, produced independently of the presence of striga. Blends of secondary metabolites produced by activities in the root exudates of *Desmodium* spp., concomitantly with striga seed germination stimulatory isoflavanones, were found to directly interfere with *Striga* parasitism. This combined effect reduced the striga seed bank in situ through efficient suicidal germination even in the proximity of graminaceous host plants (Tsanuo et al. 2003). Additional benefits were provided by *Desmodium* spp. in terms of increased availability of nitrogen and soil shading.

Controlling both stemborers and striga resulted in significant yield increases, from an average of 1–3.5 t ha^{-1} (Khan et al. 2006a; 2008). Other *Desmodium* spp. similarly controlled both stemborers and striga weeds (Khan et al. 2006b). *Desmodium* thus became the intercrop of choice for most smallholder farmers in Eastern Africa, where both constraints affect cereal production. Other African adapted *Desmodium* spp. were also tested and found to have similar effects on stemborers and striga (Khan et al. 2006b). They have been evaluated to be incorporated as intercrops in adapted push-pull systems with maize, sorghum and millets (Midega et al. 2015b). The effectiveness of push-pull technology in pest control delivered through both direct and indirect effects is described in Fig. 10.1.

Innate, direct plant defence systems are deployed by plants to protect themselves from pests. These include production of toxins, digestion inhibitors and semiochemicals repellent to herbivorous insects (Kessler and Baldwin 2001). The trap and repellent plants exert direct effects on both adult and developing stages of cereal stemborers (Midega et al. 2015a). The VOCs mediate the behaviour of gravid stemborer moths, which are diverted to the border crop thus protecting the main crop from the pest. The main physiologically active compounds responsible for attractiveness of the trap crop to the gravid stemborer moths, i.e. hexanal, (E)-2-hexenal, (Z)-3-hexen-1-ol and (Z)-3-hexen-1-yl acetate, were identified using coupled gas chromatography–electroantennography (GC–EAG) on the antennae of stemborers (Khan et al. 2000).

Molasses grass was discovered to emit active compounds similar to those that are produced by plants damaged by herbivorous insects, known as herbivore-induced plant volatiles (HIPVs) (Turlings et al. 1990a, b, 1995). Identified compounds included (E)-ocimene and (E)-4,8-dimethyl-1,3,7-nonatriene (DMNT). These compounds are associated with a high level of stemborer colonisation and also attract parasitoids, natural enemies of stemborers, and are responsible for

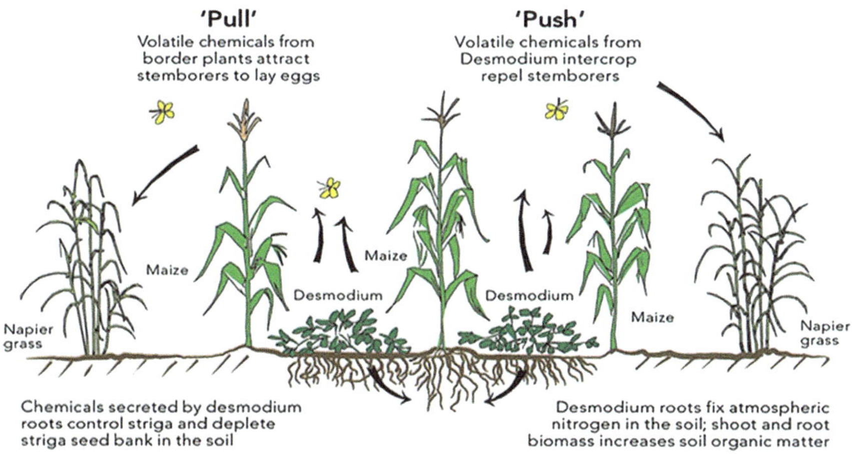

Fig. 10.1 How the push-pull system works: stemborer moths are repelled by intercrop volatiles while attracted to trap crop volatiles. Root exudates from the desmodium intercrop cause suicidal germination of striga and inhibit attachment to host cereal roots

repelling ovipositing stemborers (Khan et al. 2000). Molasses grass naturally produces high levels of HIPVs and therefore is not preferred as host. It typically produces semiochemicals emitted by a highly infested maize plant.

Similarly, *Desmodium* was also found to produce (E)-ocimene and DMNT, and large amounts of other sesquiterpenes, including alpha-cedrene (Khan et al. 2000), thus effectively repelling stemborer moths and attracting their natural enemies (Midega et al. 2009). Furthermore, a significant increase in parasitism of stemborer larvae by the indigenous parasitoid, *Cotesia sesamiae* (Hymenoptera:Braconidae), was observed in maize fields intercropped with molasses grass in western Kenya (Khan et al. 2000). It was caused by the same semiochemicals produced by molasses grass, which repelled female stemborer moths. Attraction of female parasitoid *C. sesamiae* to DMNT was confirmed in a Y-tube olfactometer bioassay, under field conditions (Khan et al. 1997b, 2000). Similarly, intercropping maize with *Desmodium* resulted in significant increase of stemborer larval and pupal parasitoid activity (Midega et al. 2014).

The push-pull cropping system seems to attract an abundance of generalist predators like ants, earwigs and spiders which prey on the stemborer immature stages (Midega et al. 2006), contributing to reduce the pest populations. The synchronized effects of the trap crops and repellent intercrops result in significant reductions in stemborer colonisation and mortality, reduced crop damage and significant improvements in grain yields. The same mechanism is thought to be responsible for reduction of FAW populations under the push-pull system (Midega et al. 2018). ICIPE is investigating the semiochemical mechanisms that prevent FAW pests from attacking crops, under the push-pull system.

In spite of Napier grass being preferred to maize for oviposition by gravid moths, it did not support the development of stemborer larvae (Khan et al. 2006a). Further investigations established that the high mortality of stemborer larvae on the grass was due to innate defence mechanism in the grass, which produces sticky sap upon injury inflicted by feeding larvae. The sap traps the young larvae and impedes their mobility, both directly exposing the larvae to the elements. Indirect mortality is caused by generalist natural enemies such as spiders that are often prevalent in Napier grass fields/strips (Midega et al. 2008). Moreover, Napier grass slows the development rates of larvae because of its poor nutritional qualities (Khan et al. 2006a).

The *Desmodium* intercrop controls the parasitic striga through a variety of mechanisms including increased availability of nitrogen, soil shading and allelopathic root exudation that happens independently of the presence of striga (Khan et al. 2002). Root exudates of different *Desmodium* legume species studied contain a complex array of plant secondary compounds that simultaneously stimulate germination and subsequently inhibit Striga radicle growth. Some of the identified compounds, including 4″,5″-dihydro-5,2′,4′-trihydroxy-5″-isopropenylfurano-(2″,3″;7,6)-isoflavanone, stimulate germination of striga seeds. At the same time, other compounds, including 4″,5″-dihydro-2′-methoxy-5,4′-dihydroxy-5″-isopropenylfurano-(2″,3″;7,6)-isoflavanone, inhibit the growth of the *Striga* root haustoria (Tsanuo et al. 2003). In maize–*Desmodium* intercrops under field conditions, both *Desmodium* and maize root exudates stimulate striga germination, but subsequent development of striga root haustoria was disrupted by a second group of compounds in the root exudates. This abortive germination is mediated by this combined allelopathic mechanism by which *Desmodium* prevents parasitism and continuously removes striga seed bank from the soil, in situ (Tsanuo et al. 2003). Multilocation field studies across several ecosystems have confirmed that *Desmodium* effectively control striga and result in significant yield increases in maize (Khan et al. 2008, 2014), sorghum (Khan et al. 2006b), finger millet (Midega et al. 2010) and upland rice (Pickett et al. 2010).

10.4 Climate Smart Push-Pull Technology

Climate change is anticipated to have far-reaching effects on sustainable agricultural development in SSA. Anticipated changes include increased frequency and intensity of plant abiotic stressors, e.g. unpredictable and erratic rainfall amounts and patterns, as well as flooding (Hoerling and Kumar 2003) and higher temperatures and droughts (Burke et al. 2009). Further and more serious land degradation, increased pest and weed pressure, increased incidences of crop failure and general increases in food and nutritional insecurity are expected for resource poor farmers, in many parts of SSA. There is therefore a need to build system adaptability to protect cropping systems from climate variability and to improve their resilience to climatic shocks. As majority of smallholder African farmers rely on rainfed agriculture, it became imperative to adapt the conventional push-pull system to withstand increasingly warmer and drier conditions. The trap and intercrops used in conventional push-pull were increasingly vulnerable to rainfall and temperature variations, as the initial system was developed for subhumid tropical conditions with an average rainfall of 800–1200 mm and moderate temperatures (15 to 30 °C).

Adaptation of the push-pull strategy for hotter and drier agroclimatic conditions has improved system resilience and adaptability needed to mitigate climate change effects (Khan et al. 2014). Using a series of studies under controlled and on-farm conditions, the push-pull technology has been adapted to drier agroecologies by selecting and incorporating new drought-tolerant trap (*Brachiaria* cv. mulato) and intercrop (drought-tolerant species of *Desmodium*, e.g. *D. intortum*) plants. These were tested and validated under farmers' conditions and incorporated into the push-pull technology.

The selection of new *Desmodium* intercrops was based on drought tolerance, appropriate chemistry to effectively control both stemborer and striga weeds, the ability to improve soil fertility and soil moisture retention as well as the ability to significantly improve yields of maize and sorghum (Khan et al. 2014, 2016; Midega et al. 2015b). *Brachiaria* cv. mulato II (*B. ruziziensis* × *B. decumbens* × *B. brizantha*) was the preferred trap plant in the climate adapted push-pull, because of its ability to control stemborers, high preference by farmers as livestock fodder and the commercial availability of its seed. In addition, drought stress significantly limited the relative attractiveness of *Brachiaria* cv mulato II to stemborer moths for oviposition

(Chidawanyika et al. 2014). Furthermore, drought stress only minimally alters secondary metabolism in the plant and does not significantly affect emission of key VOCs necessary for stemborer host location (Chidawanyika 2015). An important attribute is that *Brachiaria* spp. allow for only minimal survival of stemborer larvae (Midega et al. 2015b) and have a suitable characteristic of a border plant that supports populations of natural enemies within and out of the cereal cropping season.

The climate-adapted companion plants (both trap and repellent plants) also generate high-quality livestock fodder over long periods of drought. They also improve biodiversity, soil conservation and organic matter among other ecosystem services (Midega et al. 2015b). There is potential for further adaptation of the push-pull technology through incorporation of even more drought-tolerant African-adapted *Desmodium* species such as *D. incanum*, *D. repandum* and *D. ramosissimum*, which exhibit the right chemistry and ability to control the identified constraints and improve cereal crop yields (Hooper et al. 2015; Midega et al. 2017).

10.4.1 Climate-Adapted Push-Pull Effectively Controls Fall Armyworm

Multilocational field surveys show that the climate-adapted push-pull technology effectively controls the invasive FAW (Midega et al. 2018). The control mechanism involves the diversionary 'push and pull' tactics previously found to be effective against stemborers. Additional action by natural enemies, i.e. parasitic wasps and generalist predators, such as ants, earwigs and spiders, complements the stimulo-deterrent action to reduce FAW invasion of crops under push-pull. Studies have shown that the damage caused by fall armyworm is reduced by up to 100% with the technology, resulting in significant improvements in grain yields.

ICIPE evaluated the functionality of climate-adapted push-pull using drought-tolerant Greenleaf *D. intortum* and *Brachiaria* sp. as inter and border crops, respectively, in FAW management through direct field observations and farmers' perceptions in Kenya, Uganda and Tanzania. Data were collected from a diverse sample of farmers in different agroecologies on the number of FAW larvae on maize, the percentage of maize plants damaged by the larvae and maize grain yields. Similarly, farmers' perceptions of the impact of the technology on the pest were assessed. There were highly significant reductions in infestation by fall armyworm larvae and plant damage in climate-adapted push-pull, compared to maize monocrop plots. Reductions of 82.7% in the average number of larvae per plant and 86.7% in plant damage per plot were observed in climate-adapted push-pull, compared to maize monocrop fields. Similarly, maize grain yields were significantly higher, approximately 2.7 times, in the climate-adapted push-pull fields. Farmers rated the technology significantly effective in reducing FAW infestation and plant damage rates. These results demonstrate that the technology is also effective in controlling FAW and represents a solution that can be immediately deployed for management of the invasive pest.

10.5 New Opportunities in Exploiting Chemical Ecology and Biodiversity

In constitutive indirect plant defence, the herbivorous attack triggers production of herbivore-induced plant volatiles (HIPVs) that provide foraging cues for natural enemies, antagonistic to the pests (Turlings et al. 1990a, b). This reaction is usually triggered by pest larval stage feeding, and therefore the natural enemies are only attracted following the damage to plants. This aspect has in part limited effectiveness of biological control in reducing pest damage in farmers' fields. Ideally, responses to egg deposition on the plants should thus elicit defences that are beneficial before the larvae cause any damage to plants (Hilker and Meiners 2006; Bruce et al. 2010). Maize landraces and some selected locally adapted African open pollinated varieties (OPVs) showed promise by having this desirable trait. Oviposition by *C. partellus* on these plants was observed to induce defence responses leading to attraction of both egg and larval parasitoids, as well as their reduced attractiveness for further oviposition (Tamiru et al. 2011, 2012). Remarkably, this trait is absent in élite maize hybrids, implying that it must have been lost during the breeding processes that selected for other desirable qualities such as higher yields.

Early production of HIPVs confers adaptive value to the plant and generates also selection pressure on the parasitoids to respond to such signals, because it enhances their foraging efficiency and improvement of ecological fitness. Some selected maize varieties with the early herbivory trait have now been incorporated in the push-pull system, conferring the added benefit of providing biological control of stemborers at oviposition, the earliest stage of attack. There is further opportunity to study the molecular basis of egg-induced semiochemical production with a view to develop molecular markers that can be used in advanced selection of crop varieties and introgression of these traits into mainstream commercial hybrid maize varieties (Tamiru et al. 2015).

There is emerging evidence that plants can respond to HIPVs produced by neighbouring plants, adjusting their metabolism to increase resistance to herbivores by becoming either antagonistic to foraging herbivores or more attractive to their natural enemies (Birkett et al. 2006). Such plants have thus a higher expression of resistance genes and defence-related plant compounds (Arimura et al. 2000). In controlled experiments, maize planted next to molasses grass was found to produce similar semiochemicals as those under attack by stemborer larvae, becoming more attractive to both egg and larval parasitoids and less attractive to stemborer moths for oviposition (Midega et al. 2015a). This suggests that semiochemicals produced by molasses grass provide aerial signals inducing resistance in the neighbouring undamaged plants. Furthermore, the latter are 'primed' to respond more quickly or aggressively to future stemborers attacks. Further research is being undertaken to identify the key compounds within the HIPVs that induce and/or prime these responses in a variety of plants. Early production of HIPVs as a crop protection strategy may have potential for deployment against the newly invasive FAW in Africa.

10.6 Conclusion

The push-pull IPM system effectively exploits the science and knowledge of chemical ecology to address the key constraints to cereal production, faced by resource-poor smallholder farmers in SSA. It is an appropriate system because it uses locally available companion plants, rather than expensive external inputs. It is modelled on both the polycropping and mixed nature of smallholder farming systems, practised in Africa. It therefore allows integration with livestock through the fodder provided by the companion plants. These provide the stimulo-deterrent functionality of the technology, with trap plants attracting and trapping the gravid stemborer moths and the intercrop providing the push to the moths as well as attraction of natural enemies. The technology has been adapted for drier areas vulnerable to climate changes, by identifying and incorporating drought-tolerant trap and repellent plants. This has made the technology more resilient in the face of climate changes, as rainfall becomes increasingly unpredictable and temperatures increase. Moreover, the technology is being made 'smarter' through identification and incorporation of cereal crops with innate defence systems against stemborer pests, which include early production of HPIVs induced by egg pest deposition. Companion plants that can signal defence systems of the neighbouring smart cereals are also being identified. Deployment of inductive HPIVs as a crop protection strategy could have tremendous potential as a cost-effective solution against the more economically devastating FAW in Africa. The push-pull technology is being adopted by an increasing number of farmers in Africa, now estimated to be in excess of 170,000, because it is highly relevant to their needs and results compatible with their farming systems. Further efforts are being intensified to expand the technology to millions of farmers who need it in Africa.

Acknowledgements We gratefully acknowledge financial support to ICIPE provided by the following organizations and agencies: the European Union, Biovision Foundation, UK Department for International Development (DFID), Swedish International Development Cooperation Agency, the Swiss Agency for Development and Cooperation (SDC) and the Kenyan Government. The views expressed herein do not necessarily reflect the official opinion of these donors. The studies within the push-pull program are conducted in collaboration with Rothamsted Research of the UK, which receives grant-aided support from the UK Biotechnology and Biological Sciences Research Council (BBSRC), with additional funding provided under the Biological Interactions in the Root Environment (BIRE) initiative. Various roles played by all partners and collaborators, including farmers, are greatly acknowledged too.

References

Al-Sarar, A., Hall, F. R., & Downer, R. A. (2006). Impact of spray application methodology on the development of resistance to cypermethrin and spinosad by fall armyworm *Spodoptera frugiperda* (J.E. Smith). *Pest Management Science, 62*, 1023–1031.

Arimura, G., Ozawa, R., Shimoda, T., Nishioka, T., Boland, W., & Takabayashi, J. (2000). Herbivory-induced volatiles elicit defence genes in lima bean leaves. *Nature, 406*, 512–515.

Bebawi, F. F., & Metwali, E. M. (1991). Witch-weed management by sorghum–Sudan grass seed size and stage of harvest. *Agronomy Journal, 83*, 781–785.

Birkett, M. A., Chamberlain, K., Khan, Z. R., Pickett, J. A., Toshova, T., Wadhams, L. J., & Woodcock, C. M. (2006). Electrophysiological responses of the lepidopterous stemborers *Chilo partellus* and *Busseolafusca* to volatiles from wild and cultivated host plants. *Journal of Chemical Ecology, 32*, 2475–2487.

Bruce, T. J. A., Wadhams, L. J., & Woodcock, C. M. (2005). Insect host location: A volatile situation. *Trends in Plant Science, 10*, 269–274.

Bruce, T. J. A., Midega, C. A. O., Birkett, M. A., Pickett, J. A., & Khan, Z. R. (2010). Is quality more important than quantity? Insect behavioral responses to changes in a volatile blend after stemborer oviposition on an African grass. *Biology Letters, 6*, 314–317.

Burke, M. B., Lobell, D. B., & Guarino, L. (2009). Shifts in African crop climates by 2050, and the implications for crop improvement and genetic resources conservation. *Global Environmental Change, 19*, 317–325.

CABI. (2017). Fall armyworm: Impacts and implications for Africa. Evidence note (2), September 2017. CAB International.

Chamberlain, K., Khan, Z. R., Pickett, J. A., Toshova, T., & Wadhams, L. J. (2006). Diel periodicity in the production of green leaf volatiles by wild and cultivated host plants of stemborer moths, *Chilo partellus* and *Busseola fusca*. *Journal of Chemical Ecology, 32*, 565–577.

Chidawanyika, F. (2015). *Effects of drought on the production of ectrophysiologically active biogenic volatiles important for cereal pest management.* University of the Witwatersrand, South Africa (p. 172). Ph.D. Dissertation.

Chidawanyika, F., Midega, C. A. O., Bruce, T. J. A., Duncan, F., Pickett, J. A., & Khan, Z. R. (2014). Oviposition acceptance and larval development of *Chilo partellus* stemborers in drought-stressed wild and cultivated grasses of East Africa. *EntomologiaExperimentalis et Applicata, 151*, 209–217.

Chitere, P. O., & Omolo, B. A. (1993). Farmers' indigenous knowledge of crop pests and their damage in western Kenya. *International Journal of Pest Management, 39*, 126–132.

Cock, M. J. W., Beseh, P. K., Buddie, A. G., Cafá, G., & Crozier, J. (2017). Molecular methods to detect Spodoptera frugiperda in Ghana, and implications for monitoring the spread of invasive species in developing countries. *Scientific Reports, 7*(4103), 10.

Cook, D. R., Leonard, B. R., & Gore, J. (2004). Field and laboratory performance of novel insecticides against armyworms (Lepidoptera: Noctuidae). *Florida Entomologist, 87*, 433–439.

Cook, S. M., Khan, Z. R., & Pickett, J. A. (2007). The use of 'push–pull' strategies in integrated pest management. *Annual Review of Entomology, 52*, 375–400.

Dicke, M., & Sabelis, M. W. (1988). Infochemical terminology: Based on cost-benefit analysis rather than origin of compounds? *Functional Ecology, 2*, 131–139.

Frizzas, M. R., Neto, S. S., de Oliveira, C. M., & Omoto, C. (2014). Genetically modified corn on fall armyworm and earwig populations under field conditions. *Ciência Rural, 44*, 203–209.

Frost, D. L., Gurney, A. L., Press, M. C., & Scholes, J. D. (1997). *Striga hermonthica* reduces photosynthesis in sorghum: The importance of stomatal limitations and a potential role for ABA? *Plant, Cell & Environment, 20*, 483–492.

Gethi, J. G., Smith, M. E., Mitchell, S. E., & Kresovich, S. (2005). Genetic diversity of *Striga hermonthica* and *Striga asiatica* populations in Kenya. *Weed Research, 45*, 64–73.

Goergen, G., Kumar, P. L., Sankung, S. B., Togola, A., & Tamò, M. (2016). First report of outbreaks of the fall armyworm *Spodoptera frugiperda* (J. E. Smith) (Lepidoptera, Noctuidae), a new alien invasive pest in West and Central Africa. *PLoS One, 11*(10), e0165632. https://doi.org/10.1371/journal.pone.0165632.

Gressel, J., Hanafi, A., Head, G., Marasas, W., Obilana, A. B., Ochanda, J., Souissi, T., & Tzotzos, G. (2004). Major heretofore intractable biotic constraints to African food security that may be amenable to novel biotechnological solutions. *Crop Protection, 23*, 661–689.

Gurney, A. L., Press, M. C., & Scholes, J. D. (1999). Infection time and density influence the response of sorghum to the parasitic angiosperm *Striga hermonthica*. *New Phytologist, 143*, 573–580.

Gurney, A. L., Slate, J., Press, M. C., & Scholes, J. D. (2006). A novel form of resistance in rice to the angiosperm parasite *Striga hermonthica*. *New Phytologist, 169*, 199–208.

Hardie, J., Isaacs, R., Pickett, J. A., Wadhams, L. J., & Woodcock, C. M. (1994). Methyl salicylate and (−) -(1*R*,5S)-myrtenal are plant derived repellents for black bean aphid, *Aphis fabae* Scop. (Homoptera: Aphididae). *Journal of Chemical Ecology, 20*, 2847–2855.

Hassanali, A., Herren, H., Khan, Z. R., Pickett, J. A., & Woodcock, C. M. (2008). Integrated pest management: The push–pull approach for controlling insect pests and weeds of cereals, and its potential for other agricultural systems including animal husbandry. *Philosophical Transactions of the Royal Society B, 363*, 611–621.

Hilker, M., & Meiners, T. (2006). Early herbivore alert: Insect eggs induce plant defense. *Journal of Chemical Ecology, 32*, 1379–1397.

Hoerling, M., & Kumar, A. (2003). The perfect ocean for drought. *Science, 299*, 691–694.

Hooper, A. M., Caufield, J. C., Hao, B., Pickett, J. A., Midega, C. A. O., & Khan, Z. R. (2015). Isolation and identification of *Desmodium* root exudates from drought tolerant species used as intercrops against *Striga hermonthica. Phytochemistry, 117*, 380–387.

International Assessment of Agricultural Knowledge, Science and Technology for Development (IAASTD). (2009). *Global report: Agriculture at a crossroads*. Washington, DC: Island Press.

International Fund for Agricultural Development (IFAD). (2011). *Rural poverty report: New realities, new challenges: New opportunities for tomorrow's generation*. Rome: IFAD.

Johnson, S. J. (1987). Migration and the life history strategy of the fall armyworm, Spodoptera frugiperda in the western hemisphere. *Insect Science and its Application, 8*(4–6), 543–549.

Kessler, A., & Baldwin, I. T. (2001). Defensive function of herbivore-induced plant volatile emissions in nature. *Science, 291*, 2141–2144.

Kfir, R., Overholt, W. A., Khan, Z. R., & Polaszek, A. (2002). Biology and management of economically important lepidopteran cereal stemborers in Africa. *Annual Review of Entomology, 47*, 701–731.

Khan, Z. R., Chilishwa, P., Ampong-Nyarko, K., Smart, L. E., Polaszek, A., Wandera, J., & Mulaa, M. A. (1997a). Utilisation of wild gramineous plants for the management of cereal stemborers in Africa. *Insect Science and its Application, 17*, 143–150.

Khan, Z. R., Ampong-Nyarko, K., Chilishwa, P., Hassanali, A., Kimani, S., Lwande, W., Overholt, W. A., Pickett, J. A., Smart, L. E., Wadhams, L. J., & Woodcock, C. M. (1997b). Intercropping increases parasitism of pests. *Nature, 388*, 631–632.

Khan, Z. R., Pickett, J. A., van den Berg, J., Wadhams, L. J., & Woodcock, C. M. (2000). Exploiting chemical ecology and species diversity: Stemborer and striga control for maize and sorghum in Africa. *Pest Management Science, 56*, 957–962.

Khan, Z. R., Pickett, J. A., Wadhams, L. J., et al. (2001). Habitat management strategies for the control of cereal stemborers and Striga in maize in Kenya. *Insect Science and Applications, 21*, 375–380.

Khan, Z. R., Hassanali, A., Overholt, W., Khamis, T. M., Hooper, A. M., Pickett, A. J., Wadhams, L. J., & Woodcock, C. M. (2002). Control of witchweed *Striga hermonthica* by intercropping with *Desmodium* spp., and the mechanism defined as allelopathic. *Journal of Chemical Ecology, 28*, 1871–1885.

Khan, Z. R., Midega, C. A. O., Hutter, N. J., Wilkins, R. M., & Wadhams, L. J. (2006a). Assessment of the potential of Napier grass (*Pennisetum purpureum*) varieties as trap plants for management of *Chilo partellus. EntomologiaExperimentalis et Applicata, 119*, 15–22.

Khan, Z. R., Midega, C. A. O., Pickett, J. A., Wadhams, L. J., Hassanali, A., & Wanjoya, A. (2006b). Management of witchweed, *Striga hermonthica*, and stemborers in sorghum, *Sorghum bicolor*, through intercropping with greenleaf Desmodium, *Desmodium intortum. International Journal of Pest Management, 52*, 297–302.

Khan, Z. R., Midega, C. A. O., Wadhams, L. J., Pickett, J. A., & Mumuni, A. (2007). Evaluation of Napier grass (Pennisetum purpureum) varieties for use as trap plants for the management of African stemborer (Busseola fusca) in a push-pull strategy. *Entomologia Experimentalis et Applicata, 124*, 201–211.

Khan, Z. R., Midega, C. A. O., Amudavi, D. M., Hassanali, A., & Pickett, J. A. (2008). On-farm evaluation of the 'push-pull' technology for the control of stemborers and striga weed on maize in western Kenya. *Field Crops Research, 106*, 224–233.

Khan, Z. R., Midega, C. A. O., Bruce, T. J. A., Hooper, A. M., & Pickett, J. A. (2010). Exploiting phytochemicals for developing a push – pull crop protection strategy for cereal farmers in Africa. *Journal of Experimental Botany, 61*, 4185–4196.

Khan, Z. R., Midega, C. A. O., Pittchar, J. O., Murage, A. W., Birkett, M. A., Bruce, T. J. A., & Pickett, J. A. (2014). Achieving food security for one million sub-Saharan African poor through push–pull innovation by 2020. *Philosophical Transactions of the Royal Society B, 369*, 20120284.

Khan, Z. R., Midega, C. A. O., Hooper, A., & Pickett, J. A. (2016). Push-pull: Chemical ecology-based integrated pest management technology. *Journal of Chemical Ecology, 42*(7), 689–697.

Midega, C. A. O., Khan, Z. R., van den Berg, J., Ogol, C. K. P. O., & Pickett, J. A. (2006). Maize stemborer predator activity under 'push-pull' system and Bt-maize: A potential component in managing Bt resistance. *International Journal of Pest Management, 52*, 1–10.

Midega, C. A. O., Khan, Z. R., van den Berg, J., Ogol, C. K. P. O., Dippenaar-Schoeman, A. S., Pickett, J. A., & Wadhams, L. J. (2008). Response of ground-dwelling arthropods to a push–pull habitat management system: Spiders as an indicator group. *Journal of Applied Entomology, 132*, 248–254.

Midega, C. A. O., Khan, Z. R., Van den Berg, J., Ogol, C. K. P. O., Bruce, T. J., & Pickett, J. A. (2009). Non-target effects of the 'push–pull' habitat management strategy: Parasitoid activity and soil fauna abundance. *Crop Protection, 28*, 1045–1051.

Midega, C. A. O., Khan, Z. R., Amudavi, D. A., Pittchar, J., & Pickett, J. A. (2010). Integrated management of *Striga hermonthica* and cereal stemborers in finger millet (*Eleusine coracana* (L.) Gaertn.), through intercropping with *Desmodium intortum*. *International Journal of Pest Management, 56*, 145–151.

Midega, C. A. O., Jonsson, M., Khan, Z. R., & Ekbom, B. (2014). Effects of landscape complexity and habitat management on stemborer colonization, parasitism and damage to maize. *Agriculture, Ecosystems and Environment, 188*, 289–293.

Midega, C. A. O., Bruce, T. J. A., Pickett, J. A., & Khan, Z. R. (2015a). Ecological management of cereal stemborers in African smallholder agriculture through behavioral manipulation *Ecological Entomology, 40*(Suppl. 1), 70–81.

Midega, C. A. O., Bruce, T. J. A., Pickett, J. A., Jimmy, J. O., Murage, A., & Khan, Z. R. (2015b). Climate-adapted companion cropping increases agricultural productivity in East Africa. *Field Crops Research, 180*, 118–125.

Midega, C. A. O., Wasonga, C. J., Hooper, A. M., Pickett, J. A., & Khan, Z. R. (2017). Drought-tolerant *Desmodium* species effectively suppress parasitic striga weed and improve cereal grain yields in western Kenya. *Crop Protection, 98*, 94–101.

Midega, C. A. O., Pittchar, J. O., Pickett, J. A., Hailu, G. W., & Khan, Z. R. (2018). A climate-adapted push-pull system effectively controls fall armyworm, Spodoptera frugiperda (J E Smith), in maize in East Africa. *Crop Protection, 105*, 10–15.

Nordlund, D. A., & Lewis, W. J. (1976). Terminology of chemical releasing stimuli in intraspecific and interspecific interactions. *Journal of Chemical Ecology, 2*, 211–220.

Oswald, A. (2005). Striga control technologies and their dissemination. *Crop Protection, 24*, 333–342.

Oswald, A., Ransom, J. K., Kroschel, J., & Sauerborn, J. (2001). Transplanting maize and sorghum reduces *Striga hermonthica* damage. *Weed Science, 49*, 346–353.

Parker, C., & Riches, C. R. (1993). *Parasitic weeds of the world: Biology and control*. Wallingford: CAB International.

Päts, P. (1991). Activity of *Chilo partellus* (Lepidoptera: Pyralidae): Eclosion, mating and oviposition time. *Bulletin of Entomological Research, 81*, 93–96.

Pickett, J. A., & Khan, Z. R. (2016). Plant volatile-mediated signalling and its application in agriculture: Successes and challenges. *New Phytologist, 212*, 856–870.

Pickett, J. A., Hamilton, M. L., Hooper, A. M., Khan, Z. R., & Midega, C. A. O. (2010). Companion cropping to manage parasitic plants. *Annual Review of Phytopathology, 48*, 161–177.
Pinstrup-Andersen (Ed.). (2010). *The African food system and its interaction with human health and nutrition*. New York: Cornell University Press.
Pretty, J., Noble, A. D., Bossio, D., Dickson, J., Hine, R. E., Penning de Vries, F. W. T., & Morrison, J. I. L. (2006). Resource-conserving agriculture increases yields in developing countries. *Environmental Science and Technology, 40*(4), 1114–1119.
Pretty, J., Toulmin, C., & Williams, S. (2011). Sustainable intensification in African agriculture. *International Journal of Sustainable Agriculture, 9*, 5–24.
Rodenburg, J., Bastiaans, L., Weltzien, E., & Hess, D. E. (2005). How can selection for striga resistance and tolerance in sorghum be improved. *Field Crops Research, 93*, 34–50.
Sanchez, P. (2002). *Soil fertility and hunger in Africa. Science, 295*(5562), 2019–2020.
Solomon, D., Lehmann, J., Kinyanjui, J., Amelung, W., Lobe, I., Ngoze, S., Riha, S., Pell, A., Storer, N. P., Babcock, J. M., Schlenz, M., Meade, T., & Thompson, G. D. (2010). Discovery and characterization of field resistance to Bt maize: Spodoptera frugiperda(Lepidoptera: Noctuidae) in Puerto Rico. *Journal of Economic Entomology, 103*, 1031–1038.
Sparks, A. N. (1979). A review of the biology of the fall armyworm. *Florida Entomologist, 62*(2), 82–87.
Storer, N. P., Babcock, J. M., Schlenz, M., Meade, T., & Thompson, G. D. (2010). Discovery and characterization of field resistance to Bt maize: Spodoptera frugiperda (Lepidoptera: Noctuidae) in Puerto Rico. *Journal of Economic Entomology, 103*, 1031–1038.
Tamiru, A., Bruce, T. J. A., Woodcock, C. M., et al. (2011). Maize landraces recruit egg and larval parasitoids in response to egg deposition by a herbivore. *Ecology Letters, 14*, 1075–1083.
Tamiru, A., Bruce, T., Midega, C. A. O., et al. (2012). Oviposition-induced volatile emissions from African smallholder farmers' maize varieties. *Journal of Chemical Ecology, 38*, 231–234.
Tamiru, A., Khan, Z. R., & Bruce, T. J. A. (2015). New directions for improving crop resistance to insects by breeding for egg induced defence. *Current Opinion in Insect Science*. https://doi.org/10.1016/j.cois.2015.02.011.
Tenebe, V. A., & Kamara, H. M. (2002). Effect of *Striga hermonthica* on the growth characteristics of sorghum intercropped with groundnut varieties. *Journal of Agronomy and Crop Science, 188*, 376–381.
Todd, E. L., & Poole, R. W. (1980). Keys and illustrations for the armyworm moths of the noctuid genus Spodoptera Guenée from the Western Hemisphere. *Annals of the Entomological Society of America, 73*(6), 722–738.
Tsanuo, M. K., Hassanali, A., Hooper, A. M., Khan, Z. R., Kaberia, F., Pickett, J. A., & Wadhams, L. (2003). Isoflavanones from the allelopathic aqueous root exudates of *Desmodium uncinatum*. *Phytochemistry, 64*, 265–273.
Turlings, T. C. J., Scheepmaker, J. W. A., Vet, L. E. M., Tumlinson, J. H., & Lewis, W. J. (1990a). How contact experiences affect preferences for host-related odors in the larval parasitoid Cotesia marginiventris (Cresson) (Hymenoptera: Braconidae). *Journal of Chemical Ecology, 16*, 1577–1589.
Turlings, T. C. J., Tumlinson, J. H., & Lewis, W. J. (1990b). Exploitation of herbivore-induced plant odors by host-seeking parasitic wasps. *Science, 250*, 1251–1253.
Turlings, T. C. J., Loughrin, J. H., McCall, P. J., Rose, U. S. R., Lewis, W. J., & Tumlinson, J. H. (1995). How caterpillar-damaged plants protect themselves by attracting parasitic wasps. In *Proceedings of the National Academy of Sciences of the United States of America* (Vol. 92, pp. 4169–4174).
World Bank. (2007). World development indicators. Washington, DC.
World Bank. (2008). *World development report 2008: Agriculture for development*. Washington, DC: The World Bank.

World Resources Institute; Department of Resource Surveys and Remote Sensing, Ministry of Environment and Natural Resources, Kenya; Central bureau of Statistics, Ministry of Planning and National Development, Kenya; and International Livestock Research Institute. (2007). *Nature's benefits in Kenya: An Atlas of ecosystems and human well-being*. Washington, DC and Nairobi: World Resources Institute.

Yoder, J. I. (1999). Parasitic plant responses to host plant signals: A model for subterranean plant–plant interactions. *Current Opinion in Plant Biology, 2*, 65–70.

Young, R. (1979). Fall armyworm: Control with insecticides. *Florida Entomologist, 62*, 130–133.

Yu, S. J. (1992). Detection and biochemical characterization of insecticide resistance in fall armyworm (Lepidoptera: Noctuidae). *Journal of Economic Entomology, 85*, 675–682.

Chapter 11
Insect Resistance to Insecticides and Bt Cotton in India

Sandhya Kranthi, Keshava R. Kranthi, Chetali Rodge, Shilpa Chawla, and Sarita Nehare

Abstract Yield losses in cotton are often attributed, worldwide, to biotic factors, of which insect pests, categorized as sucking pests and bollworms, are dominant. Insecticides are the most potent tools used against them. Pest management was severely affected when bollworms developed resistance to commonly used insecticides. Transgenic Bt cotton Bollgard and Bollgard II were introduced in India in 2005 and 2006, respectively, to control lepidopteran pests. Cotton crop protection in India, supported by strategic research, made strides beginning in the late 1990s. From a less-rational, calendar-based schedule in the 1980s, today insecticide use is integrated with other components of pest management in a "windows" approach to ensure sustainability, and these programs are comparable with cotton crop protection programs across the world. This chapter briefly documents the advances made in the field of insecticide resistance, including Bt resistance. Stewardship of Bt cotton in India did not facilitate a delay in the development of resistance in target pests. Bt cotton, a promising tool in crop protection, stands threatened with the development of field-evolved resistance to both toxins, Cry1Ac and Cry2Ab, in Bollgard II, by the pink bollworm. In addition, sucking and emerging pests have been reported to limit yields of BGII cotton in specific locations. Dissemination of crop protection strategies is still inadequate. Cotton crop protection needs to be strengthened quickly through an effective network that includes stakeholders, exploiting recent technologies to ensure that rational strategies are disseminated and implemented in the right place and at the right time, for effective pest and crop management, before

---

S. Kranthi (✉) · C. Rodge · S. Chawla
Crop Protection Division, Central Institute for Cotton Research, Nagpur, Maharashtra, India
e-mail: sandhya.kranthi@gmail.com; chet.roj@gmail.com; chawlashilpa11@gmail.com

K. R. Kranthi
Technical Information Section, International Cotton Advisory Committee, Washington, DC, USA
e-mail: krkranthi@gmail.com

S. Nehare
Anand Niketan College of Agriculture, Warora, Maharashtra, India
e-mail: sarita.nehare@gmail.com

R. Peshin, A. K. Dhawan (eds.), *Natural Resource Management: Ecological Perspectives*, Sustainability in Plant and Crop Protection,
https://doi.org/10.1007/978-3-319-99768-1_11

a powerful, expensive, and useful technology such as genetically modified (GM) cotton is rendered unsustainable.

Keywords Cotton · Insecticide · Bt · Resistance · Insect pests

Abbreviations/Terms

$	US dollar
BHC(HCH)	Hexachlorocyclohexane
Bollgard II	Cotton contains two genes derived from the common soil bacterium *Bacillus thuringiensis*
Bt	*Bacillus thuringiensis*
CICR	Central Institute for Cotton Research, Nagpur, India
Cry1Ac	Cry1Ac toxin is a crystal protein produced by the bacterium Bacillus thuringiensis (Bt) during sporulation. Cry1Ac is one of the delta endotoxins produced by this bacterium which acts as an insecticide
Cry2Ab	Pesticidal crystal protein Cry2Ab
DDT	Dichlorodiphenyltrichloroethane
Dissemination	Active spread of innovations/new practices to the target audience using planned strategies
g	Gram
ha	Hectare
HDPS	High density planting system
IARI	Indian Agricultural Research Institute, New Delhi
IPM	Integrated pest management
IRM	Insecticide resistance management
LC_{50}	Lethal concentration 50 (LC_{50}).
Mt.	Metric tons
NPV	Nuclear polyhedrosis virus
RIB	Refuge in Bag, cotton hybrid seeds mixed with the Bt cotton seeds
Rs	Indian Rupee: the official currency of the Republic of India
USA	The United States of America

11.1 Introduction

Cotton is a key commercial crop, the cultivation of which is usually affected by many insect pests and diseases. Insect pests are dominant and occur throughout the season, often requiring input of intensive management, particularly in the tropics. Cotton is cultivated in an area of 12 million ha in India that constitutes 38% of the global cotton acreage. More than 90% of the cotton cultivated in the country comprises Bt cotton (Bollgard II) that expresses Cry1Ac and Cry2Ab toxins of *Bacillus thuringiensis* (Bt). Prior to the introduction of Bt cotton, whiteflies and bollworms were the major insect pests. After the introduction of Bt cotton, bollworm infestation became negligible, until 2010, during which pink bollworm developed resistance to the Cry1Ac toxin (Dhurua and Gujar 2011), but jassids, thrips, and whiteflies, however, continued to cause damage. Cotton cultivated in North India was severely affected by insecticide-resistant whiteflies that caused serious economic losses, around an estimated US$ 636 million (Kranthi 2016). Whiteflies transmit the dreaded leaf curl virus disease, which accentuates the damage. Pink bollworm *Pectinophora gossypiella* developed resistance to the two gene variants of Bt cotton, in Central and South India, where it has been causing yield losses since 2013 (Naik et al. 2018). It is now acknowledged that the pink bollworm, whitefly, and the cotton leaf curl virus disease are the major biotic factors affecting cotton production in the country.

11.2 Insecticide Use in India

Insecticide use in India started in the 1950s. Estimates indicate that about 35–50% of the annual average insecticide used was applied on cotton during 1970–2000. Insecticide usage (active ingredients) on cotton varied from 6,863 to 13,176 Mt. at an annual average of 10,665 Mt. during 1996–2004. About 58–71% of the total insecticide use during this period was for bollworm control. As the area under Bt cotton increased after 2004, insecticide usage declined to an average of 6,863 Mt. (4,623–11,598 Mt) during 2005–2013. Insecticide use on cotton during 1996–2004 was 26–40% of the total insecticides used in agriculture in India. The proportion decreased to 15–25% during 2005–2013. Insecticide usage on cotton was reported to have increased during 2014–2017, mainly due to enhanced sucking pest infestation and pink bollworm's resistance to Bt cotton.

11.3 Insecticide Influenced Changes in Insect Pest Dynamics

Prior to 1980, the pink bollworm (*Pectinophora gossypiella*), jassids (*Amrasca biguttula biguttula*), and cotton leafworm (*Spodoptera litura*) were the major insect pests of cotton in India. The main insecticides used on cotton in India during 1950–1990 were BHC, DDT, endosulfan, carbaryl, carbofuran, parathion, dimethoate, monocrotophos, acephate, triazophos, metasystox, chlorpyrifos, and quinalphos. Synthetic pyrethroids were introduced in India in 1981 and used on cotton to control pink bollworm and the cotton leafworm. Indiscriminate use of pyrethroids during the 1980s replaced these pests with the American bollworm *Helicoverpa armigera* and whitefly *Bemisia tabaci*. Outbreaks of American bollworm notably intensified over time and were confirmed in 1978, 1983, 1990, 1995, 1997, 1998, and 2001 (Dhawan et al. 2004). By the early 1990s, *H. armigera* and *B. tabaci* showed high levels of resistance to almost all insecticides recommended for their control. Efforts were stepped up to develop and implement integrated pest management (IPM) and insecticide resistance management (IRM) strategies, mainly to combat the American bollworm and the whitefly. During the mid-1990s, chloronicotinyl insecticides such as imidacloprid, acetamiprid, and thiamethoxam were introduced, initially as seed treatment and later on as foliar sprays for sucking pest control. These insecticides were found to be very effective as seed treatment in protecting seedlings against sap-sucking insects for the first 2 months and as foliar sprays for 15–20 days. Cotton yields started increasing due to the effective protection of the vegetative stage of the crop from sucking pest infestation. During the late 1990s, new chemicals such as rynaxypyr, novaluron, spinosad, indoxacarb, emamectin benzoate, and lufenuron were introduced for the control of the *H. armigera* and *S. litura*. However, with the introduction of Bt cotton in 2002 the demand for these insecticides declined.

11.4 Bt Cotton and Changes in Insect Pest Dynamics

Transgenic Bt cotton containing the crystal (Cry) toxin Cry1Ac, derived from the insect pathogenic bacterium *B. thuringiensis*, was introduced in India in 2002. Bt cotton that expressed two crystal toxins, Cry1Ac and Cry2Ab, was introduced in the country in 2006. In India, the Bt technology was introduced in only hybrid varieties and not in pure line, open-pollinated varieties. Because of their responsiveness to nitrogenous fertilizers and excessive vegetation due to hybrid vigor, hybrid cotton varieties in general were known to be susceptible to aphids, jassids, and a few other sap-sucking insect pests. Experimental evidence showed that, without seed treatment, a vast majority of the Bt cotton hybrids would not have survived the damage caused by sap-sucking insects. Thus, it is widely believed that the chloronicotinyl group of insecticides may have played a major role in the adoption of Bt cotton hybrids and their subsequent near saturation of cotton cultivated area in the country. The introduction of more than 1,000 Bt cotton hybrids, most of which were susceptible to sap-sucking insect pests, led to the emergence of new insect pests such as mirid bugs, tea mosquito bugs, flower bud maggots, thrips, and mealybugs, during 2006–2011. By the mid-2000s, mainly after 2007, the main sap-sucking pests, aphids, jassids, and whiteflies developed resistance to the chloronicotinyl insecticides and other major insecticide groups, thereby leading to enhancement in insecticide use for their control. By 2011, more than 75% of India's cotton area was covered by Bollgard II Bt cotton. The intensive selection pressure led to the development of resistance to Cry toxins in the oligophagous pest, the pink bollworm, which prompted the use of insecticides for its control on Bt cotton. Currently, Gujarat resorts to 6–7 rounds of sprays on BGII cotton.

A similar situation was seen in other countries cultivating Bt cotton. The green mirid (*Creontiades dilutus*), green vegetable bug (*Nezara viridula*), leafhoppers (*Austroasca viridigrisea* and *Amrasca terraereginae*), and thrips (*Thrips tabaci*, *Frankliniella schultzei*, and *Frankliniella occidentalis*) increased in importance in

Australia on Bt cotton (Lei et al. 2003; Wilson et al. 2006). The bug *Lygus hesperus*, which is a sucking insect pest not susceptible to *Bt* proteins, is considered to be the number one pest of cotton in Arizona on the basis of the proportion of the total insecticide sprays targeting it (Ellsworth and Jones 2001; Ellsworth et al. 2007). Wu et al. (2002) observed that populations of a complex of mirid plant bugs (*Adelphocoris suturalis*, *A. lineolatus*, *A. fasciaticollis*, *Lygus lucorum*, and *L. pratensis*) arose dramatically in association with reduced insecticide use in Bt cotton in northern China.

11.5 Insecticide Resistance in Sucking Pests

Among aphids, *Aphis gossypii* was reported to have developed high levels of resistance to the synthetic pyrethroids, organophosphates, and carbamates in China, Hawaii, Australia, France, Pakistan, and other countries (Kung et al. 1961; Furk et al. 1980; Wei et al. 1988; Robert et al. 1994; Deguine 1996; Cheng et al. 1997; Delorme et al. 1997; Zhang et al. 1997; Villatte et al. 1999, Herron et al. 2001; Nibouche et al. 2002; Ahmad et al. 2003; Herron et al. 2003, 2014; Herron and Wilson (2011); Bass et al. 2015).

The jassid *A. biguttula biguttula* was reported to have developed resistance to endosulfan and a range of organophosphate insecticides in India (Santhini and Uthamasamy 1997; Challam and Subbaratnam 1999; Jeyapradeepa 2000; Challam et al. 2001; Praveen 2003). Studies conducted from 2008 to 2017 at CICR Nagpur showed high levels of jassid resistance to neonicotinoids and organophosphates.

Whitefly *Bemisia tabaci* showed high levels of resistance to dimethoate and monocrotophos (Dittrich and Ernst 1983) to buprofezin, imidacloprid, and other organophosphate insecticides in the USA, China, Egypt, Europe, Pakistan, Sudan, and Israel (Cahill et al. 1996; Ahmad et al. 2002; El-Kady and Devine 2003; Horowitz et al. 2004; Dennehy et al. 2005; Wang et al. 2010), as well as to BHC, endosulfan, organophosphates, and carbaryl in India (Prasad et al. 1993), and to methomyl and monocrotophos, with moderate resistance to cypermethrin, in India (Kranthi et al. 2002a, b).

Resistance in the leafhopper and whitefly were quantified more recently against the commonly used insecticides. Insecticide resistance to selected organophosphates, pyrethroids, and neonicotinoids in seven Indian field populations of *B. tabaci* genetic groups Asia-I, Asia-II-1, and Asia-II-7 was reported (Naveen et al. 2017). The variability of the LC_{50} values was 7 times for imidacloprid and thiamethoxam, 5 times for monocrotophos, and 3 times for cypermethrin among the Asia-I, whereas they were 7 times for cypermethrin, 6 times for deltamethrin, and 5 times for imidacloprid within the Asia-II-1 populations. When compared with the most susceptible population, PUSA (Asia-II-7), a substantial increase in resistant ratios was observed in both the populations of Asia-I and Asia-II-1. Evidence of potential control failure was detected using probit analysis estimates for cypermethrin, deltamethrin, monocrotophos, and imidacloprid. Studies conducted at CICR (Rishi Kumar et al., unpublished) during 2014–2017 showed that whitefly populations in North India have acquired resistance to the commonly used insecticides. Resistance ratio varied from 98- to 1,400-fold for bifenthrin 10EC, 14- to 137-fold for dinotefuran 20SG, 60- to 131-fold for acephate 75SP, 21- to 331-fold for acetamiprid 20SP, 153- to 340-fold for fipronil 5SC, 371- to 2,237-fold for triazophos 40EC, 51- to 706-fold for buprofezin 25SC, 9- to 512-fold for imidacloprid 17.8SL, 40- to 347-fold for diafenthiuron, 2- to 19-fold for chlorpyrifos 20EC, 1- to 2-fold for thiamethoxam 30FS, 2- to 7-fold for clothianidin 50WDG, 2- to 23-fold for pyriproxyfen, and 1- to 6-fold for flonicamid.

Resistance ratio to imidacloprid was high, up to 2,089-fold, in leafhopper populations from Jalna, in Maharashtra (Central India), and 7,264-fold with leafhopper populations in the Haveri district of Karnataka (South India). The highest resistance ratio to thiamethoxam was 6,554-fold in the populations of leafhoppers from the Indore district of Madhya Pradesh (Central India) and 13,945-fold in the populations of leafhopper from the Haveri district of Karnataka (South India). Broadly, leafhopper populations in Central and South India were resistant to neonicotinoids, imidacloprid, and thiamethoxam as compared with that in the populations from North India (K.R. Kranthi et al., unpublished). The level of resistance in *A. biguttula biguttula* from Tamil Nadu as revealed by the percent survival, varied from 6.67 (Salem) to 15.38 (Srivilliputhur) for imidacloprid, 3.33 (Salem) to 15.09 (Srivilliputhur) for thiamethoxam, 5.00 (Bhavanisagar) to 20.00 (Srivilliputhur) for acetamiprid, and 5.00 (Bhavanisagar) to 9.09 (Srivilliputhur) for thiacloprid (Preetha et al. 2014).

11.6 Bollworm Resistance to Insecticides and Bt Cotton

The American bollworm, *H. armigera* was found to be resistant to parathion, endosulfan, DDT, organophosphates, pyrethroids, and endrin in Australia (Forrester et al. 1993; Gunning 1993); to carbamate and pyrethroids in Thailand (Ahmad and McCaffery 1988); to organophosphate insecticides (Cheng and Lieu 1996) and spinosad in China (Wang et al. 2009); to deltamethrin in South Africa (Martin et al. 2003); to cypermethrin in Turkey (Ernst and Dittrich 1992); to pyrethroids in Central Africa (Djihinto et al. 2009); and to organophosphates and pyrethroids in Pakistan (Ahmad et al. 1995, 1997). In India, *H. armigera* was reported to have developed high levels of resistance to endosulfan, carbamates, organophosphates, and pyrethroids (Armes et al. 1996; Dhingra et al. 1988; McCaffery et al. 1989; Mehrotra and Phokela 1992; Sekhar et al. 1996: Kranthi et al. 2001a, b; 2002a, b). The corn earworm *Helicoverpa zea* was reported to have developed resistance to Bt cotton in the USA (Tabashnik et al. 2008). Fourteen populations from northern China showed very strong resistance to fenvalerate (from 43- to 830-fold) and low levels of resistance to phoxim (3.0- to 8.9-fold) when compared with the susceptible SCD strain of *H. armigera*, whereas two populations from Northwestern China showed low levels of resistance to fenvalerate (3.0- and 10-fold) and no resistance to phoxim (0.7- and 0.9-fold). In comparison with the resistance in field populations before Bt cotton adoption, a maintenance of high levels of fenvalerate resistance was observed in northern China, with a reversion of phoxim resistance from high levels to low levels, in field populations of *H. armigera* (Yang et al. 2013).

Studies on insecticide resistance in cotton bollworms have been carried out since the 1990s in India (Kranthi et al. 2002a). Pyrethroid resistance was found to be high and constant throughout the cotton season in *H. armigera*. Resistance in the pest built up over the seasons to some insecticides, such as endosulfan, and was correlated with excessive use of that molecule (Kranthi et al. 2002b). Mechanisms and genetics of inheritance of resistance were worked out, and strategies of insecticide resistance management were developed and validated. Protocols were standardized for systematic studies on metabolic enzymes mediating insecticide resistance (Kranthi 2005). Robust repeatable protocols were developed for insecticides and Cry toxin bioassays and for maintaining healthy lab cultures of bollworms (Kranthi et al. 2000).

Resistance of the pink bollworm, *P. gossypiella*, to organophosphates and pyrethroids was reported in the USA (Osman et al. 1991) and China (Xianchun et al. 1997). Spotted bollworm *Earias vittella* was found to be resistant to monocrotophos in India (Kranthi et al. 2002a).

In India, resistance to Cry1Ac-based Bt cotton (Dhurua and Gujar 2011) and to Cry1Ac + Cry2Ab-based Bt cotton (Naik et al. 2018) were reported for the pink bollworm. First evidence, found in 2011, of field-evolved resistance of pink bollworm to Cry1Ac and lack of cross-resistance to Cry2Ab2 suggested that plants producing this toxin were likely to be more effective against resistant populations than plants producing only Cry1Ac. In less than 2 years, field resistance to the 2 gene Bt cotton was reported by Naik and coworkers (2018), where high PBW larval recovery on BGII in conjunction with high LC_{50} values to both Cry1Ac and Cry2Ab were recorded.

Laboratory studies showed that *H. armigera* resistance to Cry1Ac increased by 76-fold at the end of the tenth generation, whereas the unexposed population remained susceptible (Kranthi et al. 2000). Temporal and spatial variations in the expression of Cry1Ac were studied in Bollgard cotton (Kranthi et al. 2005). A decline in toxin expression was recorded as the plants aged. Also certain plant parts, such as the square bracts and boll rind, had significantly lower toxin expression as compared to expression in leaves.

F_2 screening method was used to estimate the frequency of Cry1Ac-resistant alleles in *H. armigera* populations collected from Bangalore, Dharwad, Raichur, and New Delhi (IARI). The F_2 screening results showed that the expected Bt resistance allele frequency in the collected populations was 0.085 with 95% confidence interval of 0.009–0.256, indicating that the F_2 screening method can be used to detect major alleles conferring resistance to Bt cotton in the targeted insect (Kennedy et al. 2017a, b).

A stochastic model, Bt Adapt, was developed to simulate the rate of resistance development in *H. armigera*, using genetic and ecological factors in addition to the response of *H. armigera* to Cry toxins expressed in plants (Kranthi and Kranthi 2004). Protocols were designed at CICR for field evaluation based on the concept of "refuge in bag" (Rishi Kumar et al., unpublished). The "refuge in bag" (RIB) concept was approved by the government through its official Gazette S.O. 4215(E) in 2016, with a Bt cotton RIB seed pack (475 g) with a minimum of 90% and a maximum of 95% seeds positive for each transgene. The 475 g RIB seed packet shall hence contain a minimum of 5% and a maximum 10% of non-Bt cotton seeds. The non-Bt seeds provided along with the Bt hybrid as a separate pack or as a refuge in bag shall be of a non-Bt hybrid isogenic version corresponding to the Bt hybrid or a non-Bt hybrid with similar flowering period and fiber traits as that of the Bt hybrid as per label claim. RIB was formulated to ensure the sustainability of the Bt technology to the other target insect pests.

11.7 Development of Window-Based IRM Strategies

In the late 1980s and early 1990s, cotton pest management in India was based mainly on biological control, where emphasis was placed on the multiplication and release of natural enemies (Rajendran et al. 2005) in an environment that still relied on the use of hazardous chemicals for cotton crop protection. A remarkable change was noticed in the approach to cotton pest management since 1997. IPM strategies were implemented in the village Astha, in Marathwada. Farmers were trained in the use of botanicals, NPV, reducing diversity in the varieties being cultivated in the village, with the adoption of farm operations of sowing and spraying in a synchronous manner. The model, however, did not spread after its initial adoption largely due to the non-availability of recommended inputs. Using the data generated in the lab, insecticide resistance management strategies were developed by CICR to combat insecticide resistance, first in bollworms in 1997, mealybugs in 2007, sucking pests in 2008, and whiteflies and pink bollworm in 2013. Strategies were developed based on robust scientific data on insecticide resistance monitoring and mechanisms mediating resistance, generated under various funded projects. Exploitation of host plant resistance in the first 60 days, thereby avoiding use of broad-spectrum organophosphates against sucking pests and withdrawal of pyrethroids against bollworms, was an important feature in this program. The choice of insecticides was based on their ecotoxicological profiles, ensuring minimal disruption of the cotton ecosystem. A "window" concept of pest management was introduced for the first time in the country, where emphasis was on the conservation of natural enemies, through intelligent selection and use of insecticides on the basis of economic threshold levels instead of calendar-based sprayings. Strategies were fine-tuned for compatibility with Bt cotton. A widespread use of neonicotinoid seed-treated Bt hybrids caused an upsurgence in leafhopper populations. Sucking pest management on Bt cotton, with the emergence of resistance in leafhoppers to the commonly used neonicotinoids, was also addressed. Scientific data generated in the lab, for the first time, was directly made relevant to address the issues of cotton pest management at the farmers' level. Bt resistance management strategies must be implemented henceforth on BGII cotton, to sustain the production of seed cotton.

11.8 Dissemination of Insecticide Resistance Management Strategies

The IRM dissemination program was carried out under the Technology Mission on Cotton Mini Mission II, funded by the Ministry of Agriculture and Farmers' Welfare, Government of India, and linked the Central Institute to state agricultural universities. Nine state coordinators, from national institutes or universities, worked along with the state agricultural departments to supervise the program implementation through 28 district coordinators and 56 research fellows.

The program brought about a radical change in the farmers' perception on the insect pests and native natural enemies of cotton, thereby bringing about a change in the pesticide use. The results of this farmer participatory approach were encouraging, and farmer awareness and reduction in the usage of pesticides in all the districts implementing the program were noticed. Staff for resistance monitoring were trained under the project, and they, in turn, helped farmers take guided decisions on the appropriate choice of insecticides.

In 2013–2014, IRM for high density planting system (HDPS) was introduced. HDPS is a method of growing cotton varieties with more number of plants per unit area. This is in contrast to the Bt hybrid technology, where each plant produces a higher number of bolls. HDPS is recommended for marginal soils in rain-fed regions, where the program was launched to enhance the productivity of cotton. Grown at a spacing of 45 cm or 60 cm between rows and 10 cm between plants, the systems support 1.5 lakh plants/ha, as compared with Bt (11,000 plants/ha). Use of early maturing, sucking pest-tolerant varieties such as PKV081, NH615, and Suraj helped popularize the technology in the rain-fed district of Vidarbha. Because the varieties were non-Bt, necessary bollworm protection needed to be accorded. The technology also aims to fit the boll development stage into the window of available moisture in rain-fed regions, to help overcome moisture stress at boll development stage. HDPS was demonstrated on about 2,000 acres (800 ha, 1 ha = 2.5 acres) showing that, with proper pest management and appropriate production practices, cotton yield from recommended varieties could be equal or more than that in Bt, with a dramatic reduction in input costs, especially in rain-fed regions.

During the period 2007–2011, IRM strategies were disseminated to 1,69,268 farmers in 3,33,883 hectares in 2,922 villages of 28 districts from 10 states across India. Implementation of the program resulted in yield increases estimated at a net additional benefit of ca. US $51.87 million (Rs 2593.6 million, US $1 = ca. Indian Rs. 50 at 2011 average conversion rate) and a saving due to reduction in insecticide use accounting for ca. US $15.26 million (Rs 763 million), thus adding up to a total additional benefit of ca. US $67.13 million (Rs 3356.6 million).

From 2012 to 2015, IRM strategies were disseminated to 24,613 farmers in 12,231 hectares in a total of 445 villages of 22 districts from 10 states across India. Implementation of the program resulted in yield increases estimated at a net additional benefit of ca. US $9.94 million (Rs 656 million, US $1 = ca. Indian Rs. 66 at 2015 conversion rate) and a saving due to reduction in insecticide use accounting for ca. \$1.75 million (Rs 115.6 million), thus adding up to a total additional benefit of ca. \$11.69 (Rs 771.6 million).

11.9 Pink Bollworm Management Strategies for Cotton Including BGII Cotton[1]

Pink bollworm management strategies were devised for implementation in the affected districts of India. The causes for pink bollworm outbreak on BGII cotton were determined and strategies devised for its management.

The major reasons for pink bollworm outbreak were (1) cultivation of large number of long-duration hybrids that serve as continuous hosts to the pink bollworm, (2) long-term storage of raw cotton, (3) early sowing and extended duration of the crop with supplemental irrigation, (4) noncompliance of refugia, and (5) lack of adoption of timely interventions for bollworm infestation on BGII cotton. In addition, squares and flowers have less Bt-toxin expression as compared with that in leaves and seeds. Also, the segregating seeds in bolls of F1 hybrid plants accelerate resistance development. India is the only country in the world that cultivates Bt cotton as hybrid F1 plants, harboring the F1 bolls carry seeds that segregate in the ratio of 9:3:3:1 (Cry1Ac + Cry2Ab in 9; Cry2Ab alone in 3; Cry1Ac in 3; and none in 1). Thus a spectrum of non-Bt seeds, seeds with Cry1Ac alone, seeds with Cry2Ab alone, and seeds with Cry1Ac + Cry2Ab is present in a single boll. This situation is ideal for resistance development, due to the selection for resistance to independent toxins. The segregation pattern is further affected by the unauthorized cultivation of BGII hybrids of F2 and F3 generations.

IRM programs consider fine-tuned regular pest surveillance, resistance monitoring, and monitoring of field efficacy in bollworms to ensure that Bt cotton continues to be effective for the longest possible time. Management strategies relying on the following approaches in an area-wide manner are:

(a) Regular monitoring of bollworm resistance to Bt cotton, including Bollgard II.
(b) Use of the parasitoid *Trichogramma bactriae* in Bt cotton fields, for pink bollworm management.
(c) Refugia: recommend planting of *desi* (*G. herbaceum*)/conventional non-Bt *G. hirsutum* cotton and late-planted okra as refugia crop.
(d) Timely termination of the crop latest by December, avoiding ratoon and/or extended crop.
(e) Utilization or destruction of crop residues and cotton stalks immediately after harvest.
(f) Crop rotation strongly recommended to break the pest cycle.
(g) Use of short-duration, single-pick varieties (150 days) that provide high yields in high density and escape the pink bollworm.

[1] See also http://www.cicr.org.in/pdf/Kranthi_art/Pinkbollworm.pdf.

(h) Installation of light traps (timer operated) and pheromone traps in fields during the season and also near load areas, ginning mills, market yards, etc., to trap post-season moths.
(i) Mass trapping and mating disruption using pheromone traps. Use of "pheromone traps" and "green boll dissection" for regular monitoring and initiation of control interventions, based on economic threshold levels of eight moths/trap/night and/or 10% damage in green bolls.
(j) Utilization of insecticides such as quinalphos in early stages and synthetic pyrethroids after October at economic threshold levels of damage.
(k) Strictly avoiding spraying pyrethroids before November, or acephate, fipronil, or any insecticide mixtures at any time to prevent whitefly outbreaks.
(l) Select hybrids/varieties that are tolerant to sucking pests.

The USA recently celebrated successful eradication of the pink bollworm (https://www.usda.gov/sites/default/files/documents/usda-pink-bollworm-proclamation.pdf). Impeccable planning and area-wide implementation of strategies for its management was evidenced during the entire program. The exercise was grower driven, involved extensive surveys, incorporated Bt cotton varieties as a management tool, used pheromones for monitoring, mass trapping, and mating disruption. Sterile moth release along with cultural practices such as mandatory crop termination (closed season), destruction of crop residues, shredding stalks, disking, ploughing, and winter irrigation were followed over phase I, phase II, phase IIIa, phase IIIb, and containment where phases denoted particular mapped areas where the area-wide approach was executed. The United States Department of Agriculture, state departments of agriculture in Arizona, California, New Mexico, and Texas adopted these approaches even before the pink bollworm developed resistance to the Cry toxins. With the pink bollworm having developed resistance to Cry toxins in India, area-wide management approaches ought to be focused and executed even in years of low incidence as an outbreak of resistant pink bollworms would not only impact seed cotton yield and its quality, but would also render a major tool (Bt cotton) useless in the Indian cotton scenario.

11.10 Conclusion

To summarize, cotton crop protection has considerably progressed since the early 1990s in India. The technology/strategies were not only relevant and timely but were also sustainable. Scientific lab experiments directly contributed to the evolution of sustainable strategies and technologies that were evaluated in multiple locations. The most significant achievement was the ability of cotton scientists of the country to work together as a team in funded programs in an effort to contribute to area-wide cotton crop protection, fine tuned to location-specific strategies if necessary, for the welfare of cotton farmers. Unexpected pest control failures, despite technologies and strategies being in place, were due to reasons other than failure of the technology itself.

References

Ahmad, M., & McCaffery, A. R. (1988). Resistance to insecticides in a Thialand strain of *Helicoverpa armigera* (Lepidoptera: Noctuidae). *Journal of Economical Entomology, 81*, 45–48.

Ahmad, M., Arif, M. I., & Ahmad, Z. (1995). Monitoring insecticide resistance of *Helicoverpa armigera* (Lepidoptera: Noctuidae) in Pakistan. *Journal of Economic Entomology, 88*, 771–776.

Ahmad, M., Arif, M. I., & Attique, M. R. (1997). Pyrethroid resistance of *Helicoverpa armigera* (Lepidoptera: Noctuidae) in Pakistan. *Bulletin of Entomological Research, 87*, 343–347.

Ahmad, M., Arif, I., Zahoor, A., & Denholm, I. (2002). Cotton white fly (*Bemisia tabaci*) resistances to organophosphate and pyrethroide insecticides in Pakistan. *Pest Management Science, 58*, 203–208.

Ahmad, M. M., Arif, I., & Denholm, I. (2003). High resistance of field populations of the cotton aphid *Aphis gossypii* glover (Homoptera: Aphididae) to Pyrethroid insecticides in Pakistan. *Journal of Economic Entomology, 96*(3), 875–878. https://doi.org/10.1093/jee/96.3.875.

Armes, N. J., Jadhav, D. R., & DeSouza, K. R. (1996). A survey of insecticide resistance in *Helicoverpa armigera* in the Indian subcontinent. *Bulletin of Entomological Research, 86*, 499–514.

Bass, C., Denholm, I., Williamson, S. M., & Nauen, R. (2015). The global status of insect resistance to neonicotinoid insecticides. *Pesticide Biochemistry and Physiology, 121*, 78–87.

Cahill, M., Groman, K., Day, S., Denholm, I., Elbert, A., & Nauen, R. (1996). Baseline determination and detection of resistance to imidacloprid in *Bemisia tabaci* (Homoptera:Aleyrodidae). *Bulletin of Entomological Research, 86*, 343–349.

Challam, M. S., & Subbaratnam, G. V. (1999). Insecicide resistance in cotton leaf hopper *Amrasca biguttula biguttula* (Ishida) in Andhra Pradesh. *Pest management and Economic Zoology, 7*, 105–110.

Challam, M. S., Subbaratnam, G. V., & Rao, G. R. C. (2001). Role of mixed function oxidases in imparting resistance to the cotton leaf hopper *Amrasca biguttula biguttula* (Ishida). *Pest management and Economic Zoology, 9*, 49–53.

Cheng, G., & Lieu, Y. (1996). Cotton bollworm resistance and its development in northern cotton region of China 1984–1985. *Resistant Pest Management Newsletter, 8*, 32–33.

Cheng, G. L., Liu, R., & Mingjiang, H. (1997). Comparison of cotton aphid resistance level between Xiang and Shandong populations. *Resistant Pest Management Newsletter, 9*, 10–12.

Deguine, J. P. (1996). The evolution of insecticide resistance in *Aphis gossypii* (Homoptera: Aphidae) in Cameroon. *Resistant Pest Management Newsletter, 8*, 13–14.

Delorme, R. D., Auge, M. T., Bethenod, & Villatte, F. (1997). Insecticide resistance in a strain of *Aphis gossypii* from Southern France. *Pesticide Science, 49*, 90–96.

Dennehy, T. J., DeGain, B. A., Harpold, V. S., Brown, J. K., Morin, S., Fabrick, J. A., Byrne, F. J., & Nichols, R. L. (2005). *Vegetable report*. https://arizona.openrepository.com/handle/10150/215014

Dhawan, A. K., Sohi, A. S., Sharma, M., Kumar, T., & Brar, K. S. (2004). *Insecticide resistance management on cotton*. Ludhiana: Punjab Agricultural University.

Dhingra, S., Phokela, A., & Mehrotra, K. N. (1988). Cypermethrin resistance in the populations *Heliothis armigera*. In: National Academy of Sciences. *India; Science Letters, 11*, 123–125.

Dhurua, S., & Gujar, G. T. (2011). Field-evolved resistance to Bt toxin Cry1Ac in the pink bollworm, *Pectinophora gossypiella* (Saunders) (Lepidoptera: Gelechiidae), from India. *Pest Management Science, 67*, 898–203.

Dittrich, V., & Ernst, G. H. (1983). The resistance pattern in white flies of Sudanese cotton. *Mitteilumgen der Deutschen Gesellschatt Fuir Allgemeine and Augewandte and Entomologie, 4*, 96–97.

Djihinto, A. C., Katary, A., Prudent, P., Vassal, J. M., & Vaissayre, M. (2009). Variation in resistance to pyrethoids in *Helicoverpa armigera* from Benin Republic, West Africa. *Journal of Economic Entomology, 5*, 1928–1934.

El-Kady, H., & Devine, G. J. (2003). Insecticide resistance in Egyptian populations of the cotton whitefly, *Bemisia tabaci* (Hemiptera: Aleyrodidae). *Pest Management Science, 59*(8), 865–871.
Ellsworth, P. C., & Jones, J. S. (2001). Cotton IPM in Arizona: A decade of research, implementation and education. In *Proceedings of the Beltwide cotton conference* (pp. 1088–1096). Memphis: National Cotton Council.
Ellsworth, P. C., Fournier, A., & Smith, T. D. (2007). *Arizona cotton insect losses* (Publication AZ1183). College of Agriculture and Life Sciences, Cooperative Extension, University of Arizona: Tucson. http://cals.arizona.edu/crops/cotton/insects/cil/cil.html
Ernst, G. H., & Dittrich, V. (1992). Comparative measurements of resistance to insecticides in three closely-related old and new bollworm species. *Pesticide Science, 34*, 147–152.
Forrester, N. W., Cahill, M., Bird, L. J., & Layland, J. K. (1993). Management of pyrethroid and endosulfan resistance in *Helicoverpa armigera* (Lepidoptera: Noctuidae) in Australia. *Bulletin of Entomological Research, 1*, 132p.
Furk, C., Powell, D. F., & Heyd, S. (1980). Pirimicarb resistance in the melon and cotton aphid *Aphis gossypii* Glover. *Plant Pathology, 29*, 191–196.
Gunning, R. V. (1993). Comparison of two bioassay techniques for larvae of *Helicoverpa* spp. (Lepidoptera: Noctuidae). *Journal of Economic Entomology, 86*, 234–238.
Herron, G. A., Powis, K., & Rophial, J. (2001). Insecticide resistance in *Aphis gossypii* Glover (Homoptera:Aphidae), a serious threat to Australian cotton. *Australian Journal Entomology, 40*, 85–91.
Herron, G. A., Gullick, G. C., & Powis, K. J. (2003). Cross resistance studies in cotton aphid. *Aphis gossypii Glover from Australian cotton. General and Applied Entomology, 32*, 13–20.
Herron, G. A., & Wilson, L. J. (2011). Neonicotinoid resistance in *Aphis gossypii* Glover (Aphididae: Homoptera) from Australian cotton. *Australian Journal of Entomology, 50*, 93–98.
Herron, G. A., Langfield, B. J., Bogema, D. R., & Chen, Y. (2014). Baseline susceptibility and cross resistance in *Aphis gosssypii* Glover to phorate and sulfoxaflor. *Australian Entomology, 53*, 32–35.
Horowitz, R. A., Kontsedalov, S. I., & Ishaaya, I. (2004). Dynamics of resistance to the neonicotinoids acetamiprid and thiamethoxam in Bemisia tabaci (Homoptera: Aleyrodidae). *Journal of Economic Entomology, 97*(6), 2051–2056.
Jeyapradeepa, S. (2000). Studies on insecticide resistance in cotton leaf hopper *Amrasca devastans* (Distant). M Sc. (Agri) Thesis. Tamil Nadu Agricultural University, Madurai, India. 82 pp.
Kennedy, N., Kirti, S., & Gujar, G. T. (2017a). Biological traits associated with Cry1Ac resistance in *Helicoverpa armigera* (Hübner) (Lepidoptera: Noctuidae). *Indian Journal of Experimental Biology, 55*(10), 722–727.
Kennedy, N., Kirti, S., & Gujar, G. T. (2017b). Frequency of alleles conferring resistance to Cry1Ac toxin of *Bacillus thuringiensis* in population of *Helicoverpa armigera* (Hübner) (Lepidoptera: Noctuidae) from India. *Indian Journal of Experimental Biology, 55*(10), 728–733.
Kranthi, K. R. (2005). *Insecticide Resistance – Monitoring, mechanisms and management manual*. Published by CICR, Nagpur, India and ICAC, Washington, USA.
Kranthi, K. R. (2016). Let not the whitefly see red. *Cotton News and Statistics*, 9th February 2016. http://www.cicr.org.in/pdf/Kranthi_art/whitefly_red.pdf
Kranthi, K. R., & Kranthi, N. R. (2004). Modelling adaptability of cotton bollworm *Helicoverpa armigera* (Hubner) to Bt-cotton in India. *Current Science, 87*, 1096–1107.
Kranthi, K. R., Banerjee, S. K., & Russell, D. A. (2000). Insecticide resistance management of cotton pests. *Pestology, 24*, 58–67.
Kranthi, K. R., Jadhav, D. R., Wanjari, R. R., Kranthi, S., & Russell, D. (2001a). Pyrethroid resistance and mechanisms in field strains of *Helicoverpa armigera* Hubner (Lepidoptera: Noctuidae). *Journal of Economic Entomology, 94*, 253–263.
Kranthi, K. R., Jadhav, D. R., Wanjari, R. R., Shakir Ali, S., & Russell, D. A. (2001b). Carbamate and organophosphate resistance in cotton pests in India, 1995–1999. *Bulletin of Entomological Research, 91*, 37–46.
Kranthi, K. R., Jadhav, D. R., Kranthi, S., Wanjari, R. R., Ali, S., & Russell, D. (2002a). Insecticide resistance in five major insect pests of cotton in India. *Crop Protection, 21*, 449–460.

Kranthi, K. R., Russell, D., Wanjari, R., Kherde, M., Munje, S., Lavhe, N., & Armes, N. (2002b). In-season changes in resistance to insecticides in *Helicoverpa armigera* (Lepidoptera: Noctuidae) in India. *Journal of Economic Entomology, 95*, 134–142.

Kranthi, K. R., Dhawad, C. S., Naidu, S., Mate, K., Patil, E., & Kranthi, S. (2005). Bt-cotton seed as a source of *Bacillus thuringiensis* insecticidal Cry1Ac toxin for bioassays to detect and monitor bollworm resistance to Bt-cotton. *Current Science, 88*(05), 796–800.

Kung, K. Y., Chang, K. L., & Chai, K. Y. (1961). Detecting and measuring resistance of cotton aphids to systox. *Acta Entomologia Sinica, 13*, 19–20.

Lei, T., Khan, M., & Wilson, L. (2003). Boll damage by sucking pests: An emerging threat, but what do we know about it? In A. Swanepoel (Ed.), *World cotton research conference III: Cotton for the new millennium* (pp. 1337–1344). Cape Town: Agricultural Research Council, Institute for Industrial Crops.

Martin, T., Ochou, G. O., Vaissayre, M., & Fournier, D. (2003). Organophosphorus insecticides synergize pyrethroids in the resistant strain of cotton bollworm, *Helicoverpa armigera* (Hübner) (Lepidoptera: Noctuidae) from West Africa. *Journal of Economic Entomology, 96*, 468–474.

McCaffery, A. R., King, A. B. S., Walker, A. J., & El-Nayir, H. (1989). Resistance to synthetic pyrethroids in the bollworm, *Heliothis virescens* from Andhra Pradesh, India. *Pesticide Science, 27*, 65–76.

Mehrotra, K. N., & Phokela, A. (1992). Pyrethroid resistance in *Helicoverpa armigera* (Hubner). V. Response of populations in Punjab cotton. *Pesticide Research Journal, 4*, 59–61.

Naik, V.C.B., Kumbhare,S., Kranthi,S., Satija,U and K.R.Kranthi (2018). Field-evolved resistance of pink bollworm, Pectinophora gossypiella (Saunders) (Lepidoptera: Gelechiidae), to transgenic Bacillus thuringiensis (Bt) cotton expressing crystal 1Ac (Cry1Ac) and Cry2Ab in India. *Pest Management Science 74*, 2544–2554. https://doi.org/10.1002/ps.5038.

Naveen, N. C., Chaubey, R., Kumar, D., Rebjith, K. B., Rajagopal, R., Subrahmanyam, & Subramanian, S. (2017). Insecticide resistance status in the whitefly, *Bemisia tabaci* genetic groups Asia-I, Asia-II-1 and Asia-II-7 on the Indian subcontinent. *Scientific Reports, 7*, 40634. https://doi.org/10.1038/srep40634.

Nibouche, S., Beyo, J., Brevault, T., Cretenet, M., Goze, E., Jallas., Martin, P., & Moussa, A. A. (2002, July 15–18). Cotons-Sinbad: A tool for establishing cotton bollworm economic damage thresholds. In: Villalobos, F.J., Testi, L. (Eds.*) VII congress of the European society for agronomy*, Cordoba/Seveilla, Junta de Andalucia 307–308.

Osman, A. A., Watson, T. F., & Sivasupramaniam, S. (1991). Susceptibility of field populations of pink bollworm (Lepidoptera: Gelechiidae) to azinphosmethyl and permethrin and synergism of permethrin. *Journal of Economic Entomology., 84*, 358–362.

Prasad, V. D., Bharati, M., & Reddy, G. P. V. (1993). Relative resistance to conventional insecticides three populations of cotton white fly *Bemisia tabaci* (Genn.) in Andhra Pradesh. *Indian Journal of Plant Protection, 21*, 102–103.

Praveen, P. M. (2003). *Studies on insecticide resistance in early season sucking pests of cotton in Tamil Nadu*. Ph D Thesis. Tamil Nadu Agricultural University. Coimbatore, India. 116 pp.

Preetha, G., Stanley, J., Manoharan, T., & Kuntalam, S. (2014). Baseline toxicity of chloronicotinyls against cotton leaf hopper, *Amrasca bigittula biguttula* and detection of insecticide resistance. *Archives of Phytopathology and Plant Protection, 47*(17), 2095–2105.

Rajendran, T. P., Venugopalan, M. V., & Praharaj, C. S. (2005). Cotton research towards sufficiency to Indian textile industry. *Indian Journal of Agricultural Sciences, 75*, 699–708.

Robert, G. H., Tabashnik, B. E., Ulliman, D. E., Marshal, W. J., & Messing, R. (1994). Resistance of *Aphis gossypii* (Homoptera:Aphidae) to insecticides in Hawaii, spatial patterns and relation to insecticide use. *Journal of Economic Entomology, 87*, 293–300.

Santhini, S., & Uthamasamy, S. (1997). Susceptibility of cotton leaf hopper *(Amrasca devastans*) to insecticides in Tamil Nadu. *Indian Journal of Agricultural Sciences, 67*, 330–331.

Sekhar, P. R., Venkataiah, M., Rao, N. V., Rao, B. R., & Rao, V. S. P. (1996). Monitoring of insecticide resistance in *Helicoverpa armigera* (Hubner) from areas receiving heavy insecticidal applications in Andhra Pradesh (India). *Journal of Entomological Research, 20*, 93–102.

Tabashnik, B. E., Gassmann, A. J., Crowder, D. W., & Carriere, Y. (2008). Insect resistance to Bt crops: Evidence versus theory. *Nature Biotechnology, 26*, 199–202.

Villatte, F., Auge, D., Touton, P., Delorme, R., & Fournier, D. (1999). Negative cross-insensitivity in insecticide-resistant cotton aphid *Aphis gossypii* Glover. *Pesticide Biochemistry and Physiology, 65*, 55–61.

Wang, D., Qiu, W., Ren, X., Niu, F., & Wang, K. (2009). Resistance selection and biochemical characterization of spinosad resistance in *Helicoverpa armigera. Pesticide Biochemistry and Physiology, 95*, 90–94.

Wang, Z., Yan, H., Yang, Y., & Wu, Y. (2010). Biotype and insecticide resistance status of whitefly, *Bemisia tabaci*, from China. *Pest Management Science, 66*(12), 1360–1366.

Wei, C., Rui, C. H., & Fan, X. L. (1988). A study on the resistance of cabbage aphid, *Liphaphis erysimi* to pyrethroids in Beijing area. *Plant Protection, 14*, 17–20.

Wilson, L., Hickman, M., & Deutscher, S. (2006). Research update on IPM and secondary pests. In *Proceedings of the 13th Australian cotton conference* (pp. 249–258). Gold Coast: ACGRA.

Wu, K., Li, W., Feng, H & Guo,Y (2002).Seasonal abundance of the mirids, Lygus lucorum and Adelphocoris spp. (Hemiptera: Miridae) on Bt cotton in northern China. *Crop Protection, 21*(10), 997–1002.

Xianchun, L., Yinchang, W., Qiansong, Z., Ganjun, Y., Dunyang, Z., Yantao, Y., Zhi, Z., Jianping, Z., Shoushan, L., Caixia, C., & Shiyin, D. (1997). Insecticide resistance in field strains of P*ectinophora gossypiella* (Saunders) in China and effect of synergists on deltamethrin and parathion methyl activity. *Pesticide Science, 50*, 183–186.

Yang, Y., Li, Y., & Wu, Y. (2013). Current status of insecticide resistance in *Helicoverpa armigera* after 15 years of Bt cotton planting in China. *Journal of Economic Entomology, 106*(1), 375–381.

Zhang, Y., Han, X., Zhang, W., Luo, L., & Zhou, P. (1997). An electrophysiological study on resistance to pyrethroid insecticides in *Helicoverpa armigera. Acta Entomologica Sinica, 40*, 113–121.

Chapter 12
Ecological Pest Management in the Twenty-First Century: An Analysis of Challenges and Future Strategies

Dharam P. Abrol and Uma Shankar

Abstract In modern agriculture, the role of ecological pest management (EPM) has been enhanced to address environmental and human health issues while improving farms' productivity and profitability in a sustainable way. EPM has shown promises in the reduction of pesticide pressure and in maintaining biodiversity and natural resources. Apart from these, there is a need to make location-specific changes for multiple pest control tactics, educating farmers through community approaches with the use of information and communication technology (ICT) to understand both direct and indirect benefits and risks involved in it. The present chapter addresses information on pesticide usage and its adverse impacts on safer food production for masses, leading to EPM development.

Keywords Ecological pest management · Pesticides · Environmental issue · Food safety · Multiple pest control tactics · ICT · Community participation · Rational use

Abbreviations

BIPM	Biointensive integrated pest management
Bt	*Bacillus thuringiensis*
Cry	Crystal
DFS	Diversified farming systems
EBPM	Ecologically based pest management
EPA	Environmental Protection Agency
EPM	Ecological pest management
FAO	Food and Agriculture Organization

D. P. Abrol · U. Shankar (✉)
Division of Entomology, Sher-e-Kashmir University of Agricultural Sciences & Technology of Jammu, Jammu, India
e-mail: dharam_abrol@rediffmail.com; umashankar.ento@gmail.com

R. Peshin, A. K. Dhawan (eds.), *Natural Resource Management: Ecological Perspectives*, Sustainability in Plant and Crop Protection,
https://doi.org/10.1007/978-3-319-99768-1_12

FFS	Farmer field school
GM	Genetically modified
GMOs	Genetically modified organisms
HCH	Hexachlorocyclohexane
ICPs	Insecticidal crystal proteins
ICT	Information and communication technology
IPM	Integrated pest management
NPV	Nuclear polyhedrosis viruses
NSKE	Neem seed kernel extract
ORP	Operational Research Project
Pis	Proteinase inhibitors
UC-IPM	University of California
US$	US dollar
VoIP	Voice over Internet Protocol
WHO	World Health Organization

12.1 Introduction

India achieved self-sufficiency in the mid-1960s with the advent of the Green Revolution. Its rapid pace of population explosion coupled with urbanization and degrading natural resources has put tremendous pressure on agricultural production. In spite of several initiatives for enhancing agricultural production, feeding the rapidly growing human population in the twenty-first century remains a challenging task. Eventually, there is a need to promote ecologically based pest management (EBPM) which differs significantly from integrated pest management (IPM), relying on a broader view of all pests within an agroecosystem. IPM approaches rely more on insect pests than on weeds, pathogens, and vertebrate pests (Helenius 1997). EBPM supports the natural stability of the agroecosystem while approaching the pest outbreak when pest population density is still low. It emphasizes the agroecosystem stability by achieving high biodiversity, through a combination of different crops and lands with a natural complex of plants. In this context, Ehler and Bottrell (2000, pp. 61–64) pointed out in the online periodical of the US National Academy of Sciences: "*Despite three decades of research, there is very little 'I' in IPM.*" *Instead, the vast majority of "IPM" programs are dominated by single technologies, a few of them by biological control, host plant resistance, or biopesticides that are used as replacements for synthetic chemicals.*

The use of chemical pesticides aims at a "stop or brought to low levels" approach to increase the ecosystem stability (Tshernyshei 1995; UC-IPM 2008). The use of multiple pest control tactics, a fundamental paradigm underlying IPM, presents additional levels of complications, especially when multiple pest types, such as plant pathogens, insects, mites, and nematodes, are targeted. This is particularly important because simply combining different management tactics is not sufficient for the implementation of true IPM programs (Ehler and Bottrell 2000).

There are four different ways of pest management, which include ecological pest management, pesticide-free management, the use of transgenic, and pesticide-intensive management. The first three components aim at reducing the pesticide pressure, harmonizing the ecological balance while focusing on increasing agricultural productivity. The excessive use of chemicals in agriculture in general is harmful to the agroecosystem, and careful decision-making is needed in choosing agrochemicals, particularly the synthetic and new classes of chemical insecticides. No doubt, IPM had been a powerful tool in the hands of the farming community engaged in sustainable agricultural production, enhancing crop productivity (Pretty and Bharucha 2015). But it has not been so successful in many developing countries having small areas and needing to feed an ever-growing, large population. In these circumstances, IPM tactics couldn't achieve their original mandate and must rely on the use of chemical pesticides. Similar observations have been highlighted by Pretty and Bharucha (2015), who portrayed that in spite of a large number of IPM projects implemented in different Asian countries over the past 20 years, IPM's impacts on pesticide use and yields still remain patchy.

12.2 The Pesticide Syndrome

We have travelled a long way (about six decades) since the inception of the economic threshold level proposed by Stern et al. in 1959, and at this juncture we have to redefine and analyze the challenges and strategies of IPM. This is particularly true in developing countries, to boost a safe and sustainable food production. The use of synthetic chemical insecticides increased phenomenally with the introduction of the Green Revolution in India, resulting in substantial yield increases for some time, but boomeranged with several ecological consequences, as documented in *Silent Spring* by Rachel Carson (1962). In the IPM program, chemical pesticides have been set as the last resort in IPM practices, but farmers still use them as the first resort for getting quick results.

However, pesticides proved to be more dangerous when overused, resulting in the deterioration of soil and plant health; contamination of food and the environment; resistance (Atwal and Dhaliwal 1997); pest resurgence (Chelliah 1987); loss of beneficial fauna like bees, birds, and soil microbes (Dhaliwal and Arora 1996; Muralidharan et al. 1992); pesticide poisoning; and frequent, localized pest outbreaks (Abrol and Shankar 2014). No doubt pesticides are credited to have saved millions of lives by controlling diseases, such as malaria and yellow fever which are insect borne. However, their use also causes a variety of adverse health effects and environmental pollution, as described above. Alternative pest control methods and the restricted use of pesticides can minimize the risks due to their usage (Soundararajan 2012). Pesticides can prove to be the most effective instruments in crop protection, and, if correctly used, their effect is fast and complete, which makes them applicable against nearly every pest (Oomen and Bouma 2003). Consequently, the need for IPM strategies to produce safe food and reduce negative externalities caused by pesticides arose (Kogan 1998; Wilson and Tisdell 2001; Abrol and Shankar 2012; Shankar and Abrol 2012). To overcome these challenges and to prevent the situation from further worsening, ecosystem-based IPM appears to be the only feasible solution, as described below.

12.3 Ecosystem-Based IPM

Ecosystem-based IPM involves various eco-friendly tactics to reduce the losses caused by pests to tolerable levels. It involves management programs based on sound ecological principles such as the knowledge of the interactions between the fauna (pests, pollinators, and their natural enemies) and its environment. Gliessman et al. (1981) discussed the perspective ecological basis for the application of traditional agricultural technology in the management of tropical agroecosystems and focused on sustaining yields on a long-term basis, rather than maximizing them in the short-term.

Therefore, it is important to develop sustainable production systems that require low inputs of fertilizer and pesticides and provide incentives for the selection of high-yield and resistant crops that can help alleviate these ecological and economical problems. By understanding the interactions among fertilizers, pesticides, rotations, and other cultural practices, and how they influence the growth and yields of crops, appropriate systems can satisfy human needs for food and also mitigate the impacts of the inputs on the environment. Well-designed systems and other alternatives can contribute to the success of these approaches, such as crop rotation with legumes, use of organic waste either from plants or animals, IPM, use of organic and mechanical weed control and conservation tillage, intercropping or double-row cropping, strip cropping, and undersowing (Edwards 1989). Consequently, focusing on a diverse sustainable agriculture has led to the development of Diversified Farming Systems (DFS) described by Kremen et al. (2012).

12.4 IPM Scenario in India

In India, IPM has remained an exercise, and its adoption among the farming community has not achieved the desired level. This has mostly been due to the lack of proper knowledge among farmers and because of low-level exposure of IPM and pressing needs of farmers to produce more food. This way it has become a sort of "integrated pesticide management" rather than an "integrated pest management" (Peshin and Zhang 2014). The IPM strategy and its implementation and evaluation have always struggled with interpretation and true progress (Peshin and Pimentel 2014). There are several challenges in the successful implementation and adoption of IPM programs, which are discussed in the following subsections.

12.4.1 Scenario of Pesticide Usage in India

In India, the consumption of pesticides is very low, about 0.5 kg/ha of pesticides against 6.6 and 12.0 kg/ha in Korea and Japan, respectively. However, there has been a widespread contamination of food commodities with pesticide residues, basically due to a non-judicious use or application. According to an estimate by Gupta (2004), 51% of food commodities are contaminated with pesticide residues, of which 20% were above the maximum residue levels, on a worldwide basis. Among the various pesticides used in India, more than 40% belong to the organochlorine class (FAO 2005; Agrawal et al. 2010) followed by organophosphorus, another major category of pesticides. It is further strange to find out that, of the technical pesticides used in India, more than 56% are insecticides, followed by herbicides and fungicides. This is in contrast to the pesticide use at the global level, where herbicides are the leading category, followed by insecticides and fungicides. In India, the share of fungicide and herbicide markets has increased by 23% and 18%, respectively, until 2011 (Peshin and Zhang 2014). Further, Abhilash and Singh (2009) observed several, constraints such as the lack of training, potential dangers to health and environment, poor literacy, inappropriate mixing and application methods, repeated and excessive dosage of pesticides to their usage, and issues on safety measures.

Sustainable IPM systems are extremely rare globally, and pesticide use has increased in recent years as the farmers are easily influenced by the pesticide industry. The farmers are driven to apply inexpensive, generic compounds available, rather than scouting their fields (Abrol and Shankar 2012). Interestingly, the use of insecticides, in proportion to all used pesticides, is much higher in developing countries than in developed ones, because of the higher prevalence of herbicide-tolerant transgenic crops (Abrol and Shankar 2012; Peshin and Zhang 2014).

12.4.2 Environmental Degradation/Losses

Indiscriminate and excessive use of pesticides has resulted in the degradation of the environment, adverse effects on human health and other beneficial organisms, development of resistance, and insect pest resurgence. Reliance on single control tactics has seriously impaired the sustainability of our ecosystem, resulting in the contamination of food, problems on pesticide residues, and resistance in target species (Vega et al. 2009; Miller et al. 2010). Therefore, a system based on sound ecological principles is desired for sustainable agricultural production with a low impact on the natural balance (Overton 1996; Lewis et al. 1997; Kennedy and Sutton 2000).

12.4.3 Lack of Proper Taxonomy of Pests and Biocontrol Agents

Identification or proper diagnosis is the prerequisite in IPM programs. Numerous examples can be cited where wrong classification has led to misinterpretations. For instance, *Salvinia molesta* (kariba weed), native of Brazil, is an aquatic fern and one of the world's worst weeds. It causes extensive environmental damage by choking lakes, reservoirs, slow-moving rivers, irrigation systems, rice paddies, fishponds, etc. with continuous meter-thick mats of dense vegetation. In addition to rendering the water useless for normal use, its presence can lead to the breeding of mosquitoes. It was initially identified as *Salvinia auriculata*. A weevil, *Cyrtobagous singularis*, from Trinidad was used in Africa to control it, but the efforts failed. Later, it was identified as *S. molesta*, whose growth in Queensland was controlled by *Cyrtobagous* sp. from Brazil. It is evident from these examples as to how effective control or mitigation measures could be implemented, depending on an appropriate taxonomy (Abrol 2013).

12.4.4 Problems in Pest Management in the Context of Climate Change

Climate change is no longer a matter of opinion or speculation, almost worldwide. Of concern now is the assessment of the extent of the changes and their potential impacts. Climate change, food insecurity, and energy demand are major concerns for modern agriculture, and their impact is increasing rapidly. The last decade has seen new developments in food production: the genetic engineering of organisms and the organic chemical-free agriculture. Biotechnology and the release of genetically modified organisms (GMOs), such as engineered soybean (*Glycine max*), colza (*Brassica napus* subsp. *napus*), maize (*Zea mays*), and tomatoes (*Lycopersicon esculentum*), promise a solution to food security needs and nutritional problems (Khush 2002). Most biotechnology companies claim that their GMOs may be resistant to insect pests, molds, frost, dry conditions, etc., and it could be instrumental in producing a new revolution in agriculture (Pingali and Traxler 2002). For instance, soybean and other plants were modified to be tolerant to glyphosate, which not only enabled crops to fight weeds but also produce much higher yields. However, because the weeds become increasingly resistant to this herbicide, the use of these genetically modified (GM) plants renders the farmers dependent on the use of more and more glyphosate.

12.4.5 Lack of Information and Communication Technology (ICT) in IPM

Practically, any IPM is concerned with most constitutional levels present in the agroecosystems, from populations and commodities down to individual viruses or genotypes, genomes and genes, as well as up to the levels of landscape and global ecosystems. In fact, IPM practices are involved in a complex course responding to climatic change, soil dynamics, vegetation evolution, and human activities.

Radio, television, and print media are vital in many developing countries. Over the last decade, "new" ICTs, such as mobile phones and Internet-associated applications such as Voice over Internet Protocol (VoIP), have become easily available to people worldwide. Developing countries face challenges when harnessing the potential of ICT for economic development (Michelle and Fong 2009). The potential of ICT for the speedy dissemination of information to farmers needs to be felt (Meera et al. 2004), and up to a certain extent, they have been exploited for pest and disease diagnosis and forecasting, management options, and marketing information, in more accurate ways. In modern days, ICT has proved to be a powerful tool in pest forecasting as a propeller giving priority to prevention. Pest forecasting involves data acquisition, processing, and information dissemination. ICT can also be very helpful in terms of enforcing IPM (Shen et al. 2012).

12.4.6 IPM Implementation and Adoption

In the 1970s and 1980s, the first IPM program under the Operational Research Project (ORP) focused on pilot programs using a prescriptive approach to demonstrate IPM practices in cotton and rice crops in a cluster of villages in seven states. The government of India adopted IPM as the main strategy for plant protection in 1985. In the early 1990s, the farmer field school (FFS) model was adopted to implement IPM by educating farmers and extension workers. Between 1990 (before many ad hoc IPM programs began) and 2002 (when Bt cotton was introduced), the use of pesticides active ingredient (a.i.) by weight decreased by 35% (Peshin et al. 2014). This was mainly due to the ban of hexachlorocyclohexane (HCH) in 1997, accounting for 30% of the total pesticides; consequently, low-dosage pesticides were introduced (Peshin et al. 2014). Only about 2–4% of the total cultivated area, including only 5% of farmers, however, is covered under IPM programs, so whether IPM has reduced the overall pesticide use in Indian agriculture is debatable. Although the introduction of Bt cotton has reduced the insecticide use in cotton by almost 50%, the mass pesticide use in Indian agriculture overall has increased by 9% since 2002 (Peshin et al. 2014).

Bt crops are compatible with IPM strategies, but they are not sustainable on their own. Overreliance on transgenic crops has already led to weed and insect resistance, which may lead farmers into a transgenic-cum-pesticide treadmill (Peshin et al. 2007). It is evident from the published literature that transgenic insecticidal crops derived from the genes of the bacterium *Bacillus thuringiensis* (Bt) are the most important technological advancement in insect pest management since the development of synthetic insecticides (Vaeck et al. 1987; Koziel et al. 1993; Perlak et al. 1990, 1993). At least 18 transgenic insecticidal crops have been field-tested in the USA, and three (corn, cotton, and potato) have been widely planted (Federici 1998; Gould 1998). But as the commercial availability of these crops has grown, so too has the controversy over how to assess and manage the risks posed by this method of control. The widespread planting of a million hectares of transgenic crops with high levels of insecticidal proteins raises concerns that pest populations might develop resistance to Bt toxins and that food webs might be disrupted (Gould 1998; McGaughey et al. 1998; Marvier 2001). Indeed, the US Environmental Protection Agency (EPA) requires the industry to maintain populations of susceptible (nonresistant) insect pests to slow down the development of resistant populations.

Experiences with the implementation of pesticide action plans and IPM programs around the world confirm that reduction in pesticide use by mass is not a robust indicator to measure the success of IPM (Sharma et al. 2015). Low-volume pesticides propelled the pesticide use reduction in many countries. The pesticide treatment frequency index and the field use environmental impact quotient are better evaluation indicators to measure the impact of IPM programs (Peshin et al. 2014; Sharma et al. 2015). How can the experiences with IPM technology and extension, documented in this book, be bracketed successful and viable? In many instances, IPM technologies developed at the research level have not been effectively scaled up to industry-wide practices, because of the lack of a well-conceived and evaluated extension process. Different extension approaches are needed in different situations for greater adoption of IPM by farmers. IPM practices in most cases are tested for success at pilot scale but fail when facing the constraints, mainly the IPM attributes, for replication on a larger scale (Peshin and Pimentel 2014). Some of the strategies which need to be adopted to make an IPM program successful are included below.

12.5 Ecological Pest Management Program

Levins and Wilson (1980) have raised certain issues related to the ecological theory and pest management, to avoid the narrow theoretical base of contemporary pest management practices. The main drivers of ecological approach are processes used and manipulated in conjunction with traditional methodologies, as an alternative for maintaining the high productivity and harvest quality of an agroecosystem. This approach is more environmentally compatible and requires less input of energy and resources than conventional pest management (Brown 1999). The aim of this new approach is to shift management strategies so that they have less reliance on chemicals and more on the biology of pests and their interactions with crops. Thus, ecologically based IPM combining all approaches – physical, cultural, chemical, and biological – is the only option for sustaining productivity and maintaining the health of ecosystems (Kennedy and Shutton 2000).

12.5.1 Biointensive Integrated Pest Management (BIPM) Program

BIPM is the recent trend to reduce pesticide pressure and to enhance farmers' interest to fetch higher remuneration for their produce. It begins with steps to diagnose the nature and source of pest problems accurately and then relies on a range of preventive tactics and biological controls to keep pest populations within acceptable limits. Reduced-risk pesticides are used if other tactics have not been adequately effective, as a last resort, and are used with care to minimize risks. BIPM also opens new horizons to opt for a better choice, such as biological control and the use of bio-rational products, which are less toxic and only affect the target pest. It also includes biopesticides derived from microbials, parasitoids, predators, botanicals, and all conventional, nonchemical methods of pest management. It refers to the more dynamic and ecologically informed approach to IPM that considers farms as a vital part of an agroecosystem (Kaul et al. 2009).

12.5.2 Ensuring Availability of Biological Control Agents, Microbial and Botanical Pesticides, and IPM Devices

A sensible use of these biocontrol agents can contribute substantially to stable pest management. Hoy and Herzog (1986) have clearly documented the role of biological control in agricultural IPM system. However, several issues such as high costs involved in production, poor persistence requiring more frequent release, poor quality of bioproducts sold in the Indian market, and short shelf-life should be addressed to promote biocontrol programs in India. Insect predators and parasitoids act as biological pest control agents in agroecosystems and thus provide valuable ecosystem services (Ramsden et al. 2014). Microbial pathogens such as viruses, bacteria, and fungi are potential bio-rational pesticides. The usefulness of nuclear polyhedrosis viruses (NPV) in the management of polyphagous pests such as *Helicoverpa armigera* and *Spodoptera litura* on several crops has been demonstrated (Rabindra 1998). It is possible to isolate strains of baculoviruses with increased virulence and genetically improve them further (Rabindra 2000). The major advantages with viral insecticides are their specificity and high levels of safety. Insects also do not develop resistance to baculoviruses.

Bacillus thuringiensis is widely used for lepidopteran pests. The plant kingdom is a rich source of insecticidal components, and several hundred plants that have been identified possess pesticidal action. They can be used as insecticide, acaricide, nematicide, and fungicide. Neem (*Azadirachta indica*) seeds contain several toxic properties, including antifeedant and insecticidal compounds. Though several commercial formulations are available in the market, the most effective is the neem seed kernel extract (NSKE).

12.5.3 Cropping System Approach

Our ancestors have long been managing pests through trial and error, learning the best crops to grow, the best sowing times, and the best crop combinations and spacings that would reduce the insect pest share of the harvest. They selected seeds of plants that were tolerant to the pests in their fields. Cropping systems such as intercropping or mixed cropping, strip cropping, etc., when carefully selected, can reduce pest incidence by (1) acting as a barrier; (2) acting as alternate hosts, diverting the pest away from the crop at risk; and (3) helping natural enemies of the pest or minimizing risks involved in monocultures. Therefore, crop diversification is one of the best drivers to mitigate pests' pressure and to maintain the biodiversity of beneficial insects, naturally suppressing the pest population.

12.5.4 Habitat Manipulation or Ecological Engineering

Ecological engineering has recently emerged as a paradigm for considering pest management approaches that are based on cultural practices informed by ecological knowledge, rather than those based on high-technology approaches such as synthetic pesticides and genetically engineered crops (Gurr et al. 2004). This requires that the functional mechanisms that allow some components of the local species biodiversity to suppress pest activity are better understood and exploited. Pest suppression via ecological engineering is placed in the broader context of "ecosystem services" provided by farmland biodiversity including nitrogen fixation and the conservation of pollinator species and wildlife. Habitat manipulation aims to provide the natural enemies of pests with resources such as nectar (Baggen and Gurr 1998), pollen (Hickman and Wratten 1996), physical refugia (Halaji et al. 2000), alternative preys (Abou-Awad et al. 1998), alternative hosts (Viggiani 2003), and lekking sites (Sutherland et al. 2001). Habitat manipulation approaches provide the above-listed resources and operate to reduce pest densities via enhancement of natural enemies (Landis et al. 2000) (Fig. 12.1a–d).

It is evident that crop diversification tends to increase natural enemy abundance and diversity, providing a system more resilient to pest population increase. Overall, farming diversity within the agroecosystem may also affect biological control by

Fig. 12.1 Habitat manipulation with flowering resources for natural enemies (**a**, **b**), diversified farming garden pea and linseed (**c**), and strips of flowering plants in the vicinity of an onion crop (**d**)

natural enemies, due in part to a wider range of flowering plants that provide nectar (carbohydrate) and pollen (protein) resources to insects during longer times of the growing season. Thus, pest outbreaks tend to be less common in polycultures than in monocultures (Andow 1991).

12.5.5 Farmers' Demonstration Through Community Participation for Enhancing the Adoption of IPM

The problem of acute pesticide poisoning (Bhanti et al. 2004) is well established among the Indian farmers, and its long-term exposure (Kamel et al. 2003) results in many neurological diseases (Richter 2002; Michael et al. 2004; Kamel et al. 2005; Costa et al. 2008; Mostafalou and Abdollahi 2013) in the near future. Evidently, the current need is to address the awareness for long-term exposure and the usage of new pesticide spraying technologies among farmers. Besides these, the existing traditional or indigenous technology needs to be blended with the new one for better implementation and adoption by farmers (Pina and Forcada 2004; Falconer and Hodge 2000). In this context, farmer field schools have proved to be a very effective method for creating learning opportunities for farmers and to look for solutions to their problems. Gaining knowledge and practical experience is necessary for farmers to manage their farms successfully.

12.5.6 Avoidance and Excessive Use of Pesticides

Currently, pesticides are being used on 25% of cultivated area, and their consumption is showing a slight declining trend, probably due to farmers' shift toward biopesticides, natural plant sources, and other alternative methods. Despite such large amount of consumption of pesticides, crop losses have declined from 23.3% to 17.5% in post–Green Revolution India. In monetary terms, these losses amount to 19,283 million US dollars (Rs. 8,63,884, 1 US dollar = Rs 44.80 at 2010 rates) per year, due to insect pests (Dhaliwal et al. 2010). It has been observed that long-term, low-dose exposures are increasingly linked to human health effects such as immune suppression, hormone disruption, diminished intelligence, reproductive abnormalities, and cancer. In the light of this, pesticide safety, regulation of application, and use of biotechnologies, biopesticides, and products obtained from natural plant sources such as neem extracts are some of the future strategies expected to minimize human exposure to pesticides.

12.5.7 Use of Insect-Resistant Crop Varieties

Host plant resistance is an important component around which an IPM program should be built. Use of resistant varieties is in harmony with nature. While exploiting this valuable component of host plant resistance, entomologists should be aware of the pest dynamics resulting in the development of biotypes, as in the case of BPH and gall midge. Horizontal or field resistance, which is polygenic, should be preferred rather than vertical resistance, in breeding. Developments of molecular tools such as marker-aided selection for identification of resistant sources and wide hybridization and embryo rescue techniques, with insect-resistant genes, are landmark achievements in the development of insect-resistant crop plants. Any insect-resistant variety, however, should be compatible with other components of IPM, particularly biocontrol agents such as parasites, predators, and pathogens. Development of such genotypes possessing good agronomic traits, i.e., increased yield and quality, will have greater acceptance and application in pest management programs.

12.5.8 Rational Use of New Interventions and Biotechnological Approaches

New technological interventions such as transgenics and the offering of bio-rational compounds may become the more suitable choice for the farming community to maintain the balance in any sustainable agroecosystem. These new interventions may prove to be environmentally safe, economically sound, and socially viable in the long run. Major technological advances in biochemistry, molecular biology, genetics, and biotechnology have facilitated in the designing of plants with improved resistance to insect pests. The potential of genetic transformation technology has been widely recognized over the last two decades. Recent advances in molecular biology, plant tissue culture, and genetic engineering have clearly demonstrated the possibility of incorporating foreign genes for desired characteristic s while preserving the existing traits of improved genotypes. Bt is a Gram-positive, aerobic, sporulating bacterium that synthesizes crystalline proteins during sporulation. These crystalline (Cry) proteins are highly insecticidal at very low concentrations to lepidopterans, coleopterans, dipterans, lice, mites, and nematodes, depending on the bacterium isolate, and are nontoxic to mammals and other organisms.

By using biotechnological approaches, it has become possible to use Bt more effectively and rationally by introducing the insecticidal crystal proteins (ICPs) of Bt into crop plants. The ICPs are classified on the basis of their amino acid sequence homology. The first transgenic tobacco plants using cry genes were developed against tobacco hornworm in 1987. The commercialization of Bt transgenic crops started in 1996 in the USA with the introduction of bollworm-resistant cotton (Bollgard), carrying the Bt toxin gene, while Bt cotton was commercialized in India in 2002.

Other advanced approaches consider a wide array of defense proteins that plants possess, including the PIs and lectins induced in response to insect attack.

12.5.9 Organic Farming: A Safer Way to Get Remunerative Foods

Organic farming is a production system which eliminates the use of synthetically compounded fertilizers, pesticides, growth regulators, and livestock feed additives. It relies on crop rotation and residues, animal manures, legumes, green manures, organic wastes, mechanical cultivation, and aspects of biological pest control to maintain productivity, to supply nutrients, and to control weeds and other insects. The principles of organic food productions aim at encouraging and enhancing biological cycles within the farming system, to maintain and increase long-term fertility of soil, minimizing different forms of pollution caused by synthetic chemicals (fertilizers and pesticides), and producing foods of high quality, in sufficient amounts. It has been reported that more than 75% of consumers are very concerned with pesticide residues in food. All these issues have a direct bearing on the marketing of the products (Hallman et al. 2003). However, this process is costly, quite labor-intensive, and, in some cases, ineffective (Abhilash and Singh 2009).

Sikkim became the first state in India to have gone organic, which portrays the sound picture of the state to promote a pesticide-free agri-horticulture produce and simultaneously agritourism, to provide job opportunities to the rural masses. The organic scenario will definitely assist in maintaining the balance of an ecosystem, promoting safe and residue-free produce and thereby fetching enhanced remuneration from their organic certification.

12.5.10 Crop Protection Models

Pest models help to know how key components and processes affect pests' development, crop damage, and the effectiveness of management. Using an optimum quantity of eco-friendly chemicals is more important in the context of health hazard, environmental aspects minimizing the resistance development, cost of chemicals, etc. Crop protection models play an important role in the estimation of more accurate economic injury level and also assist in forecasting and forewarning pest population insurgence to the farming community.

12.6 Conclusion

Keeping the above facts in mind, it is imperative to educate farmers to use pesticides judiciously. Misuse and overuse of pesticides result in the contamination of food, degradation of environment, and health hazards to humans. Other alternatives such as the use of biotechnology, biopesticides, and pesticides obtained from natural products such as neem, pongamia oil, etc. should be encouraged. Moreover, the government should frame a policy to propagate and strengthen the dissemination and the adoption of IPM by farmers through demonstration of technologies, which will not only ensure the production of safe food but also reduce severe environmental and economic losses.

References

Abhilash, P. C., & Singh, N. (2009). Pesticide use and application: An Indian scenario. *Journal of Hazardous Materials, 165*, 1–12.

Abou-Awad, B. A., El-Sherif, A. A., Hassan, M. F., & Abou-Eleila, M. M. (1998). Studies on development, longevity, fecundity and predation of *Amblyseius olivi* Nasr & Abou-Awad (Acari: Phytoseiidae) on various kinds of prey and diets. *Zeitschrift fur Pflanzenkrankheiten und Pflanzenschutz, 105*(5), 538–544.

Abrol, D. P. (2013). Future of taxonomy in the 21st century – Whither or wither. *Current Science, 104*(12), 1594.

Abrol, D. P., & Shankar, U. (2012). History, overview and principles of ecologically-based pest management. In D. P. Abrol & U. Shankar (Eds.), *Integrated pest management: Principles and practices* (pp. 1–26). Wallingford: CAB International.

Abrol, D. P., & Shankar, U. (2014). Pesticides, food safety and integrated pest management. In D. Pimentel & R. Peshin (Eds.), *Integrated pest management- pesticide problems* (Vol. 3, pp. 167–199). New York: Springer Dordrecht.

Agrawal, A., Pandey, R. S., & Sharma, B. (2010). Water pollution with special reference to pesticide contamination in India. *Journal of Water Resource and Protection, 2*, 432–448.

Andow, D. A. (1991). Vegetational diversity and arthropod population response. *Annual Review of Entomology, 36*, 561–586.

Atwal, A. S., & Dhaliwal, G. S. (1997). *Agricultural pests of South Asia and their management.* Ludhiana: Kalyani Publishers.

Baggen, L. R., & Gurr, G. M. (1998). The influence of food on *Copidosoma koehleri* (Hymenoptera: Encyrtidae), and the use of flowering plants as a habitat management tool to enhance biological control of potato moth, *Phthorimaea operculella* (Lepidoptera: Gelechiidae). *Biological Control, 11*, 9–17.

Bhanti, M., Shukla, G., & Taneja, A. (2004). Contamination levels of organochlorine pesticides and farmers' knowledge, perception, practices in rural India–a case study. *Bulletin of Environmental Contamination and Toxicology, 73*, 787–793.

Brown, M. W. (1999). Applying principles of community ecology to pest management in orchards. *Agriculture, Ecosystems and Environment, 73*, 103–106.

Carson, R. (1962). *Silent spring*. Boston: Houghton Mifflin.

Chelliah, S. (1987). Insecticide induced resurgence of rice brown plant hopper, *Nilaparvata lugens* (Stal) in resurgence of sucking pests. In S Jayaraj (Ed.) (p. 114). Coimbatore: TNAU Coimbatore.

Costa, L. G., Giordano, G., Guizzetti, M., & Vitalone, A. (2008). Neurotoxicity of pesticides: A brief review. *Frontiers in Bioscience, 13*, 1240–1249.

Dhaliwal, G. S., & Arora, R. (1996). *Principles of insect pest management*. Ludhiana: National Agricultural Technology Information Centre.

Dhaliwal, G. S., Jindal, V., & Dhawan, A. K. (2010). Insect pest problems and crop losses: Changing trends. *Indian Journal of Ecology, 37*(1), 1–7.

Edwards, C. A. (1989). The importance of integration in sustainable agricultural systems. *Agriculture, Ecosystems and Environment, 27*, 25–35.

Ehler, L. E., & Bottrell, D. G. (2000). The illusion of integrated pest management. *Issues in Science and Technology, 16*(3), 61–64.

Falconer, K., & Hodge, I. (2000). Using economic incentives for pesticide usage reductions: Responsiveness to input taxation and agricultural systems. *Agricultural Systems, 63*, 175–194.

FAO. (2005). *The state of food insecurity in the world 2005* (Economic and social department). Rome: Food and Agriculture Organization of the United Nations. http://www.fao.org/docrep/008/a0200e/a0200e00.htm. Accessed 30 May 2013.

Federici, B. A. (1998). Naturally occurring baculoviruses for insect pest control. In F. R. Hall & J. J. Mens (Eds.), *Biopesticides: Use and delivery* (pp. 301–320). Totowa: Humana press.

Gliessman, S. R., Garcia, E. R., & Amador, A. M. (1981). The ecological basis for the application of traditional agricultural technology in the management of tropical agro-ecosystems. *Agro-Ecosystems, 7*, 173–185.

Gould, F. (1998). Sustainability of transgenic insecticidal cultivars: Integrating pest genetics and ecology. *Annual Review of Entomology, 43*, 701–726.

Gupta, P. K. (2004). Pesticide exposure – Indian scene. *Toxicology, 198*, 83–90.

Gurr, G. M., Wratten, S. D., & Altieri, M. A. (2004). *Ecological engineering: Advances in habitat manipulation for arthropods*. Collingwood: CSIRO Publishing.

Halaji, J., Cady, A. B., & Uetz, G. W. (2000). Modular habitat refugia enhance generalist predators and lower plant damage in soybeans. *Environmental Entomology, 29*, 383–393.

Hallman, W. K., Hebden, W. C., Aquino, H. L., Cuite, C. L., & Lang, J. T. (2003). *Public perceptions of genetically modified foods: A national study of American knowledge and opinion.* Publication No. 1003–004. http://ideas.repec.org/p/ags/rutfwp/18174.html

Helenius, J. (1997). Spatial scales in Ecological Pest Management (EPM): Importance of regional crop rotations. *Biological Agriculture & Horticulture: An International Journal for Sustainable Production Systems, 15*(1–4), 162–170.

Hickman, J. M., & Wratten, S. D. (1996). Use of Phacelia tanacetifolia strips to enhance biological control of aphids by hoverfly larvae in cereal fields. *Journal of Economic Entomology, 89*, 832–840.
Hoy, M. A., & Herzog, D. C. (1986). *Biological control in agricultural IPM systems*. Orlando: Academic.
Kamel, F., Rowland, A., Park, L., Anger, K., Baird, D., Gladen, B., Moreno, T., Stallone, L., & Sandler, D. (2003). Neurobehavioral performance and work experience in Florida farmworkers. *Environmental Health Perspectives, 111*(14), 1765–1772.
Kamel, F., Engel, L., Gladen, B., Hoppin, J., Alavanja, M., & Sandler, D. (2005). Neurologic symptoms in licensed private pesticide applicators in the agricultural health study. *Environmental Health Perspectives, 113*(7), 877–882.
Kaul, V., Shankar, U., & Khushu, M. (2009). Bio-intensive integrated Pest management in fruit crop ecosystem. In R. Peshin & A. K. Dhawan (Eds.), *Integrated pest management: Innovation development process* (Vol. 1, pp. 631–666). Berlin: Springer.
Kennedy, G. G., & Sutton, T. B. (2000). *Emerging technologies for integrated pest management: Concept research and implementation*. St Paul: APS Press.
Khush, G. S. (2002). The promise of biotechnology in addressing current nutritional problems in developing countries. *Food and Nutrition Bulletin, 23*(4), 354–357.
Kogan, M. (1998). Integrated Pest management: Historical perspectives and contemporary developments. *Annual Review of Entomology, 43*, 243–270.
Koziel, M., Beland, G. L., Bowman, C., Carozzi, N. B., Crenshaw, R., Crossland, L., Dawson, J., Desai, N., Hill, M., Kadwell, S., Launis, K., Lewis, K., Maddox, D., Pherson, K. M., Meghji, M., Merlin, E., Rhodes, R., Warren, G. W., Wright, M., & Evola, S. (1993). Field performance of elite transgenic maize plants expressing an insecticidal protein derived from *Bacillus thuringiensis*. *Biotechnology, 11*, 194–200.
Kremen, C., Iles, A., & Bacon, C. (2012). Diversified farming systems: An agroecological, systems-based alternative to modern industrial agriculture. *Ecology and Society, 17*, 44.
Landis, D. A., Wratten, S. D., & Gurr, G. M. (2000). Habitat management to conserve natural enemies of arthropod pests in agriculture. *Annual Review of Entomology, 45*, 175–201.
Levins, R., & Wilson, M. (1980). Ecological theory and pest management. *Annual Review of Entomology, 25*, 287–308.
Lewis, W. J., van Lenteren, J. C., Phatak, S. C., & Tumlinson, J. H. I. I. (1997). A total system approach to sustainable pest management. *Proceedings of the National Academy of Sciences USA, 94*, 12243–12248.
Marvier, M. (2001). Ecology of transgenic crops. *American Scientist, 89*, 160–167.
McGaughey, W. H., Gould, F., & Gelernter, W. (1998). Bt resistance management: A plan for reconciling the needs of the many stakeholders in Bt-based products. *Nature Biotechnology, 16*, 144–146.
Meera, S. N., Jhamtani, A., & Rao, D. U. M. (2004). *Information and communication technology in agricultural development: A comparative analysis of three projects from India*. AgREN Network Paper No. 135, ODI, January 2004. 20 p. Available at: www.odi.org.uk/agren/papers/agrenpaper_135.pdf
Michael, C. R., Hoppin, A. J. A., & Kamel, F. (2004). Health effects of chronic pesticide exposure: Cancer and neurotoxicity. *Annual Review of Public Health, 25*, 155–197.
Michelle, W., & Fong, L. (2009). Digital divide: The case of developing countries. *Issues in Informing Science and Information Technology, 2009*, 471–478.
Miller, J. R., McGhee, P. S., Siegert, P. Y., Adams, C. G., Huang, J., Grieshop, M. J., & Gut, L. J. (2010). General principles of attraction and competitive attraction as revealed by large-cage studies of moths responding to sex pheromone. *Proceedings of the National Academy of Sciences USA, 107*, 22–27.
Mostafalou, S., & Abdollahi, M. (2013). Pesticides and human chronic diseases: Evidences, mechanisms, and perspectives. *Toxicology and Applied Pharmacology, 268*, 157–177.

Muralidharan, S., Regupathy, A., & Sundaramoorthy, T. (1992). *Organochlorine residues in the eggs of selected colonial water birds breeding at Keoladeo National Park, Bharatpur, India* (NHRI symposium series 7, pp. 189–195). Saskatoon: Environment Canada.

Oomen, P. A., & Bouma, E. (2003). *Role of pesticides in IPM*. Wageningen: Plant Protection Service.

Overton, J. (1996). *Ecologically based Pest management: New solutions for a new century*. Washington, DC: National Academy Press.

Perlak, F. J., Deaton, R. W., Armstrong, T. A., Fuchs, R. L., Sims, S. R., Greenplate, J. T., & Fischhoff, D. A. (1990). Insect resistant cotton plants. *Biotechnology, 8*, 939–943.

Perlak, F. J., Stone, T. B., Muskopf, Y. M., Peterson, L. J., Parker, G. B., Mcpherson, S. A., Wyman, J., Love, S., Reed, G., Biever, D., & Fischhoff, D. A. (1993). Genetically improved potatoes: Protection from damage by colorado potato beetles. *Plant Molecular Biology, 22*, 313–321.

Peshin, R., Dhawan, A.K., Vatta, K., & Singh, K. (2007). Attributes and socio-economic dynamics of adopting Bt cotton. *Economic and Political Weekly, 45*(51), 73–80.

Peshin, R., & Pimentel, D. (2014). *Integrated pest management experiences with implementation, global overview* (Vol. 4). Dordrecht/New York: Springer.

Peshin, R., & Zhang, W. (2014). Integrated pest management and pesticide use. In D. Pimentel & R. Peshin (Eds.), *Integrated pest management- pesticide problems* (Vol. 3, pp. 1–46). Dordrecht/New York: Springer.

Peshin, R., Kranthi, K. R., & Sharma, R. (2014). Pesticide use and experiences with integrated pest management programs and Bt cotton in India. In R. Peshin & D. Pimentel (Eds.), *Integrated pest management-experiences with implementation, global overview* (Vol. 4, pp. 269–306). Dordrecht/New York: Springer.

Pina, C. M., & Forcada, S. A. (2004). Effects of an environmental tax on pesticides in Mexico. *Industry and Environment, 27*, 34–38.

Pingali, P. L., & Traxler, G. (2002). Changing locus of agricultural research: Will the poor benefit from biotechnology and privatization trends. *Food Policy, 27*, 223–238.

Pretty, J., & Bharucha, Z. P. (2015). Integrated pest management for sustainable intensification of agriculture in Asia and Africa. *Insects, 6*, 152–182.

Rabindra, R. J. (1998, November 12–13). Development of microbial pesticides: Present and future prospects. In: *Abstract of the national symposium on development of microbial pesticides and insect pest management* (p. 10). Pune: Research and Development Centre, Hindustan Antibiotics Ltd.

Rabindra, R. J. (2000). Genetic improvement of baculoviruses for insect pest management. In *Proceedings of national symposium on microbials in insect pest management*. February 24–25. Entomology Research Institute, Loyola College: Chennai.

Ramsden, M. W., Menéndez, R., Leather, S. R., & Wackers, F. (2014). Optimizing field margins for biocontrol services: The relative role of aphid abundance, annual floral resources, and overwinter habitat in enhancing aphid natural enemies. *Agriculture, Ecosystems and Environment, 199*, 94–104.

Richter, E. D. (2002). Acute human pesticide poisonings. *Encyclopedia of Pest Management*, 3–6.

Shankar, U., & Abrol, D. P. (2012). Role of integrated pest management in food and nutritional security. In D. P. Abrol & U. Shankar (Eds.), *Integrated pest management: Principles and practices* (pp. 408–432). Wallingford: CAB International.

Sharma, R., Peshin, R., Shankar, U., Kaul, V., & Sharma, S. (2015). Impact evaluation indicators of an integrated Pest management program in vegetable crops in the subtropical region of Jammu and Kashmir, India. *Crop Protection, 67*, 191–199.

Shen, Z.-R., Cheng, D.-F., Zhang, Y.-H., & Gao, L.-W. (2012). Role of information and communication technology in integrated pest management. In D. P. Abrol & U. Shankar (Eds.), *Integrated pest management: Principles and practices*. Wallingford: CAB International.

Soundararajan, R. P. (2012). *Pesticides – advances in chemical and botanical pesticides*. Rijeka: InTech Janeza Trdine. https://doi.org/10.5772/2609. Accessed 27 Mar 2013.

Stern, V. M., Smith, R. F., den Bosch, V., & Hagen, K. S. (1959). The integration of chemical and biological control of the spotted alfalfa aphid, I: The integrated control concept. *Hilgardia, 29*, 81–101.
Sutherland, J. P., Sullivan, M. S., & Poppy, G. M. (2001). Distribution and abundance of aphidophagous hoverflies (Diptera: Syrphidae) in wildflower patches and field margin habitats. *Agricultural and Forest Entomology, 3*, 57–64.
Tshernyshei, W. B. (1995). Ecological Pest management (EPM): General approaches. *Journal of Applied Entomology, 119*, 379–381.
UC-IPM. (2008). *What is IPM?* Statewide IPM program, University of California, Agriculture and Natural Resources. Available at: www.ipm.ucdavis.edu/WATER/U/ipm.html. Accessed 26 March 2018.
Vaeck, M., Reynaerts, A., Höfte, H., De Beuckeleer, M., Jansen, S., Dean, C., Zabeau, M., van Montagu, M., & Leemans, J. (1987). Transgenic plants protected from insect attack. *Nature, 328*, 33–37.
Vega, F. E., Infante, F., Castillo, A., & Jaramillo, J. (2009). The coffee berry borer, *Hypothenemus hampei* (Ferrari) (Coleoptera: Curculionidae): A short review, with recent findings and future research directions. *Terrestrial Arthropod Reviews, 2*, 129–147.
Viggiani, G. (2003). Functional biodiversity for the vineyard agroecosystem: Aspects of the farm and landscape management in Southern Italy. *Bulletin OILB-SROP, 26*(4), 197–202.
Wilson, C., & Tisdell, C. (2001). Why farmers continue to use pesticides despite environmental, health and sustainability costs. *Ecological Economics, 39*, 449–462.

Chapter 13
Understanding the Diversity of Lac Insects of *Kerria* spp. in India and the Nature of Insect-Host Plant Interaction

Kewal K. Sharma

Abstract Lac insects are phytosuccivorous and prefer certain species of plants called lac hosts. They remain attached to the host plant throughout their life, except for crawler and adult male stages. Due to their unique biology, host preference and dispersal mechanisms, lac insects are expected to differentiate locally, forming geographic and host races without adequate morphological differentiation. Phloem sap is the sole source of nutrition for the lac insect. The factors governing the selection of only a few host taxa by the lac insect in general and preference of specific host species by different lac insect strains and species needed to be investigated. Detailed studies aimed at detection, identification and characterization of endosymbionts are desirable as they are likely to play an important role in host metabolism, reproduction as well as biosynthesis of the lac constituents. An understanding of lac insect biodiversity and insect-host plant relationship required for lac production is far from complete without the knowledge on the above aspects. Therefore, a study was undertaken for the better understanding of lac insect biodiversity and lac insect host plant interaction. The characterization parameters, namely morphology, karyology, biochemical and molecular profiles, variations in the resin and dye produced by these insects were studied and are discussed in this chapter.

Keywords Lac insect · *Kerria* spp. · Diversity · Insect-host plant interaction

Abbreviations

COI	Cytochrome c oxidase I
EF-1-alfa	Elongation factor 1-alpha
DNA	Deoxyribonucleic acid
EPIC	Electronic Privacy Information Center
GC-MS	Gas chromatography–Mass spectrometry

K. K. Sharma (✉)
Indian Institute of Natural Resins and Gums (IINRG), Ranchi, India
e-mail: kewalkks@gmail.com

R. Peshin, A. K. Dhawan (eds.), *Natural Resource Management: Ecological Perspectives*, Sustainability in Plant and Crop Protection,
https://doi.org/10.1007/978-3-319-99768-1_13

HPLC	High-performance liquid chromatography
HPTLC	High-performance thin-layer chromatography
IINGR	Indian Institute of Natural Resins and Gums
ISSR	Inter-simple sequence repeat
MS	Mass spectrometry
mtDNA	Mitochondrial DNA
NJ	Neighbour-Joining
NMR	Nuclear magnetic resonance
PCR	Polymerase chain reaction
RAPD	Random amplified polymorphic DNA
rDNA	ribosomal DNA
SSR	Simple-sequence repeat

13.1 Introduction

India is the foremost lac-producing country in the world with an annual production of about 20,000 tons and home to the richest biodiversity of economically important lac insects (Sharma et al. 2006). Lac in commerce is derived from a few species belonging to the genus *Kerria*. Lac insects yield three basic components of economical value, i.e. resin, wax and dye. Lac insects belong to the family Tachardiidae (Kerriidae) of the order Homoptera. They are phytosuccivorous and prefer certain species of plant hosts. They remain attached to the plant throughout their whole life, with exception of crawler and adult male stages. Lac insects prefer warm climates and are, therefore, distributed in the tropical and subtropical regions between the latitudes 40°N and 40°S (Varshney 1977). Lac production is restricted to a few South, East and Southeast Asian countries like India, Thailand, China, Indonesia, Bangladesh, Myanmar, Laos and Vietnam.

Due to their unique biology, host preference and dispersal mechanisms, lac insects are expected to differentiate locally, forming geographic and host races without adequate morphological differentiation. This has been substantiated through recent studies using random amplified polymorphic DNA (RAPD) and inter-simple sequence repeat (ISSR) primers on insects from 48 geographic races and other cultivated *Kerria* spp. It appeared, therefore, important to develop more reliable markers for an easy characterization of such variations in the insect collections. Plant sap is the sole source of nutrition for lac insects. The factors governing the selection of only few host plant taxa by lac insects in general, and the preference of specific host species by different insect strains and species, needed to be investigated. The presence of endosymbiont *Wolbachia* spp. was suggested by preliminary investigations (Vashishtha et al. 2011). Detailed studies aimed at detection, identification and characterization of endosymbionts were desirable, as they are likely to play an important role in host metabolism, reproduction as well as biosynthesis of the lac constituents. Therefore, an understanding of insect-host plant relationship required for lac production was far from complete, without the knowledge on endosymbionts.

In this study, we employed a multifaceted approach to analyse and document lac insect germplasm collections (Sharma and Ramani 2014). The aspects studied included the characterization of parameters which encompassed morphology, karyology as well as biochemical and molecular profiles and of economic parameters, i.e. qualitative and quantitative variations in resin, wax and dye produced by these insects.

13.2 Approach to Research

Lac cultures were maintained on potted plants of *bhalia* (*Flemingia macrophylla*), which is a common host for the lines under study. The samples were collected at required intervals to study all the aspects considered.

13.2.1 Morphology

A morphometric study-based approach is generally adopted for the characterization of the genetic diversity of the lac insect populations and strains occurring in India (Mishra et al. 1998). This approach was authenticated through a suitable documentation involving digitization and statistical analyses. The female lac insects, which are conventionally used for such studies, were investigated. Besides, the adult male and immature stages were also studied for complementing the taxonomic keys. Taxonomic keys were made more robust with SEM studies of various glands, brachia and anal tubercle openings, for finer variations (Jena et al. 2011). This approach was authenticated through suitable documentation involving digitization and statistical analyses.

13.2.2 Chemistry of Economic Products

The insect produces three economic products, namely resin, dye and wax. The lac and the insect body were obtained from fully mature females and subjected to standard extraction procedures to isolate resins, waxes and pigments. The lac resin molecule is an ester complex of straight-chain fatty and sesquiterpenic acids. Qualitative and quantitative differences in the constituents, such as hard and soft resin fractions and also constituent acids, were studied for any distinct variation of industrial significance.

The lac pigments in resin and body are derivatives of anthraquinones. They were separated using high-performance liquid chromatography (HPLC) and subsequently subjected to further analytical procedures through mass spectrometry (MS) and nuclear magnetic resonance (NMR). The knowledge about the content of lac dye and resin pigment is very important from an industrial point of view, as the former has high economic importance, whereas the latter plays an important role in definition of the resin market price. A methodology recently developed for anthraquinones, involving reversed-phase liquid chromatography with diode array detection, was used for the study of colouring pigments. Spectrophotometric estimations of lac dye and resin content from all insect lines were done by checking the absorption spectra. As a difference in the spectral property of both dye and resin pigment among the selected insects was expected, full absorption spectra rather than routine measurements of OD values at specified wavelengths were used. For high-performance thin-layer chromatography (HPTLC) analysis of body pigments, from all the collected insect lines, a method was standardized for dye extraction and HPTLC analysis. The dye was extracted per insect body weight for all the lines, for uniform sampling. A study on dye and resin yields with respect to insect size and of crop growth season was undertaken from few selected insect lines.

13.2.3 DNA and Protein Profiling

Lac insects show considerable intraspecific variations due to geographic origin and host plant adaptations. Therefore, other tools were needed to differentiate such populations. In lac insects, the ecology is further complicated due to endosymbionts and other parasites/predators. Therefore, molecular variations measured at the DNA sequence level and total proteins of the different lac insect lines analysed served as a supplementary tool integrating the classical taxonomic parameters to distinguish insect populations. For DNA-based studies, the sequence polymorphism of conserved mitochondrial genes such as cytochrome c oxidase I (COI) and ribosomal genes was used for the comparison of genetic polymorphism among the lac insect taxa. Initial polymerase chain reaction (PCR) amplification of COI gene from insect genomic DNA was performed with the universal primers proposed by Folmer et al. (1994). Considering the limitations of mitochondrial DNA (mtDNA) in precise estimation of genetic diversity, a microsatellite-based approach like simple-sequence repeat (SSR) markers helped in covering a greater amount of the nuclear genome to derive population genetic diversity data among insect lines. The sequence data from related species were employed to mine SSRs and their applicability, validated for studying insect diversity. To cover most of the genome for assessing genetic diversity, more number and types of markers were essential. In general, the introns are more variable than the coding regions and are characteristic for eukaryotes. Considering the association of bacterial endosymbionts with the lac insect and the admixture of endosymbiont DNA with the insect genome, the exon primed, intron-crossing PCR (EPIC-PCR) tool (Palumbi and Baker 1994) was explored for analysing the DNA sequence diversity.

13.2.4 Biology

The lac insect's developmental rate is influenced by the climate as well as genetic composition, as illustrated by the seasonal differences between the *kusmi* and *rangeeni* strains of *K. lacca*. Therefore, key biological parameters were studied under uniform conditions for all lines, to enable reliable parameter comparisons across them.

Synthesis of Data of Different Aspects The information collected from all the above studies was superimposed to obtain a comprehensive profile of each line. This was useful for the biodiversity documentation as well as their application in insect improvement programmes.

13.2.5 Lac Insect and Host Plant Interaction

This was addressed by studying the cellular details of tissues directly involved in proboscis insertion and sucking the plant sap, using histological approaches. A comparative study on preferred and non-preferred plant taxa and parts was carried out using light microscopy. Different parameters characterizing cell types, structures, composition, etc. were studied. The phloem sap was collected from the different host taxa and preferred and non-preferred plant parts. The exudates were analysed using HPLC and GC-MS spectrometry for sugars and amino acid, respectively. These biochemical analyses were carried out on the plant sap as well as insect body fluid during different stages in the insect life cycle. Attempts to correlate the information obtained with yield and quality data were made.

Electron microscopy-based methods were used to detect the presence of endosymbionts during the key events in the life cycle of the lac insect. Molecular techniques such as 16S DNA primer-based amplification were used to detect the presence of endosymbionts. Specific sequences were then used to confirm and characterize the endosymbionts detected.

13.3 Major Innovations Tried

Innovations sought included (i) taxonomy of lac insects using novel characters and stages, karyology, digitization and statistical analyses; (ii) population-level study of variation in chemistry of commercial constituents; (iii) new molecular approaches for characterization and development of molecular markers, such as mtDNA, SSRs, EPIC-PCR and protein profiling; (iv) insect vs host interactions through phloem and insect biochemistry, diversity and role of endosymbionts; and (v) a comprehensive biosystematic study on lac insects using a large number of specimens collected from different parts of the country (Fig. 13.1) (Ramani et al. 2009).

13.4 Significant Insights

13.4.1 Morphology

- Description of six new species from collections made by the Indian Institute of Natural Resins and Gums (IINRG), namely *K. manipurensis*, *K. maduraiensis*, *K. thrissurensis*, *K. pennyi*, *K. dubeyi* and *K. varshneyi* with an updated key for the species of *Kerria* (Ahmad et al. 2013a, b).

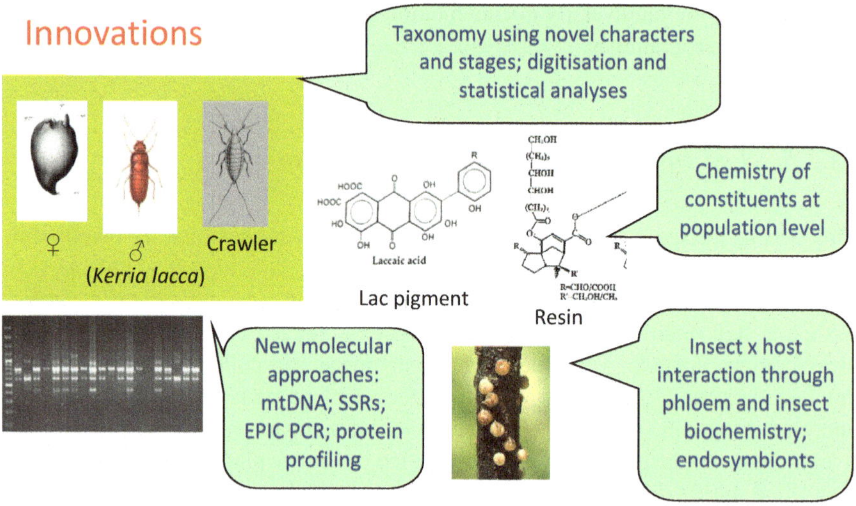

Fig. 13.1 First multifaceted study on a large number of lac insect collections from different parts of India, employing novel approaches and parameters

Fig. 13.2 Mouthpart of first-instar nymphs of *Kerria lacca* showing the labium with sensilla

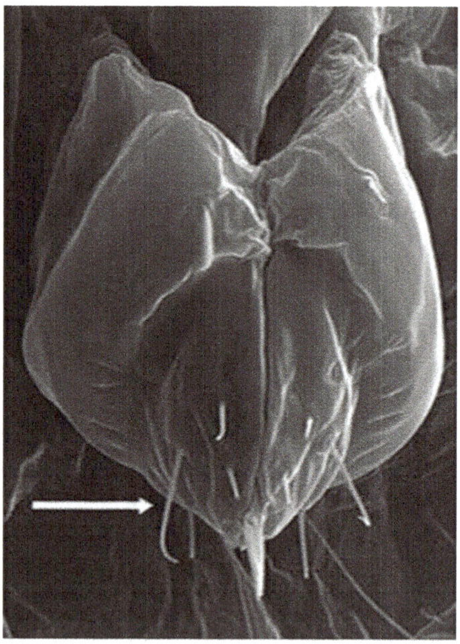

- SEM studies on lac crawlers revealed the presence of dorsal setae on the head and setae on the antennae, bordering brachial plate setae and pores, labial sensilla, anal fringes, etc. not reported earlier; SEM studies on females revealed the presence of a six-sectored anal ring, rudimentary legs and labial sensilla (Fig. 13.2); dorsal duct cluster and preanal plate are additions to the existing knowledge on the insect's description.

- Principal component analysis resulted in clustering of lines studied (Fig. 13.3); discriminant analysis assessed the utility of characters for classification with 78% of correct classification of specimen; and canonical discriminant analysis showed a clustering similar to PCA, providing a picture for possible relationships among lines (Ahmad et al. 2014).
- Templates (Fig. 13.4) for taxonomy of lac insects using mature females, adult males and crawlers developed based on conventional and new characters—60 for females, 151 for males and 70 for crawlers. Morphometric studies on female lines with characterization and illustrations lead to the characterization of 15 distinct species groups.

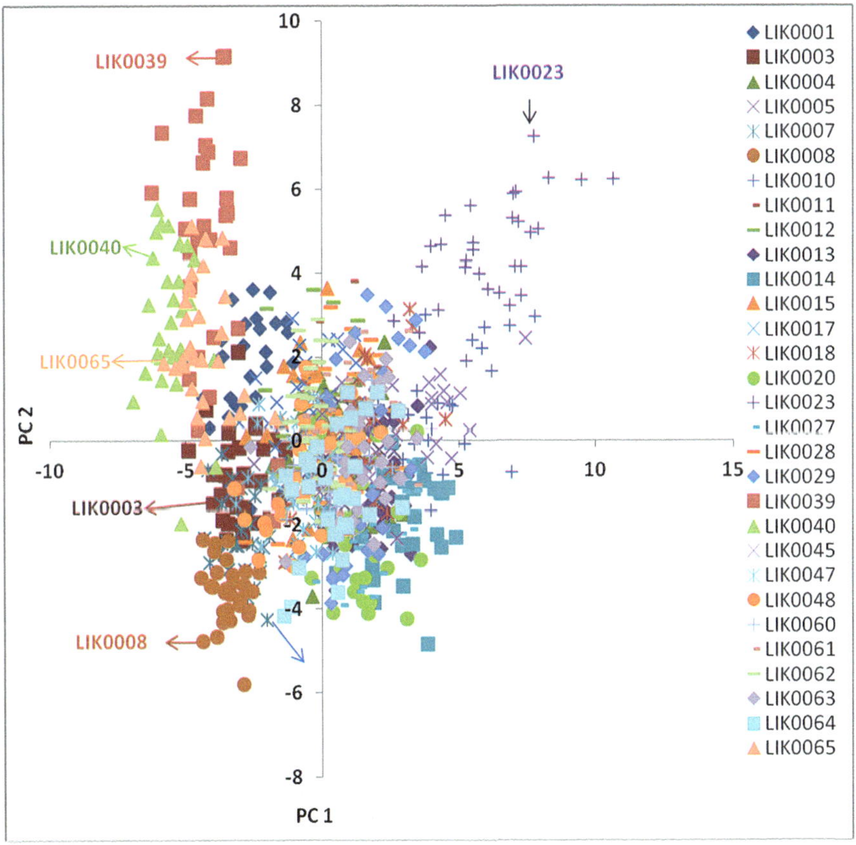

Fig. 13.3 Scatter plot of principal components 1 and 2 showing distinct grouping or clustering of lines. To statistically analyze the data, MS Excel 2007 and SAS/STAT9.1.3 were used. The morphometrics of the adult female lac insect lines were explored using single factor ANOVA (analysis of variance) and several multivariate analyses, such as principal component analysis (PCA), discriminant function analysis (DFA) and canonical discriminant analysis (CDA). Wilk's lambda statistic was used to test the significant differences occurring between the groups. PC1: Number of ducts in marginal duct clusters; PC2: Length of anal tubercle, length of pre-anal plate, distance of anterior spiracle from crater rim, length of brachia, pedicel length and total length of dorsal spine (Reproduced from: Sharma and Ramani 2014)

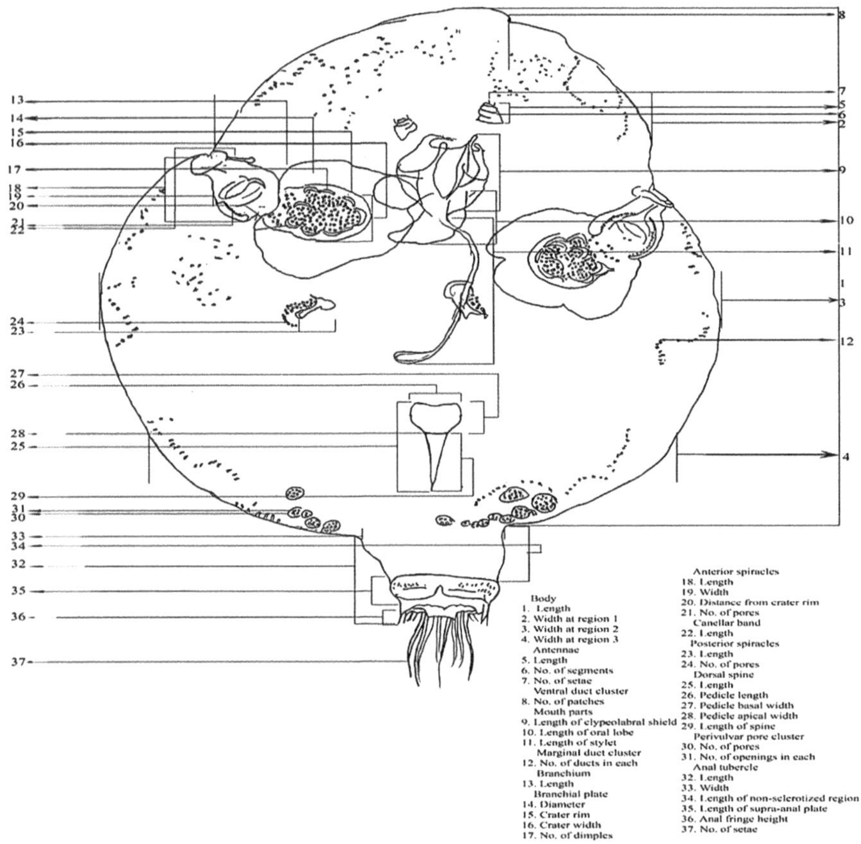

Fig. 13.4 Template for morphometrics of female lac insect (Reproduced from: Sharma and Ramani 2014)

13.4.2 Chemistry of Economic Products

- HPTLC profiling of qualitative and quantitative variations of lac dye components in the insect lines (Fig. 13.5a, b); variation in hard and soft resin fractions, using HPLC, and of lac wax components, using gas chromatography–mass spectrometry (GC-MS).

13.4.3 DNA and Protein Profiling

- Insect transcriptome and molecular characterization based on housekeeping genes and EPIC-PCR provided important leads. Data obtained from transcriptome sequencing will be used for the identification of genes responsible for the synthesis of economically important products. Intraspecific variations found will help in marker-assisted selection, in future breeding programmes.

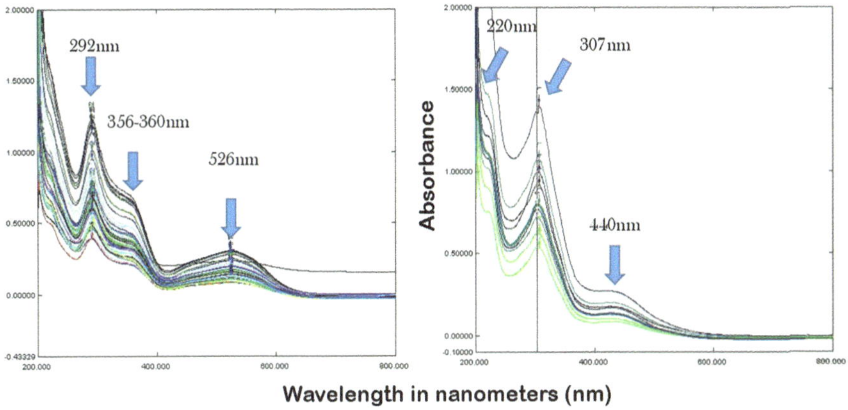

Fig. 13.5 Overlay diagram of all spectra for crimson or yellow lines

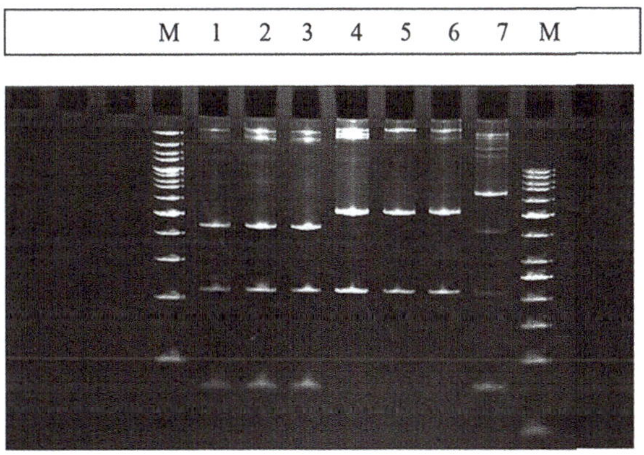

Fig. 13.6 PCR-RFLP for *kusmiand rangeeni Kerria* strains. Lane M: 100 bp ladder; lanes 1-3: *kusmi*; lanes 4-6: *rangeeni*

- Patents: (i) Group-specific primers for identification of *Kerria* species and (ii) novel molecular methods using PCR-RFLP to differentiate infra-subspecific forms, *kusmi* and *rangeeni*, of Indian *Kerria lacca* (Fig. 13.6).
- COI-based barcoding of lines, with 650 bp region amplified with patented primers developed through the analysis of sequence variations, in *Kerria* (Fig.13.7), using neighbour-joining (NJ) method (Saitou and Nei 1987), and distances computed using maximum composite likelihood method (Tamura et al. 2004). Sequence data deposited in the GenBank include three other housekeeping genes also used in diversity analysis: elongation factor 1-alpha (EF-1-alfa), 18S and 28S-D2; intron length polymorphism (EPIC-PCR) studied in the insect lines for five genes (myosin, S7 ribosomal protein, ATPS alpha, cathepsin and opsin).
- Group-specific markers were developed, based on de novo transcriptome sequencing of *K. lacca* using Roche 454 GS FLX (> 22,000 transcript contigs) and Illumina (> 31,900 contigs); 24 microsatellite markers and 126 SNPs.
- Eight different bacterial species were isolated from lac insects and categorized as endosymbionts, gut or subsurface bacteria, based on biochemical tests and 16S ribosomal DNA (rDNA) profiling.

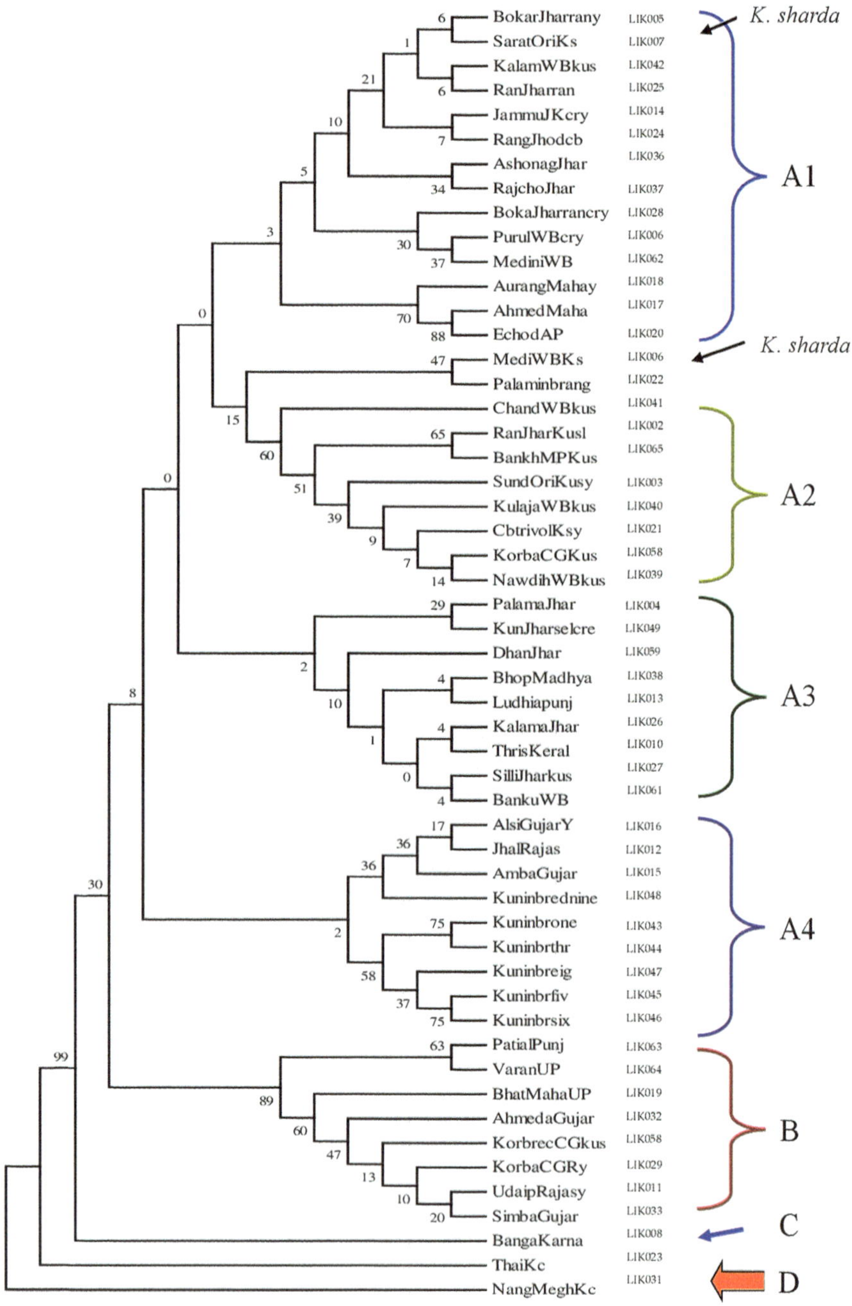

Fig. 13.7 Cluster analysis of similarity between the *Kerria* lines, on the basis of COI sequences (648 bp) using neighbourhood joining method (Saitou and Nei1987). Capital letters on the right side denote different geographical lines of lac insects under different groups: A1, A2, A3, A4 and B are subgroups of *K. lacca*. C represents *K. chinensis* and branches with arrow signs are two lines of *K. sharda* from different locations forming distant part of A1 and A2. Numbers on branches are genetic distance computed using maximum composite likelihood method (Tamura et al. 2004)

13.4.4 Biology

- Survey and collection of lac insects from the peripheral regions of 49 districts in 12 states/union territories: northeastern, northern and southern states. Surveys were undertaken in Assam (2), Manipur (9), Meghalaya (2), Mizoram (3), Nagaland (2), Tripura (4), Himachal Pradesh (9), Haryana (4), Punjab (2), Tamil Nadu (9), Kerala (2) and Andaman (1). Collection and identification of ant species associated with lac insects from Manipur. Eight new lines from Assam, Kerala, Manipur, Nagaland, Punjab and Tamil Nadu added to the existing germplasm collection (Fig. 13.8).
- Unique lac insect and ant associations were found, with ant nest covering the insect colonies recorded (Fig. 13.9). *Crematogaster flava*, *C. rogenhoferi*, *C. rothneyi* and *Technomyrmex albipes* ants were identified in symbiotic relationship with lac insect from Manipur.
- Variation in six key biological attributes documented for lac insect lines, during both generations (summer and rainy/winter) of the year (Fig. 13.10).

13.4.5 Lac Insect and Host Plant Interaction

- Precise documentation of feeding site through localization of proboscis (Fig. 13.11) by SEM and fluorescent microscopy; the ultrastructure of mouthparts was studied for first instars. Finer details of the stylet penetration path in the host plant revealed phloem feeding and adaptation of mandibular and maxillary stylets and of salivary sheath.

Fig. 13.8 Lac insect on *Amhertsia nobilis* in Thrissur, Kerala (India)

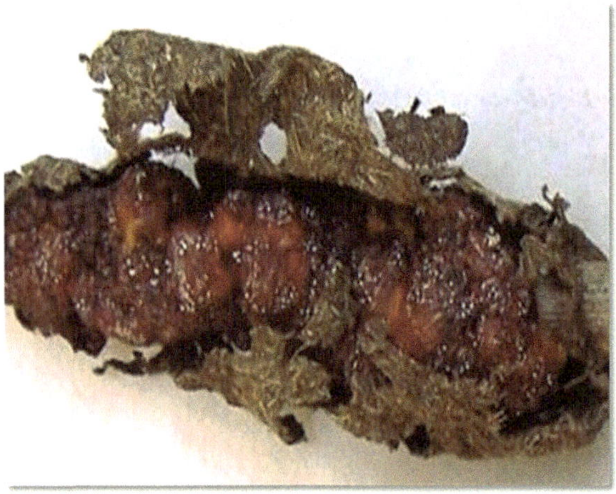

Fig. 13.9 Ant nest around lac encrustation

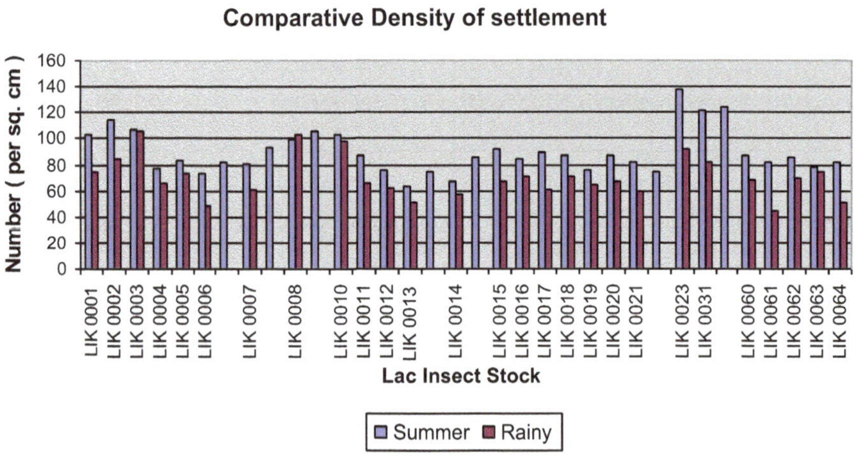

Fig. 13.10 Density of settlement of various lac insect stocks during summer and rainy seasons

- Anatomical studies were carried out for variation in feeding site distance of lac host plant species differing for performance, to understand the basis of insect preference. Phloem distance did not affect preference, but moderate secondary growth was a preferred stage, revealing the importance of age of shoot for lac infestation. Bark phytochemicals did not appear to influence the settlement of lac crawlers on the host plant.
- Phytochemical investigations in 11 host plant species of *K. lacca*, with analysis of amino acids in four host plant species, namely, *Flemingia semialata*, *F. macrophylla*, *F. bracteata* and *F. chappar*. Quantitative and qualitative studies were carried out on essential and nonessential amino acids by applying MS-MS.

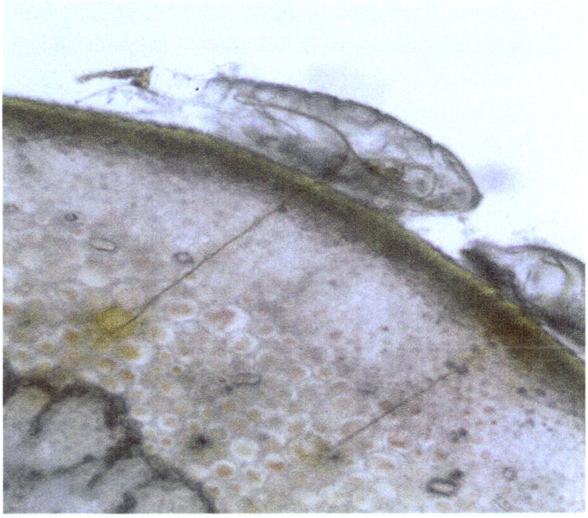

Fig. 13.11 Stylet penetration path of lac crawler in the host plant, revealing phloem feeding

- The detection of *Wolbachia* (natural bacteria) in males, crawlers and adult females was confirmed on the basis of specific primers; detection of *Wolbachia* sp. in lac insect strains also occurred using 16S *Wolbachia*-specific primers; detection and characterization of *Wolbachia* for classifying into groups using A- and B-specific primers. Multi-locus sequence typing for *Wolbachia* and two new alleles for *Wolbachia* specific to *K. lacca* were reported for the first time.
- Detection of *Wolbachia* phage using ORF7 primers was reported for the first time in *K. lacca* unravelling the multipartite association of the insect, molecular detection of a bacteriophage associated with *Wolbachia* in *Kerria* spp.; bacteriome localization was identified using fluorescent microscopy and in situ localization of *Wolbachia*.

The generation of a complete set of informations for the study encompassing morphology, cytology, biochemistry of economic products, molecular variations, insect-host interactions and endosymbiont diversity led to the production of a comprehensive document on the above aspects. The study has made an indelible mark by developing a number of novel methodologies/products and employing unconventional approaches leading to:

- Improved and reliable basis for the description of commercially important lac insect (*Kerria*) species and enhanced exploitation of wild genetic resources
- Development of molecular tools for the characterization of lac insects especially at the intraspecific level (morphologically indistinguishable) and markers to facilitate genetic improvement programmes
- Better understanding and documentation of qualitative and quantitative variation of economic products derived by lac insects, for enhanced exploitation of genetic resources

- Evolution of a scientific approach for manipulation of insects, for better performance on a host base, with knowledge generated and comprehensive analysis of biochemical composition of the phloem sap; detection and characterization of endosymbionts associated with lac insects
- Augmentation of the lac insect germplasm through extensive surveys into the country, with special emphasis on peripheral areas that could provide distinct populations
- Availability of integrated knowledge based on characterization and economic parameters of lac insect germplasm, allowing a more effective breeding for their economic products

The multipronged approach herein described has helped in developing a holistic picture of lac insect biodiversity and a better understanding of their interactions with the host plants for exploiting the applied aspect of lac insect variability.

References

Ahmad, A., Ramamurthy, V. V., Sharma, K. K., Mohansundaram, A., Vidyarthi, A. S., & Ramani, R. (2013a). Three new species of *Kerria* (Hemiptera: Coccoidea: Tachardiidae) from India. *Zootaxa, 3734*(4), 442–452.

Ahmad, A., Sharma, K. K., Ramamurthy, V. V., Vidyarthi, A. S., & Ramani, R. (2013b). Three new species of *Kerria* (Hemiptera: Sternorrhyncha: Coccoidea: Tachardiidae), a redescription of *K. yunnanensis* Ou & Hong, and a revised key to species of *Kerria*. *Zootaxa, 3620*(4), 518–532.

Ahmad, A., Ramani, R., Sharma, K. K., Vidyarthi, A. S., & Ramamurthy, V. V. (2014). Distinction of Indian commercial lac insect lines of *Kerria* spp. (Homoptera: Coccoidea) based on their morphometrics. *Journal of Insect Science, 14*(263). https://doi.org/10.1093/jisesa/ieu125.

Folmer, O., Black, M., Hoeh, W., Lutz, R., & Vrijenhoek, R. (1994). DNA primers for amplification of mitochondrial cytochrome *c* oxidase subunit I from diverse metazoan invertebrates. *Molecular Marine Biology and Biotechnology, 3*, 294–299.

Jena, A., Vidyarthi, A. S., Ramani, R., Sharma, K. K., & Ramamurthy, V. V. (2011). An illustrated diagnostics of lac insect *Kerria lacca* (Kerr) as exemplified through advanced staining and microscopy techniques. *Indian Journal of Entomology, 73*(3), 235–242.

Mishra, Y. D., Sushil, S. N., Bhattacharya, A., & Sharma, K. K. (1998). Morphometric differences between strains of Indian Lac insect, *Kerria lacca* (Kerr.) (Homoptera: Tachardiidae). *Journal of Insect Science, 11*(2), 170–171.

Palumbi, S. R., & Baker, C. S. (1994). Contrasting population structure from nuclear intron sequences and mtDNA of humpback whales. *Molecular Biology and Evolution, 11*, 426–435.

Ramani, R., Anees, K., Sharma, K. K., Saha, D., & Baboo, B. (2009). *To understand the nature of diversity in lac insects of Kerria spp. in India and the nature of insect x host interaction*. Indian Institute of Natural Resins and Gums, Ranchi. 12 pp.

Saitou, N., & Nei, M. (1987). The neighbour-joining method: A new method for reconstructing phylogenetic trees. *Molecular Biology and Evolution, 4*, 406–425.

Sharma, K. K., & Ramani, R. (2014). *To understand the nature of diversity in lac insects of Kerria spp. in India and the nature of insect x host interaction. Final Report of NAIP sub-Project.* 136 pp. Indian Institute of Natural Resins and Gums, Namkum, Ranchi - 834 010 (Jharkhand), India.

Sharma, K. K., Jaiswal, A. K., & Kumar, K. K. (2006). Role of lac culture in biodiversity conservation: Issues at stake and conservation strategy. *Current Science, 91*(7), 894–898.

Tamura, K., Nei, M., & Kumar, S. (2004). Prospects for inferring very large phylogenies by using the neighbour-joining method. *PNAS, (USA), 101*, 11030–11035.

Varshney, R. K. (1977). Taxonomic studies on lac insects of India. *Oriental Insects*, Supplement No. 5 (1976), 97 pp.

Vashishtha, A., Sharama, K. K., & Lakhanpaul, S. (2011). Co-existence, phylogeny and putative role of *Wolbachia* and Yeast-Like Symbiont (YLS) in *Kerria lacca* (Kerr). *Current Microbiology, 63*, 206–212. https://doi.org/10.1007/s00284-011-9961-x.

Chapter 14
Diffusion and Adoption: Factors Impacting Adoption of Sustainable Agricultural Practices

Rajinder Peshin, Fatima Bano, and Raj Kumar

> *Diffusion of innovations has the status of a bastard child with respect to the parent interests in social and cultural change: too big to ignore but unlikely to be given full recognition.*
>
> Fliegal and Kivlin (1966)

Abstract Managing natural resources, and at the same time increasing productivity in agriculture, is the thrust of research and extension all over the world. Achieving long-term food security without depleting natural resources can be achieved by adoption of sustainable agricultural practices. Sustainable agricultural practices, or agriculture per se, seek the wider adoption of practices that are ecologically sound and maintain the long-term ecological and biological integrity of natural resources. Sustainable practices in agriculture include integrated pest management, integrated nutrient management, soil conservation and water management, among others. Much of the dynamics of the diffusion process of sustainable agricultural practices, for adoption or rejection of these practices, can be analysed and understood on the basis of sound understanding of diffusion and adoption theory. The diffusion researchers have mostly analysed farmers' differences in analysing adoption or rejection of innovation/technology. Very little attention has been paid to technology attributes or technology inappropriateness. In this chapter, we have analysed factors impacting diffusion and adoption of sustainable natural resource management practices in agriculture and different models that can be employed to predict adoption or rejection, in future times. The researchers involved in an innovation development process should consider factors propelling adoption or rejection before commercialisation of a technology. Diffusion researchers need to employ alternative field

R. Peshin (✉) · F. Bano · R. Kumar
Division of Agricultural Extension Education, Sher-e-Kashmir University of Agricultural Sciences and Technology of Jammu, Chatha Jammu-180009, Jammu and Kashmir, India
e-mail: rpeshin@rediffmail.com; fatimabano36@gmail.com; rkchallgotra@gmail.com

R. Peshin, A. K. Dhawan (eds.), *Natural Resource Management: Ecological Perspectives*, Sustainability in Plant and Crop Protection,
https://doi.org/10.1007/978-3-319-99768-1_14

experimental before-after designs, in which data are gathered at different points in time rather than post hoc data collection, to overcome farmer blame and pro-technology biases. We are of the firm belief that diffusion of innovation research must be given full recognition by research and development, and change agencies involved in agricultural sciences.

Keywords Sustainable agricultural practices · Diffusion and adoption · Factors affecting adoption · Models for forecasting diffusion and adoption

Abbreviations

ETL	Economic threshold level
FAO	Food and Agriculture Organisation
IBM	International Business Machines
IPM	Integrated pest management
IRM	Insecticide resistance management
PAU	Punjab Agricultural University
SDGs	Sustainable Development Goals

14.1 Introduction

Agriculture has over time witnessed a great shift from the traditional practices to external, input-intensive and modern technologies, covering the mid-twentieth-century period through the 1960s, 1970s and 1980s, and then to sustainable agricultural practices, in the last quarter of the twentieth century through the current century. Sustainable agriculture farming is based on ecosystem analysis and conservation of natural resources. Sustainability, however, means different things to different people, based on their perspectives of development. Sustainable agriculture has been holistically defined by Zhao et al. (2008) to include sustainability of the rural economy, ecological and environmental sustainability within agricultural systems and sustainability of rural society. Sustainable agricultural practices, or agriculture per se, seek a wider adoption of practices that are ecologically sound and maintain the long-term ecological and biological integrity of natural resources. Managing natural resources and, at the same time, increasing productivity in agriculture are the thrusts of research in many countries. Achieving long-term food security, without depleting natural resources, for example, land and water resources, and biodiversity can be achieved by adoption of sustainable agricultural practices. These include integrated pest management, integrated nutrient management, soil conservation and water management, among others. The sustainable development agenda of the 193 member states of the United Nations that was adopted in 2015 includes 17

Sustainable Development Goals (SDGs), with sustainable agriculture as one of the primary targets of SDGs (FAO 2017).

Adoption of sustainable agricultural practices at farm, village and regional levels is required to achieve a sustainable growth in agriculture. Researchers, policy makers, extension services and private industry must have synergy to achieve sustainable development and adoption of these sustainable agricultural practices, maintaining in the long-term ecological and biological integrity of natural resources. Much of the dynamics of diffusion process of sustainable agricultural practices, adoption or rejection of these practices and their consequences can be analysed and understood on the basis of sound understanding of diffusion and adoption processes (Roling 1988). There is not a single technological innovation where the inventor-organisation/company that commercialises a new technology/product is not interested to know whether the technology is diffused and adopted/purchased by the consumer or not as well as, if not adopted, the reasons for the non-adoption thereof.

Before discussing "why some technologies are adopted and others are not", a definition of diffusion and adoption is imperative. Evert M. Rogers (1962, p. 13), in his classical diffusion of innovations theory, defined diffusion as a process by which innovation spreads. He later on expanded this definition in the third edition of his book *Diffusion of Innovations* published in 1983 and defined diffusion as "*the process by which an innovation is communicated through certain channels over time among the members of the social system*" (Rogers 1983, p.5). Diffusion and adoption are interrelated concepts, but they are distinct in many respects. Diffusion/dissemination[1] refers to the spatial and temporal spread of the new technology among different economic units. Adoption commonly refers to the decision to use a technology or practice by the economic units, on a regular basis. With the credit of being the most researched topic, diffusion and adoption research has achieved a prominent position in the field of agricultural extension, anthropology, sociology, public health, medical sociology, education, communication, marketing and management, economics, public administration, industrial engineering, geography and many other disciplines.

14.2 Diffusion and Adoption of Innovations

The four elements of diffusion are (*i*) innovation (technology), (*ii*) communication channels (interpersonal and mass), (*iii*) time and (*iv*) social system. Adoption is a decision to make full use of an innovation/product/technology as the best course of action available, whereas rejection is the decision not to adopt a technology (Rogers 1962, 1983, 1995, 2003). Ryan and Gross in 1943 first indicated the identification of adoption as a process (Rogers 1962, p. 79). The diffusion and adoption of sustainable agricultural practices are not uniform for all innovations introduced in the social system. Some innovations take a few years from first introduction/commercialisation

[1] "Diffusion" includes both the panned and the spontaneous spread of innovations and "dissemination" is directed and managed spread of innovations (Rogers 1983). In this chapter, we use the word "diffusion" and "dissemination" interchangeably.

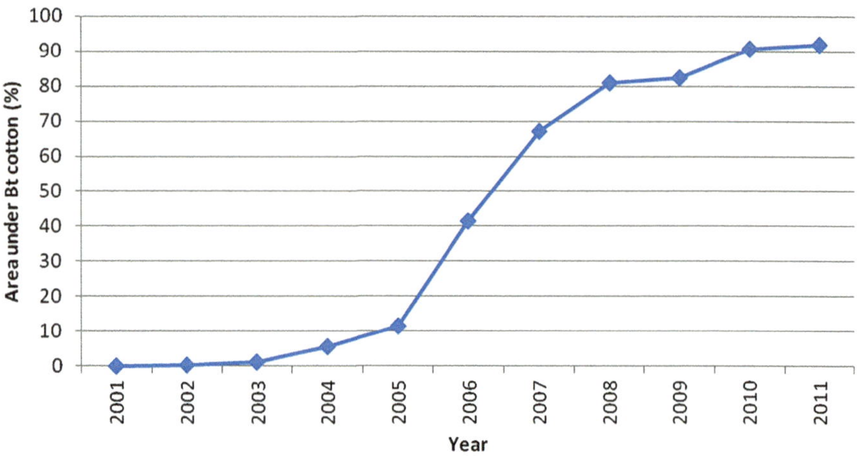

Fig. 14.1 Rate of adoption of Bt cotton in India. It fits typical Rogers' "S"-shaped rate of adoption curve. (Modified after: Peshin et al. 2014)

Table 14.1 Recommendations and extent of adoption of sustainable agricultural technologies by Punjab farmers (K. Singh 2015, unpublished data)

Innovation/technology	Year of recommendation	Adoption status (area)
Tensiometer	2006	Less than 1%
Laser land leveller	2006	70–80%
Leaf colour chart	2006	Less than 1%
Happy seeder	2006	1%

to widespread use. For example, Bt cotton diffused from first introduction in Indian agriculture in 2002 to reach more than 90% of adoption by 2010 (Fig. 14.1) (Peshin et al. 2014). Another technological innovation may not take off to gain critical mass and may level off at less than 1%, for example, tensiometer (a water-saving device that measures the moisture and water potential to save water use in rice crops) (Table 14.1) (K. Singh 2015, unpublished data). There are many studies that prove that different technologies are not diffused and adopted uniformly.

An analysis of diffusion and adoption of sustainable agricultural technologies recommended by the state agricultural universities in North India showed that some technologies have been widely adopted, namely Bt cotton and laser land leveller, among others. In the case of Bt cotton, the adoption was 72% in 2004 (prior to the government approval for the cultivation of Bt crops in Punjab, though it was approved for cultivation in western and southern states of India in 2002) (Peshin 2005), and it has reached 95% in 2016 (Peshin et al. 2017). In 12 years since its approval for cultivation in Punjab (2005), around 92% of cotton area is under Bt cotton. Laser land leveller (water-conserving technology) introduced in 2006 is also widely adopted by 97% rice growers (Peshin et al. 2017) (Table 14.2). The widespread adoption of these sustainable agricultural technologies can be compared with the adoption of mobile phones by the farming community, which is more than 90%

Table 14.2 Extent of adoption of sustainable agricultural practices by Punjab farmers, India

Practice	Project areas[1] (%)	Non-project areas (%)
Soil testing	28	15
Leaf colour chart	12	0
Laser land leveller 2016	96	98
Soil health cards	8	4
Not burning rice stubbles	67	68

Source: Peshin et al. (2017)

[1]Locations where the Punjab Agricultural University (Ludhiana, India) implemented projects to educate farmers

Table 14.3 Examples of high rates of adoption in North India

Technology	Adoption
Bt cotton (% farmers / % cotton area)	95 (farmers)/92 (cotton area)
Mobile phones (% farmers)	93
Laser land leveller (% farmers)	97

Sources: Peshin et al. (2017) and Nanda et al. (2017)

in the states of Jammu and Kashmir and Punjab (Peshin et al. 2017). In the Jammu region, it ranges between 90 and 94%, in the Kashmir Valley about 95%, in Ladakh 86% and in Punjab 93% (Peshin et al. 2017; Nanda et al. 2017) (Table 14.3). There are technologies whose adoption is moderate, like timely transplanting of rice and cotton crops and use of pesticide treated seeds sold in the market (Peshin 2013; Peshin et al. 2017).

However, there are sustainable agricultural technologies which have either been rejected or their adoption has been low. Examples are seed treatment/dressing with pesticides done by farmers (5%), sampling for economic threshold level of insect pests (7%), soil testing for determining fertilizer recommendations (15–28%), leaf colour chart – used to determine the N fertiliser needs of rice crop (project area of PAU 12% and other areas 0%) – and soil health cards, introduced in India in February 2015 (4–8%) (Peshin 2005, 2013; Peshin et al. 2017). Further, if we analyse the adoption of sustainable water management technologies, the adoption of tensiometer is less than 1% that of laser land leveller. The pertinent question that needs to be answered is, "Why does the same group of farmers exhibit different adoption decisions?"

Diffusion research traditions in disciplines such as sociology, education, public health, marketing and management, economics, agricultural extension education/rural sociology and others studied socio-personal differences in people/farmers, determining the adoption or rejection. Agriculture economists mostly concentrated on socioeconomic drivers of adoption of innovations and diffusion cycle (Feder and Umali 1993; Knowler and Bradshaw 2007) and not on technology attributes. This led to the farmer/individual-blame bias in diffusion research. *Farmer blame bias* is the tendency of agricultural scientists to side with the change agencies that promote innovations rather than the individuals who are the potential adopters (Rogers 2003, p. 106). The other shortcoming of diffusion research is the contribution of a successful diffusion of hybrid maize among Iowa farmers (Ryan and Gross 1943), having high degree of benefits being a profitable innovation. This led to the *pro-innovation bias* which is the greatest implication in the diffusion research such that “an innovation should be diffused and adopted by all members of a social system, that it should be diffused more rapidly, and that the innovation should be neither re-invented nor rejected” (Rogers 2003, p. 118). Though several critics have reported about pro-innovation bias (Downs and Mohr 1976), it continues to be the failure of diffusion researchers in particular and agricultural researchers in general, in order to know and learn about the important aspect of a technology rejection or discontinuance. The agricultural scientists involved in innovation development processes do not involve diffusion researchers in the process. Despite that being the case, diffusion researchers side with the promoters of technology. Evaluation of the technologies for their adoptability does not happen. Technology attributes responsible for non-adoption have been ignored by the majority of diffusion researchers. For example, adoption of economic threshold levels (ETL) of pests, recommended by agricultural scientists for managing pests, is patchy all over the world (Norris et al. 2003; Peshin et al. 2009a; Sharma 2011; Peshin 2013), as the use of pesticides according to good agricultural practices is limited (Litsinger et al. 2009; Peshin et al. 2009b; Hashemi et al. 2014). The biological scientists and diffusion researchers blamed it on farmers’ lack of knowledge, large landholdings and other socio-economic attributes of farmers. Much less attention has been paid to (*i*) analysing technology differences or inappropriateness, to overcome farmer blame bias, and (*ii*) changing the post hoc data gathering by time series with multipoint data gathering in time, after the commercialisation of a technology (Fig. 14.2). Another area of diffusion research can be the technology abandonment, called discontinuance by Rogers (1983), after its adoption.

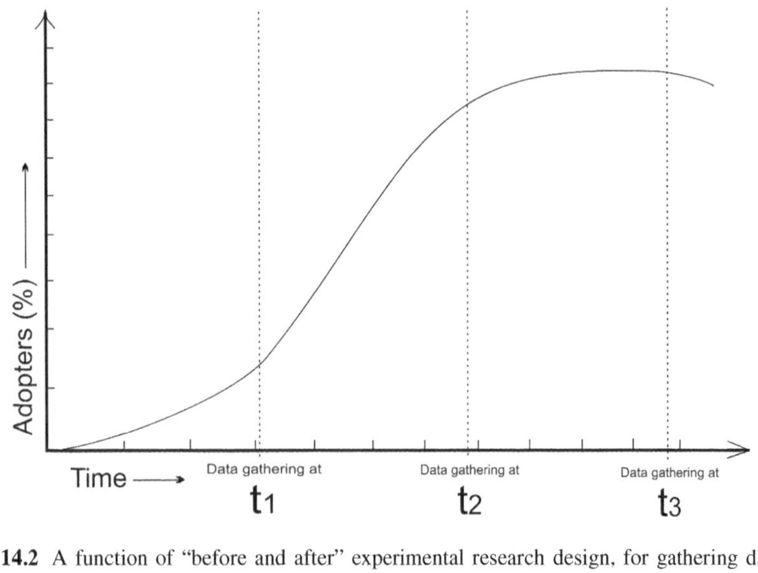

Fig. 14.2 A function of “before and after” experimental research design, for gathering data at several points in time after an innovation has been commercialised. (Source: Rogers 1995)

14.3 Factors Affecting Adoption/Non-adoption

Diffusion researchers have identified the factors driving the adoption of innovations. Rogers (1983) identified the factors impacting the rate of adoption or adoption per se, based on different diffusion traditions. Adoption of an innovation is not a one-step process or single factor driven. Major diffusion research traditions have worked on *(i)* rate of adoption (early sociology, rural sociology, education, etc.), (*ii*) characteristics of adopter categories, i.e. social, personal and economic characteristics of members of social system affecting adoption (early sociology, education, communication channels, etc.), (*iii*) stages in the innovation decision process and communication channels by stages (communication, public health, medical sociology and management) and (*iv*) desired or undesired consequences of innovations (anthropology, economics). Rogers in 1962 described a general diffusion model that brought diffusion research in such prominence that his book, and editions thereof, became the second highest cited social science work with 99052 citations. In his diffusion of innovation theory, he generalised the variables impacting adoption of an innovation. They are:

1. **Perceived attributes of innovation**, namely *relative advantage* (economic benefits, social prestige, initial cost, saving of time and effort, immediacy of reward), *compatibility* (perceived as consistent with the existing values, past experience and needs of potential adopters), *complexity* (the degree to which an innovation is perceived as relatively difficult to understand and use), *trialability* (the degree to which an innovation may be experimented with on a limited basis) and *observability* (the degree to which the results of an innovation are visible to others). While the attributes of relative advantage, compatibility, trialability and observability positively affect adoption of an innovation, i.e. they are directly proportional to adoption, at the same time complexity is negatively related, i.e. inversely proportional, to adoption. The greater the perceived complexity of an innovation, the lesser would be its adoption (Peshin 2013).

   Research has shown that most of the variance in the adoption of technologies (from 49 to 87%) is explained by these five attributes (Rogers 1995). Farmers adopt a technology keeping in view its particular, relative economic advantage (Peshin 2013). If a technology satisfies this requirement of economic profitability, the farmers will adopt it. Diffusion scholars have found relative advantage to be one of the strongest predictors of an innovation's rate of adoption (Rogers 2003).

   Few diffusion researchers have worked on the attributes of innovation. Fliegal and Kivlin (1966) have worked on the technological attributes, although in their ex post study, they determined the farmers' perception of attributes to explain the rate of adoption. They broadly classified the attributes of modern technologies into *cost attributes, returns, efficiency, risk and uncertainty, communicability of the innovation and its effects and congruence.*

   Menzel (1960) made a very insightful analysis of several attributes of innovations in the field of medicine and relied on designations of innovations as more or less communicable, risky or pervasive in their consequences, from the researchers' point of view. Diffusion of innovation studies in health care has identified ten factors impacting adoption, namely relative advantage; trialability; observability; compatibility (attribute of innovation); communication channels; homophilous groups; pace of innovation; norms, roles and social networks; opinion leaders; and infrastructure (Cain and Mittman 2002). Peshin et al. (2009c) identified integrated pest management (IPM) technology attributes which influence the adoption of IPM practices. These included five attributes generalised by Rogers (discussed above) and (*i*) *perceived risk* (the degree to which an innovation is perceived with uncertainty about its outcome among the potential adopters), (ii) *laborious* (the degree to which an innovation is perceived as more labour-intensive and time-consuming than the innovation it supersedes) and (*iii*) *communicability/non-communicability* (the ease with which know-how and usefulness of an innovation can be communicated). An example of perceived risk and complexity of an innovation is the case of tensiometer, whose adoption has been a failure compared to laser land leveller. Though both technologies have proven relative advantage as use of tensiometer helps in optimum scheduling and reducing water use by 22% (relative advantage) (Sidhu et al. 2011), its adoption is low (<1%) (Singh 2015) because it requires periodic surveillance, is laborious and is perceived risky by the farmers. While laser land leveller is reported to have no initial cost and can be hired by farmers, tensiometer has high initial cost, and farmers have to install it in their fields. The success of laser land leveller has been attributed to the observability and communicability of the relative advantages of the technology in saving water (as reported by farmers). Laser land leveller adoption is very high (97%) (Peshin et al. 2017), with an estimated water-saving capacity of 36.19 cm, and yields increase of 0.78 t/ha (Sidhu et al. 2010). But in the case of tensiometer, the nonavailability of timely assured irrigation and the complexity of the tool have resulted in rejection of the technology by farmers.

Another example of failure of an agricultural innovation is the N-Track soil nitrogen testing kit, used to reduce the overapplication of nitrogen fertiliser by Iowa farmers in the United States of America (USA). The failure of this innovation is analysed as having little relative advantage, being labour-intensive, time-consuming, incompatible with use of anhydrous ammonia nitrogen fertiliser and with a low observability of results, for other farmers (Kremer et al. 2000). In case of knowledge-intensive IPM practices, like sampling for determining ETL of pests besides complexity, the communicability and observability of the benefits are low, compared to calendar-based pesticide applications, besides being labour-intensive and time-consuming (Peshin et al. 2009c). As diffusion/dissemination is a special type of communication, in which the messages are about a new idea (Rogers 1983), the communicability of the above referred IPM practice is difficult, and it calls for complex behavioural changes. Certain IPM practices are in fact difficult to use (complexity) and more laborious than routine calendar-based pesticide applications (Rogers 1995; Peshin et al. 2009b). Besides, policy decisions are required as the sales people of pesticide industry are formidable opponents of IPM and are the main source of informations about pest management (Sharma and Peshin 2016).

Thus, technology attributes have a significant impact on the decision to adopt innovations.

2. **Communication channels**: mass media (for creating awareness knowledge) or interpersonal channels (in order to develop favourable and unfavourable attitudes towards a technology, which are determined by an innovation attribute).
3. **Nature of social system**: a social system is the patterned network of relationships constituting a coherent whole that exists among individuals living in an area, group of farmers and institutions. The nature of social system sets the norms (modern or traditional), the degree of communication and the integration between individuals, groups and institutions, among others.
4. **Types of innovation decisions:**

   (a) *Optional innovation decision*, which is the choice to adopt or reject an innovation by an individual, independent of the decision by other members of a system. This type of innovation decision is free from the control of an authority, any individual being free to decide on a technology, for example, adoption of Bt cotton and mobile phones, among others.

   (b) *Collective innovation decision*, which is the choice to adopt or reject an innovation, made by consensus among the members of a system. This type of innovation decision is made collectively by an organisation or community as a whole, for example, implementing IPM at village level is an example of collective innovation. Farmers at village level deciding not to burn rice crop stubbles, thereby not polluting the environment, is also an example of collective innovation decision.

   (c) *Authority innovation decision*, which is the choice to adopt or reject an innovation made by relatively few individuals in a system who possess power, high social status or technical expertise (Rogers 2003). For example, collective and authority decisions for managing wetlands (See Chap. 1), stopping farmers from rice stubble burning and eliminating the use of extremely hazardous pesticides in agriculture, are required. Burning of rice stubbles was recommended under the rice IPM programme to destroy overwintering eggs or larvae of stem borers by the entomologists in the early 1990s, but scientists from soil and environmental sciences considered it harmful for the soil and the environment. Thus, this practice is now banned in North India. Harvesting of rice and wheat crops by mechanised combined harvesters has led to farmers burning the stubbles of these crops after harvesting, as ploughing becomes impossible in combined harvested fields, the undesired consequences of mechanised harvesting. Due to environmental concerns, the burning of stubbles was banned by the Supreme Court of India in 2017. Despite this, a good percentage of farmers burn stubbles (around 38% rice farmers) (Peshin et al. 2017).

A policy decision by authorities/governments to ban the use of extremely hazardous pesticides detrimental to human health and the environment is required. At the time of writing this chapter, the European Union has approved the ban on use of three neonicotinoid pesticides (clothianidin, imidacloprid and thiamethoxam) on flowering crops that attract bees, which will come into force by end of 2018. This group of pesticides is reported to be responsible for *colony collapse disorder* (a phenomenon which results in a rapid loss of adult worker bees), posing a serious threat to the honeybees. High amount of pesticides are used in apple orchards in the Kashmir Valley (more than 23.750 kg/ha), where application of mancozeb, a cancer-causing pesticide, is very high (7.428 kg/ha) (Peshin et al. 2017). The decision to ban these pesticides requires authority adoption decision. More recently, Punjab state government took a decision on October 23, 2018 to ban the use of cancer-causing herbicide glyphosate in Punjab.

5. **Socio-personal and economic attributes of the potential adopters:** this includes *preference proxies* (education, age, gender and social status), *resource endowments* (income, assets, labour and credit/savings), *market incentives* (potential income gain, distance to market and price), *biophysical factors* (soil, slope, plot size and irrigation) and *risk and uncertainty* (tenure, experience, extension, membership) (Pattanayak et al. 2003).
6. **Extent of change agents' promotion efforts:** a change agent is an individual who influences clients' innovation decision in a direction deemed desirable by a change agency (Rogers 2003). Change agents are consultants, extension workers, sales people, etc. Usually, a change agent seeks to secure the adoption of new ideas, but in certain cases of innovations with undesirable effects, he/she may also attempt to slow the diffusion process and prevent the adoption of the same. The promotion efforts of a change agent are thus crucial for adoption/rejection of a technology.

However, much of the diffusion research has analysed personality variables/socio-personal/socioeconomic variables (58%) of potential adopters, and very meagre (1%) research has been conducted on other factors (as enlisted above) impacting adoption (Rogers 2003). Factors driving adoption of an innovation, modified after Rogers (1995), are given in Fig. 14.3.

Thus, before developing and disseminating a technology, it is pertinent for scientists to have basic knowledge about diffusion and adoption theory. This will help to explain why a sustainable agricultural innovation/technology is adopted or rejected and will help to shift the onus of non-adoption from farmers to the basic intricacies of the particular technology, which may be responsible for its low adoption or rejection in total. Scientists should also conduct adoptability (i.e. the future adoption of a particular technology) pilot studies before recommending the same to farmers. Farmers' society, economy and the environment they live in have to be taken into consideration by change agents before introducing sustainable agricultural innovations. Brand ambassadors/celebrities, who enjoy public recognition, are hired to create nation-wide mass awareness campaigns about consumer goods through mass media, for example, endorsement by celebrities for sports shoes and Swachh Bharat campaign (clean India campaign launched in 2014). Empirical evidence shows that celebrity endorsements help in disseminating innovations (Till and Shimp 1998; Addo 2016). To create mass awareness about adverse consequences/impacts of extremely hazardous pesticides on the environment, farm worker and consumer brand ambassadors/celebrities can be hired.

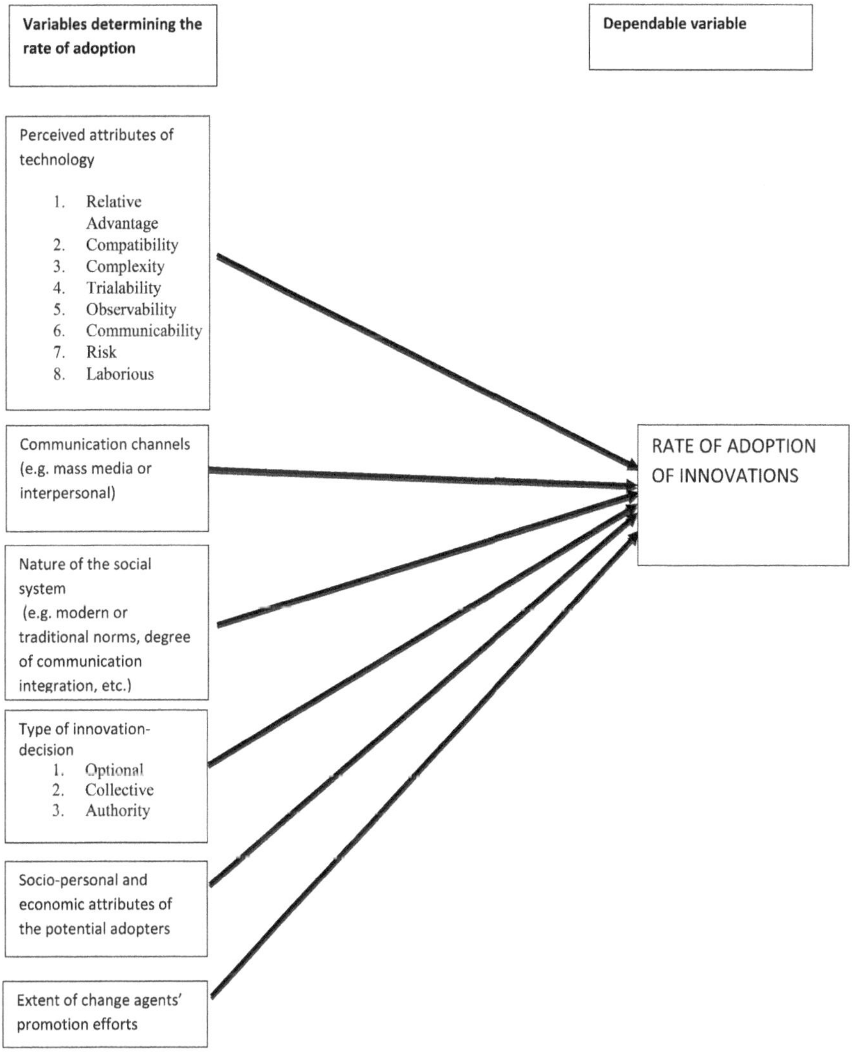

Fig. 14.3 Factors impacting the rate of adoption. (Modified after Rogers 1983)

14.4 Models Explaining Factors Responsible for Adoption/Non-adoption

14.4.1 Singh's Four-Factor Model of Adoption

In order to explain why an agricultural innovation is not adopted, Galjart (1971) developed a three-factor model which consisted of the three factors of ignorance, low motivation and inability. This model was further enlarged by Singh in 1979, by adding the element of inappropriateness vis-a-vis the characteristics of particular innovations. Thus, Singh laid emphasis on the innovation attributes that impact adoption or rejection of an innovation. These factors are:

1. *Inappropriateness*, i.e. negative technical characteristics or non-relevance of the innovation, due to incompatibility as well as complexity of the technology. The adoption of economic threshold level (ETL) of pests by farmers is low because of negative technology characteristics, making the technology inappropriate for wider adoption (Norris et al. 2003; Peshin et al. 2009a, c; Peshin 2013).
2. *Ignorance* on the part of the adopters, which is due to a lack of knowledge and skills. Majority of the Indian farmers do not wear protective clothing at the time of application of pesticides due to ignorance.
3. *Unwillingness*, which includes lack of incentives, low net returns, unremunerative prices, too much risk, inequitable land relations and psychological low aspirations. Many Punjab and Haryana (India) farmers are unwilling to stop rice stubble burning because of lack of incentives provided by the governments.
4. And finally *inability*, i.e. the lack of resources including land, labour, capital including cash, input supplies and equipment. For example, stubble burning is low-cost straw-disposal practice to reduce the turnaround time between harvesting of rice crop and sowing of wheat.

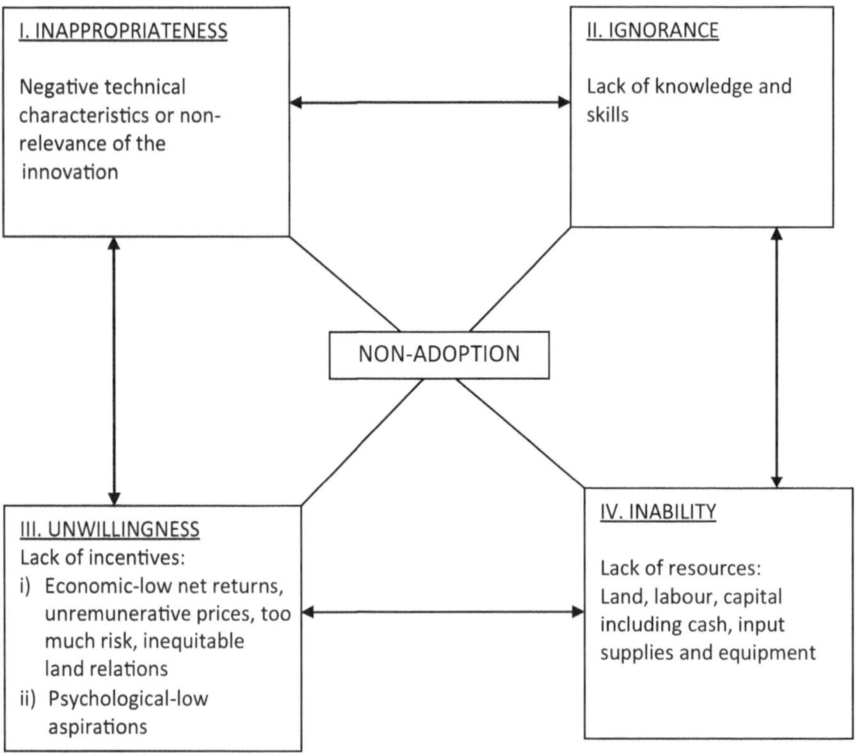

Fig. 14.4 Singh's four-factor model for non-adoption of innovation. (Based on Galjart 1971)

The four-factor model along with its operational indicators and their structure of relationships is presented in Fig. 14.4. As explained by Singh (1979), the four factors determining non-adoption of an innovation have been specified and operationally defined in the interconnected set of boxes. It implies that four factors of non-adoption have to be transformed into their opposite condition for changing the problematic state of non-adoption to the solution state of innovation adoption (Bhatia 1990; Singh 1979). It is further added that such a transformation requires:

- The generation of appropriate technologies by the research system with proven techno-economic viability
- A strong diffusion/dissemination effort focused on non-formal education and training of farmers and farm women, by the extension system including mass media
- A large-scale adoption support from infrastructural agencies concerned with supplies of inputs and services
- A favourable socioeconomic milieu, consisting mainly of policies and programmes of incentives and equitable land tenure relations

The model also emphasised that the beginning point in the adoption process is the availability of appropriate technology (here emphasis is on "appropriate") and role of other factors and institutions, contingent on the fulfilment of this condition.

14.4.2 Peshin Model of Adoptability

As discussed in Sect. 14.3 of this chapter, the attributes of a technology as perceived by farmers are considered important for determining the rate of adoption. Peshin (2013) developed a statistical model for predicting the adoptability of agricultural technologies, developed by scientists at experimental stations/universities, which may help to take into account the farmers' perceptions about attributes of an innovation, causing variance in the rate of adoption of a technology. A Likert-type scale was constructed to measure the farmers' attitude towards IPM practices disseminated under the insecticide resistance management (IRM) programme (Peshin 2013). The model included scale items to measure the five attributes: relative advantage, compatibility, observability, trialability and complexity (Table 14.4). The predicted adoption (adoptability) and actual adoption in 2004 and 2016 are highly correlated (Spearman's ρ, at $p < 0.001$). Besides, cotton farmers' attitude towards different IPM practices disseminated under the IRM programme (Table 14.4) was elicited on open-ended questions. Additional attributes of "risk" and "laborious" in adoption of complex IPM practices were reported by cotton growers. Trialability attribute was not perceived as an important characteristic of cotton IPM by farmers, as there was no variance between groups of cotton growers in the 15 villages. The adoptability indices can be calculated by Eqs. 14.1, 14.2 and 14.3. To improve Peshin's model of adoptability, a procedure followed by Kearns (1992) can be followed to elicit the important characteristics for a particular technology perceived by a sample set of respondents, before constructing the scale items. Besides, the other variables impacting adoption, namely the change agents' efforts in disseminating

Table 14.4 Adoptability and adoption of selected IPM practices

Practice	Indices of attributes[a]						Actual adoption[b]	
	Relative advantage (ap_1)	Compatibility (ap_2)	Observability (ap_3)	Trialability (ap_4)	Complexity (aq_1)	Adoptability index (AI)	2004	2016[c]
Timely sowing	0.94	0.58	0.93	1.00	0.00	0.86	74	74
Recommended resistant varieties (other than Bt cotton)	0.32	0.37	0.34	1.00	0.00	0.51	29	–
Bt cotton	0.99	0.65	0.99	1.00	0.00	0.91	72	92
Seed dressing	0.58	0.50	0.43	1.00	0.73	−0.10	05	–
Treated seed	0.58	0.50	0.43	1.00	0.00	0.63	72	–
Economic threshold level	0.54	0.29	0.45	1.00	0.55	0.02	07	19
Insecticide resistant management	0.66	0.60	0.66	1.00	0.17	0.56	42	–

[a] ap_1, ap_2, ap_3, and ap_4 = positively rated attribute indices; aq_1 = negatively rated index (Sources: Peshin 2005, 2013)
[b] % of farmers;
[c] R. Peshin, unpublished data, 2016

IPM and resistance to change from calendar-based pesticide use, should also be considered for predicting future adoption of a technology, on a pilot basis:

$$\text{Index of positively related innovation attribute}\left(\text{ap}_\text{i}\right) = \frac{\text{Sum of score of n respondents}}{\text{Maximum score obtainable}} \tag{14.1}$$

$$\text{Index of negatively related innovation attribute}\left(\text{aq}_\text{i}\right) = \frac{\text{Sum of score of n respondents}}{\text{maximum score obtainable}} \tag{14.2}$$

The index of attribute can range from 0 to 1:

$$AI = \frac{\sum_{i=1}^{n1} \text{ap}_\text{i}}{n_1} - \frac{\sum_{i=1}^{n2} \text{aq}_\text{i}}{n_2} \tag{14.3}$$

where, AI is the adoptability index of a technology Y and can range from −1 to 1, ap_i, positively related attributes of a technology; aq_i, negatively related attributes of a technology; n_1, number of positively related attributes; and n_2, number of negatively related attributes.

14.4.3 Bass Model

In marketing diffusion, the Bass forecasting model is the most significant tool to predict the diffusion of new consumer products (Bass 1969). The Bass forecasting model reduced the uncertainty associated with the introduction of a new product in the marketplace and was used by Kodak, IBM and other large US corporations. The Bass model has also been used to predict the diffusion of educational ideas (Lawton and Lawton 1979), pesticide use in coconut (Akinola 1986) and pest management (Rebaudo and Dangles 2011). The Bass model made a contribution to forecast rate of adoption at future time periods, based on mass media, interpersonal communication channels and time. The model is one of the most cited empirical generalisations in marketing; as of December 2014, its publication in *Management Science* showed (approx.) 5740 citations in Google Scholar:

$$\frac{dN(t)}{dt} = \frac{q}{m}.N(t)\left(m - N(t)\right) + p.(m - N(t) \tag{14.4}$$

where, $N(t)$ is the cumulative number of adopters, $dN(t)/dt$ the variation of the cumulative number of adopters as a function of time t and p and q the coefficients of mass media and word-of-mouth influence, with m the number of potential adopters.

Key elements in Frank Bass's forecasting model are (*i*) adopters due to mass media message (p, innovators), (*ii*) adopters due to interpersonal communication channels (q, imitators) and (*iii*) an index of market potential (m) for a new product (Fig. 14.5).

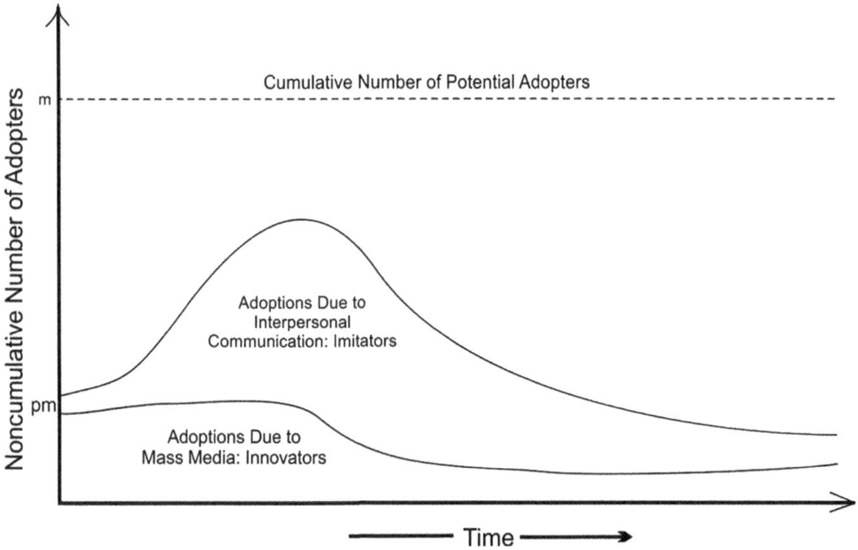

Fig. 14.5 The Bass model for forecasting rate of adoption of a new product. (Modified after Mahajan et al. 1990)

Rogers' (1962) contribution is widely utilised by the private agribusiness and other companies to sell their products. These companies have well-established marketing departments for diffusing new products and market research that investigate the diffusion of products to aid the company marketing efforts (Bass 1969, 2004). As already discussed above, the Bass model has been used in different sectors. However, it does not account for social networks through which the innovation is often diffused (Rebaudo et al. 2014; Flower and Christakis 2010). It also does not consider the attributes which are the main drivers of innovation adoption. But the significant contribution of the Bass model in marketing, to forecast future adoption of innovation/new products, has been the most significant to overcome pro-innovation bias (Rogers 1995).

Analyses of the models of innovation diffusion clearly depict that, for the diffusion of a technology and its adoption thereafter, a researcher should identify/consider the important factors first, trying to minimise the effect of attribute factors on adoption. Clearly, a technology will be adopted by an end user (e.g. farmer) when its benefits overcome its perceived negative attributes, which can only be possible with the timely awareness-knowledge and interest development in the recipient individual. Technological attributes are the strong predictors impacting adoption or rejection of a technology (Peshin 2013).

14.5 Conclusion

As a concluding remark, it is imperative to mention the importance that diffusion and adoption theory hold in the dissemination of innovations, be it a technology, idea or object. Scientists have to get acquainted with the intricacies of diffusion research so that, instead of farmers' blame and innovation biases, the researchers would focus on innovation inappropriateness. The attributes of a technology, as perceived by farmers, determine the rate of adoption. The research that studies the effect of the attribute variables on the rate of adoption or adoption per se has been, however, meagre (1%) compared to socio-personal/socioeconomic variables (58%) (Rogers 2003). Thus, it can be inferred that a group of farmers having the same socio-personal and economic attributes will exhibit different adoption behaviours for different innovations. Therefore, agricultural innovations should not be treated as equal for analysis. Researchers should take the innovation attributes and constraints faced by farmers into consideration, before recommending technologies which are partially compatible with the farming system.

We should employ path analysis of the socio-personal, economic variables of potential adopters, together with perceived attributes of innovation and communication variables, to find out their contributions in rate of adoption or adoption per se. The decision to adopt a sustainable innovation/technology is a dynamic process, with interactions among socioeconomic factors, communication behaviour, efforts of the change agency, social system, contextual factors and attributes of innovation. For that reason, Singh's four-factor model of non-adoption, Peshin model of adoptability and Bass model need to be analysed to develop a composite model for predicting future adoption of innovations, at the time of any technology commercialisation.

Acknowledgements We are grateful to Rizwan Jeelani, Ph.D student at the Division of the Veterinary Extension Education of Sher-e-Kashmir University of Agricultural Sciences and Technology of Jammu for Figs. 14.3 and 14.5.

References

Addo, J.O. (2016). Moderating factors of celebrity endorsement on diffusion. *Global Journal of Management and Business Research: E Marketing, 16*(1),19–31.

Akinola, A. A. (1986). An application of Bass model in the analysis of diffusion of cocoa spraying among Nigerian cocoa farmers. *Journal of Agricultural Economics, 37*(3), 395–404. https://doi.org/10.1111/j.1477-9552.1986.tb01607.x.

Bass, F. M. (1969). A new product growth model for consumer durables. *Management Science, 13*(5), 215–227.

Bass, F. M. (2004). Comments on a new product growth for model consumer durables. *Management Science, 50*(12), 1833–1840.

Bhatia, B. S. (1990). *Adoption of farm mechanization in a developing economy*. New Delhi: Daya Publishing House.

Cain, M., & Mittman, R. (2002). *Diffusion of innovation in health care*. Oakland: California Health Care Foundation.

Downs, G. W., & Mohr, L. B. (1976). Conceptual issues in the study of innovation. *Administrative Science Quarterly, 21*, 700–714.

FAO. (2017). *Food and agriculture: Driving action across the 2030 agenda for sustainable development*. Food and Agriculture Organisation of the United Nations. 40p. http://www.fao.org/publications/card/en/c/b31f7040-a387-466d-99e2-04896d421fca/. Accessed 12 May 2018.

Feder, G., & Umali, D. (1993). The adoption of agricultural innovations: A review. *Technological Forecasting and Social Change, 43*(3–4), 215–239.

Fliegal, F. C., & Kivlin, J. E. (1966). Attributes of innovations as factors in diffusion. *American Journal of Sociology, 72*(3), 235S–248S.

Flower, J. H., & Christakis, N. A. (2010). Cooperative behaviour cascades in human social networks. *Proceedings of the National Academy of Sciences, 107*(12), 1287–1300.

Galjart, B. (1971). Rural development and sociological concepts: A critique. *Rural Sociology, 36*(1), 31–41.

Hashemi, S. M., Peshin, R., & Feola, G. (2014). From the farmers' perspective: Pesticide use and pest control. In D. Pimentel & R. Peshin (Eds.), *Integrated pest management pesticide problems* (Vol. 3, pp. 409–432). New York/Heidelberg/Dordrecht/London: Springer.

Kearns, K. P. (1992). Innovations in local government: A socio-cognitive network approach. *Knowledge and Policy, 5*(2), 45–67.

Knowler, D., & Bradshaw, B. (2007). Farmers' adoption of conservation agriculture: A review and synthesis of recent research. *Food Policy, 32*, 25–48.

Kremer, K. S., Crolan, M., Gaiteyer, S., Tirmizi, S. N., Korsching, P. F., Peter, G., & Tong, P. (2000). Evolution of an agricultural innovation: The N-track soil nitrogen test: Adopt, and discontinuance, or reject? *Technology in Society, 23*, 93–108.

Lawton, S. B., & Lawton, W. H. (1979). An autocatalytic model for diffusion of educational innovations. *Educational Administration Quarterly, 15*(1), 19–53.

Litsinger, J. A., Libetario, E. M., & Canapi, B. L. (2009). Eliciting farmer knowledge, attitudes, and practices in the development of integrated pest management programs for rice in Asia. In R. Peshin & A. K. Dhawan (Eds.), *Integrated pest management: Dissemination and impact* (Vol. 2, pp. 119–274). Dordrecht: Springer. https://doi.org/10.1007/978-1-4020-8990-9_5.

Mahajan, V., Muller, E., & Bass, F. M. (1990). New product diffusion models in marketing: A review and direction for research. In N. Nakicenovic & A. Grubler (Eds.), *Diffusion of technologies and social behaviour* (pp. 125–177). New York: Springer.

Menzel, H. (1960). Innovation, integration, and marginality: A survey of physicians. *American Sociological Review, 25*, 704–713.

Nanda, R., Peshin, R., Sharma, L. K., Sharma, R., Sharma, P., & Liyaqat, A. (2017). *Determinants of income diversification of family farm household for livelihood security in J&K*. Jammu: Sher-e-Kashmir University of Agricultural Sciences and Technology of Jammu, India.

Norris, R. F., Caswell-Chen, E. P., & Kogan, M. (2003). *Concept in integrated pest management*. New Delhi: Prentice-Hall of India Private Ltd.

Pattanayak, S. K., Mercer, D. E., Sills, E., & Yang, J. C. (2003). Taking stock of agroforestry adoption studies. *Agroforestry Systems, 57*, 173–186.

Peshin, R. (2005). *Evaluation of dissemination of insecticides resistance management technology in cotton crop in Punjab*. PhD dissertation, Punjab Agricultural University, Ludhiana.

Peshin, R. (2013). Farmers' adoptability of integrated pest management of cotton: Revealed by a new methodology. *Agronomy for Sustainable Development, 33*(3), 563–572. https://doi.org/10.1007/s13593-012-0127-4.

Peshin, R., Dhawan, A. K., Kranthi, K. R., & Singh, K. (2009a). Evaluation of the benefits of an insecticide resistance management programme in Punjab in India. *International Journal of Pest Management, 55*(3), 207–220. https://doi.org/10.1080/09670870902738786.

Peshin, R., Bandral, R. S., Zhang, W. J., Wilson, L., & Dhawan, A. K. (2009b). Integrated pest management: A global overview of history, programs and adoption. In R. Peshin & A. K. Dhawan (Eds.), *Integrated pest management: Innovation-development process* (Vol. 1, pp. 1–49). Dordrecht: Springer. https://doi.org/10.1007/978-1-4020-8992-3_1.

Peshin, R., Jayaratne, J., & Singh, G. (2009c). Evaluation research: Methodologies for evaluation of IPM programmes. In R. Peshin & A. K. Dhawan (Eds.), *Integrated pest management: Dissemination and impact* (Vol. 2, pp. 31–78). Dordrecht: Springer.

Peshin, R., Kranthi, K. R., & Sharma, R. (2014). Pesticide use and experiences with integrated pest management programs and Bt cotton in India. In R. Peshin & D. Pimentel (Eds.), *Integrated pest management, experiences with implementation: Global overview* (Vol. 4, pp. 269–306). Dordrecht: Springer.

Peshin, R., Hansra, B. S., Nanda, R., Singh, K., & Sharma, R. (2017). *Impact evaluation of integrated pest management technologies*. Division of Agricultural Extension Education, Sher-e-Kashmir University of Agricultural Sciences and Technology of Jammu, India.

Rebaudo, F., & Dangles, O. (2011). Coupled information diffusion—pest dynamics models predict delayed benefits of farmer cooperation in pest management programs. *PLoS Computational Biology, 7*(10), 1–10., E1002222. https://doi.org/10.1371/journal.pcbi.1002222.

Rebaudo, F., Carpio, C., Crespo-Perez, V., Herrara, M., Scurrah, M. M. D., Canto, R. C., Montanez, A. G., Bonifacio, A., Mamani, M., Saravia, R., & Dangles, O. (2014). Agent-based models and integrated pest management: Diffusion in small scale farmer communities. In R. Peshin & D. Pimentel (Eds.), *Integrated pest management, experiences with implementation: Global overview* (Vol. 4, pp. 367–383). Dordrecht: Springer.

Rogers, E. M. (1962). *Diffusion of innovations*. New York: The Free Press.

Rogers, E. M. (1983). *Diffusion of innovations* (3rd ed.). New York: The Free Press.

Rogers, E. M. (1995). *Diffusion of innovations* (4th ed.). New York: The Free Press.

Rogers, E. M. (2003). *Diffusion of innovations* (5th ed.). New York: The Free Press.

Roling, N. (1988). *Extension science*. Cambridge: Cambridge University Press.

Ryan, B., & Gross, N. (1943). The diffusion of hybrid seed corn in two Iowa communities. *Rural Sociology, 8*(1), 15–24.

Sharma, R. (2011). *Evaluation of the vegetable integrated pest management farmer field school programme in Jammu region*. Ph.D. dissertation, Sher-e-Kashmir University of Agricultural Sciences and Technology of Jammu, India.

Sharma, R., & Peshin, R. (2016). Impact of integrated pest management of vegetables on pesticide use in subtropical Jammu, India. *Crop Protection, 84*, 105–112 https://doi.org/10.1016/j.cropro.2016.02.014.

Sidhu, R. S., Vatta, K., & Dhaliwal, H. S. (2010). Conservation agriculture in Punjab: Economic implication of technologies and practices. *Indian Journal of Agricultural Economics, 65*(3), 413–427.

Sidhu, R. S., Vatta, K., & Lall, U. (2011). Climate change—Its impact on agriculture productivity and livelihood: The policy response-climate change impact and management strategies for sustainable water-energy-agriculture outcomes in Punjab. *Indian Journal of Agricultural Economics, 66*, 328–339.

Singh, R. (1979). *An extension strategy for scientific storage of food grains at the farm level*. Department of Extension Education, Punjab Agricultural University Ludhiana, India (Technical paper).

Singh, K. (2015). *My experiences with KVK Jalandhar. Invited lecture delivered in the department of extension education*. Punjab Agricultural University Ludhiana, India (Unpublished data).

Till, B. D., & Shimp, TA (1998). Endorsers in advertising: The case of negative celebrity information. *Journal of Advertising, 27*(1), 67–82.

Zhao, J., Luo, Q., Deng, H., & Yan, Y. (2008). Opportunities and challenges of sustainable agricultural development in China. *Philosophical Transactions of the Royal Society B: Biological Sciences, 363*(1492), 893–904.

Index

R. Peshin, A. K. Dhawan (eds.), *Natural Resource Management: Ecological Perspectives*, Sustainability in Plant and Crop Protection,
https://doi.org/10.1007/978-3-319-99768-1

B

I

J

K

Printed by Printforce, the Netherlands